ROUTLEDGE LIBRARY EDITIONS:
POLLUTION, CLIMATE AND CHANGE

Volume 10

THE UNCERTAINTY BUSINESS

THE UNCERTAINTY BUSINESS

Risks and Opportunities in Weather and Climate

W. J. MAUNDER

LONDON AND NEW YORK

First published in 1986 by Methuen

This edition first published in 2020
by Routledge
2 Park Square, Milton Park, Abingdon, Oxon OX14 4RN

and by Routledge
52 Vanderbilt Avenue, New York, NY 10017

Routledge is an imprint of the Taylor & Francis Group, an informa business

British Library Cataloguing in Publication Data
A catalogue record for this book is available from the British Library

ISBN: 978-0-367-34494-8 (Set)
ISBN: 978-0-429-34741-2 (Set) (ebk)
ISBN: 978-0-367-36269-0 (Volume 10) (hbk)
ISBN: 978-0-367-36273-7 (Volume 10) (pbk)
ISBN: 978-0-429-34500-5 (Volume 10) (ebk)

Publisher's Note
The publisher has gone to great lengths to ensure the quality of this reprint but points out that some imperfections in the original copies may be apparent.

Disclaimer
The publisher has made every effort to trace copyright holders and would welcome correspondence from those they have been unable to trace.

The Uncertainty Business

Risks and opportunities in weather and climate

W. J. MAUNDER

with a Foreword by John R. Mather

Methuen
London and New York

To Melva
in appreciation of her
help and support

First published in 1986 by
Methuen & Co. Ltd
11 New Fetter Lane, London EC4P 4EE

Published in the USA in 1987 by
Methuen & Co.
in association with Methuen, Inc.
29 West 35th Street, New York NY 10001

Photoset by Rowland Phototypesetting Ltd,
Bury St Edmunds, Suffolk
Printed in Great Britain at
The University Press, Cambridge

British Library Cataloguing in Publication Data
Maunder, W. J.
The uncertainty business: risks and
opportunities in weather and climate.
1. Economic development 2. Man ——
Influence of climate
I. Title
330.9 HD75.6
ISBN 0-416-36100-5

Library of Congress Cataloging in Publication Data
Maunder, W. J.
The uncertainty business.

1. Climatology – Social aspects. 2. Weather
forecasting – Social aspects. 3. Planning. I. Title
QC981.M447 1986 304.2′5 86-16266
ISBN 0-416-36100-5

Contents

Figures and map

Figures

Map

Tables

Acknowledgements

It is a pleasure to acknowledge the substantial help I have had from many people and organizations in the preparation of this book. In particular, a special word of appreciation is extended to Dr W. R. D. Sewell, of the University of Victoria (Canada), for his initial encouragement to me to complete a new book, rather than a second edition of *The Value of the Weather*, and to Dr J. R. Mather of the University of Delaware (USA) who critically read all the draft text. Dr Sewell was particularly helpful in suggesting the format for the book and the approach I eventually took, while Dr Mather made many very useful suggestions for improvements during his reading of the draft text. I hope that the final text reflects his wise counsel.

Many other people also read parts of the draft text and I would like in particular to thank my colleagues in the New Zealand Meteorological Service for their valuable criticism and assistance. Special thanks in this regard are due to Mr J. S. Hickman, Director of the New Zealand Meteorological Service for providing facilities for the preparation of several sections of the book, some parts of which have been previously published in reports issued by the New Zealand Meteorological Service. In addition I would like to acknowledge the assistance of I. P. Brown, N. T. Challands, S. W. Goulter, M. J. Salinger, J. Sanson, G. S. M. Smith, R. M. Smith, C. S. Thompson and G. F. A. Ward of the New Zealand Meteorological Service for their specific constructive comments on various sections of the book. In addition, the administrative support provided by W. S. Jamison, I. D. McMillan, D. A. Hogben, Mrs J. N. Barton, Miss T. Peleseuma, Ms D. Myer, Miss M. H. Pinfold and Miss K. Tredrea of the New Zealand Meteorological Service, and the computer programming support of Mr J. M. H. Waller, of the New Zealand Meteorological Service, are acknowledged with grateful thanks. The very helpful co-operation and guidance of Mary Ann Kernan, Anna Fedden, Eleanor Rivers and Catherine Stewart of Methuen's London office is also very much appreciated. The cover design by Stonecastle Graphics was based in part on an idea of my daughter Denise.

Acknowledgement is also made to several granting agencies and organizations who directly or indirectly have supported my endeavours in the economic and social aspects of meteorology and climatology over the last twenty-five years. Special thanks in this regard are due to the University of Otago, the New Zealand Meteorological Service, the University of Victoria (Canada), the University of Missouri (Columbia), the University of Delaware, the (then) Meteorological Branch of the Canadian Department of

Transport, the World Meteorological Organization (WMO), the United Nations Environment Programme (UNEP), the Food and Agriculture Organization (FAO), the Scientific Committee on Problems of the Environment (SCOPE) of the International Council of Scientific Unions (ICSU), the Meteorological Society of New Zealand, the American Meteorological Society, the New Zealand Geographical Society, the New Zealand Institute of Agricultural Science, the (United States) National Science Foundation, the University of Oklahoma and in particular the Oklahoma Climatological Survey, the Atmospheric Environment Service (of Canada), A. G. B. McNair Surveys Limited, the (United Kingdom) Meteorological Office, the Canadian Climate Centre, the International Institute of Applied Systems Analysis (IIASA), the (United States) National Center for Atmospheric Research, the New Zealand Dairy Board, the New Zealand Meat and Wool Boards' Economic Service, the Reserve Bank of New Zealand, the Swedish Meteorological and Hydrological Institute (SMHI), the Assessment and Information Services Center (AISC) and the Center for Climatic and Environmental Assessment (CCEA) of the United States Department of Commerce, and the United States Department of Agriculture.

I would also like to thank the following societies, editors of publications, organizations and individuals for permission to reproduce tables or figures, or to use specific textual quotations. (Specific acknowledgement is given at the appropriate place in the text.)

(1) *Societies*
American Meteorological Society, Meteorological Society of New Zealand, New Zealand Geographical Society, Royal Meteorological Society, Royal Society of New Zealand.

(2) *Publications*
Britannica Book of the Year, *Business Week*, *Christchurch Star*, *The Economist*, *Encyclopaedia Britannica*, *Evening Post*, *Financial Times*, *Monthly Weather Review*, *New Zealand Agricultural Science*, *New Zealand Geographer*, *Newsweek*, the *Press*, *Time*, *Tokyo Newsletter* (Mitsubishi Corporation), *US News and World Report*, *Wall Street Journal*, *Weather*, *Weather and Climate*, and *Weatherwise*.

(3) *Organizations*
Australian Bureau of Meteorology, Department of Scientific and Industrial Research (NZ), Department of Statistics (NZ), East Coast Fertiliser Company, Food and Agriculture Organization, Her Majesty's Stationery Office, Lincoln College, National Academy of Science, National Defense University, New Zealand Dairy Board, New Zealand Institute of Economic Research, New Zealand Meat Producers Board, New Zealand Meat and Wool Boards' Economic Service, New Zealand Meteorological Service, New Zealand Wool Board, Oklahoma Climatological Survey, Reserve Bank of

New Zealand, United Nations University, United States Department of Transportation, World Bank, and the World Meteorological Organization.

(4) *Individuals*
J. A. Ausubel, E. A. Bernard, Francie Brentzel, R. D. Cess, P. W. Dini, A. Eddy, J. M. Elsom, E. S. Epstein, L. W. Gandar, M. H. Glantz, H. L. Halanger, L. D. B. Heenan, J. S. Hickman, Joan Hoak, S. B. Idso, S. R. Johnson, J. Joyce, R. W. Kates, G. W. Kearsley, W. W. Kellogg, J. W. Kidson, H. H. Lamb, Kitty Liebhart, Sir John Mason, J. R. Mather, G. A. McKay, Mrs J. D. McQuigg, J. M. Miller, J. K. Monteith, Sir Robert Muldoon, A. H. Murphy, Jean Palutikof, Sir Brian Pippard, W. R. D. Sewell, L. P. Smith, B. F. Taylor, J. A. Taylor, N. W. Taylor, J. C. Thompson, J. H. Troughton, G. B. Tucker, S. Unninayar, C. Wahlbin, M. J. Walsh, R. A. Warrick, and R. M. White.

A special word of appreciation is offered to the many colleagues I have been associated with outside New Zealand over the years, and who in many ways are the silent authors of this book. Specifically I would like to thank Oliver Ashford, Jessie Ausubel, Wolfgang Baier, E. A. Bernard, Reid Bryson, Juan Burgos, Michael Connaughton, Wayne Decker, Jim Dooge, Amos Eddy, Harry Foster, Bill Gibbs, Michael Glantz, Ken Hare, Stan Johnson, Slavka Jovičić, Bob Kates, Will Kellogg, Sharon Le Duc, Adopho Mascarenhas, (Sir) John Mason, Russ Mather, Gordon McKay, John Monteith, Allan Murphy, Lars Olsson, Martin Parry, David Phillips, Tom Potter, Frank Quinlan, Jim Rasmussen, Bill Riebsame, George Robertson, Clarence Sakamoto, Steve Schneider, Derrick Sewell, Frank Singleton, Frances Slater, Norton Strommen, Morley Thomas, Louis Thompson, Sushel Unninayer, Peter Usher, Jan De Vries, C. C. Wallén, Henry Warren, Bob White, Tom Wigley, and John Zillman.

Finally the inspiration offered to me over many years by the late Jim McQuigg of Columbia, Missouri, and the late Stuart Hurnard of the New Zealand Meteorological Service is gratefully acknowledged. I am sure that without their guiding hands this book would not have been published.

January 1986

New Zealand Meteorological Service,
Box 722
Wellington
New Zealand

W. J. Maunder
31 Bosun Terrace
Whitby
Wellington
New Zealand

Foreword

JOHN R. MATHER
Center for Climatic Research, Department of Geography
University of Delaware, Delaware, USA

Weather and climate are known to influence human activities in many pervasive ways. Our vulnerability to climatic fluctuations has become more evident in recent years as world population has increased and competition for available resources has intensified.

National political and business leaders, planners and economists are being and will continue to be forced to make decisions involving climate-sensitive activities. Thus, climatologists concerned with the economic value of weather information must be prepared to provide these leaders and planners with the guidance needed to manage such economic activities as national food production and energy consumption as well as human health and well-being. Managed properly, climate can become a resource of great value, improving the economic competitiveness of a nation.

The need to recognize climate as a resource to be managed is one of the fundamental precepts developed by John Maunder in this study. He further points out that we must achieve a greatly improved understanding of the many interrelationships between climate and society, and especially a better knowledge of how changes in one may produce shifts in the other. Finally, we must recognize that many issues of concern to the meteorological and climatological community have potential, if not real, political implications. If we are to live within the limit of our 'climatic income' or our 'élite atmospheric resource', appropriate meteorological and climatological consultation must be involved at highest levels of national planning.

The call is made for climatologists to formulate precise relationships between climate factors and various aspects of production and consumption to provide the basis needed for operational decision-making. The real climate sensitivity of a nation as well as of different commodities must be assessed within the existing economic, social and political milieu and presented by climatologists in terms of production figures, costs, alternate choices, or other similar measures that can be used directly by decision-makers. Education of both the producers of weather and climate information and the potential users of these products is a necessary first step. The politician, the planner, the economist, the manager must all become more weather and climate oriented through closer interaction with climatologists.

Maunder has provided a realistic vision of a new relationship that will allow the utilization of our climatic resource for the betterment of all nations.

Preface

A PERSONAL COMMENT

From the many comments I have received since 1970 – when *The Value of the Weather* was first published – it is clear that many of the 'ideas' expressed in that book were several years ahead of their time. However, during the last decade there has been an increasing amount of activity in what some people would call the 'non-scientific' aspects of meteorology/climatology. These include the data, application, and impact components of the WMO World Climate Programme, and the growing recognition by both the more influential newspapers and journals, and by the more influential politicians, of the real role of weather and climate in economic, social, strategic, and political activities.

I understand that the readership of *The Value of the Weather* was mainly outside the education field. That is, few university departments prescribed the book for specific courses, but many of them suggested the book as 'additional reading'. However, a large number of copies of *The Value of the Weather* were purchased by libraries and individuals, and it is believed that the book has had an influence on meteorological and climatological thinking. When *The Value of the Weather* was written, it was, among other things, a reaction against the considerable emphasis on 'meteorological processes' which I considered were more than adequately covered in many books. The book was also written to emphasize the political, planning, legal, sociological, and overall 'value' of the weather, and in particular the value of *information* about the weather. Fifteen years on I still consider that the meteorological processes are adequately covered in various books, but the 'non-scientific' aspects are still neglected by many meteorologists and climatologists (but not the politicians!), as well as by many economists, agriculturalists, etc. Accordingly I consider that the economic, social, political, planning, and legal aspects still need to be very much emphasized – even though many advances have been made in the area (particularly by the UN agencies, some national meteorological services, and many meteorological consultants).

Since *The Value of the Weather* was published in 1970, the acceptance of the atmosphere as a variable and 'élite' resource that may be despoiled, modified, monitored, ignored, used, and forecast, has become a reality. Nevertheless, this realization exists only in certain nations, communities, institutions, and companies. Indeed, there remain many people – both within and external to the field of weather and climate – who still consider that the weather and climate resource is devoid of value, and therefore has no place in operational decision-making. This present book, *The Uncertainty Business: Risks and opportunities in weather and climate*, goes far beyond the pioneer 1970 volume. While the latter provided for the first time an overview of the 'economic dimensions' of climate and human activities, it stopped short of addressing the manner in which decisions involving the atmosphere are *actually* made. The new volume provides fresh insights into this key issue. Specifically it considers how the variable nature of a particular and élite resource so often taken for granted – the atmosphere – must be accepted as an integral part of the management package.* The book consists of eight chapters.

Chapter I, 'The atmosphere as an élite resource', considers the development of the concept that the atmosphere is a variable élite resource. The nature of this élite resource is first examined from the viewpoint of understanding its utility in human terms. An in-depth examination of the changing definitions and concepts relating to climate, climatology, and meteorology during the past 100 years is then presented. This review includes the reasons why the economic, social, and political aspects of the subject are now receiving so much attention. The chapter then looks at the changing ideas associated with the terms 'weather' and 'climate' especially in the operational decision-making area, and considers the various aspects of the weather/climate 'game' as played by its many participants. This is followed by an examination of the atmospheric part of what is commonly called the 'economic' climate, with particular reference to food supplies, and the important 'near average' conditions in which most people and nations live. The economic part of the atmospheric resource, however, is not the only aspect that is significant. Overall human well-being is perhaps even more critical, and the chapter concludes by giving specific attention to the serious nature of the weather and climate game.

The management of national resources takes place within an institutional framework of laws, policies, and administrative arrangements. These are all conditioned by traditions and other cultural attributes which vary considerably from country to country. Chapter II, 'The institutional framework', addresses first the matter of human responses and public policies to the

* It should be emphasized that this book does not cover all aspects of applied meteorology and climatology. In particular it does not include several of the topics discussed at length in *The Value of the Weather*, nor does it discuss many of the 'traditional' applied aspects of meteorology and climatology which are usually included in 'standard' textbooks. For example, topics such as hurricanes, urban climates, human biometeorology, weather modification, technological developments in meteorology, numerical weather prediction, and floods, are not discussed in any detail, and for these and many other similar topics the reader is referred to *The Value of the Weather* or other relevant textbooks. Rather, the purpose of this book is to provide a basis for lateral thinking that is necessary if a true appreciation of the value and use of weather and climate information is to be realized.

weather and climate, and the management of the atmospheric resource. This is followed by a discussion of specific public (state) educational and private (commercial) sector responses of several nations to these questions. The international institutional framework is considered next with an emphasis on the changing and significant role of the World Meteorological Organization (WMO), and its various Commissions and Expert Working Groups, together with the supporting role of FAO, Unesco, WHO, UNEP and ICSU. Specific attention is focused on the importance of the WMO World Climate Programme (WCP) and the functions of its four parts (Data, Applications, Research, and Impact), as well as on the long-term plans of WMO. The strategic aspects of monitoring the various spatial and secular dimensions of the global weather and climate are then discussed – emphasizing the now current realization within most of the international community of the necessity to understand *both* the weather *and* the climate. This, it should be noted, is in contrast to the World Weather Watch approach of the 1960s and 1970s when only 'synoptic' weather monitoring was considered to be necessary, such measurements being restricted in the main to supporting purely 'meteorological' activities such as marine and aviation forecasts.

Chapter III, 'Events and information', focuses on marketing and communicating the weather/climate package. The chapter opens with a discussion of weather and climate information in the context of the wider issue of both forecasts and information. This distinction is important because the public at large, as well as individual decision-makers, characteristically regard accurate, reliable, and timely forecasts as being the prime product – and in some cases the only product – of a national meteorological service. Further, they believe that forecasts are all that is required to make sound judgements. This, however, is far from the truth. What is needed, in fact, is information about the *past* and the *present*, as well as the *future*. All are equally important, and in some circumstances a knowledge of past and current conditions may be of considerably more value than the forecast conditions, irrespective of the accuracy of such predictions. As an illustration, a forecast that there is a 50 per cent probability of precipitation of a certain magnitude in the next twenty-four hours is by itself insufficient information for a manager of an electric power utility to decide whether or not to release water from a reservoir. Similarly, such a forecast would be inadequate for a farmer to decide to move stock to a market, or to plant corn instead of wheat. In both of these cases the decision-maker would need to know, among other things, how much water or soil moisture was already available, whether such conditions were normal or otherwise, how such conditions compared with those in 'competitor areas', as well as what the longer-term prospects were. Indeed, the wider issues of weather and climate *information* provide the theme for the rest of the book. The value and costs of both weather and climate information – particularly as they relate to the observation and transmission of basic data – are also discussed. In addition, the dissemination of processed meteorological and climatological products (such as maps, forecasts, data sets, and application summaries, etc.) to users, by both the public and private media, is examined. An important consideration in this regard is the extent to which a public meteorological agency should charge for services which it provides to the public at large and to specific users. This discussion of 'what should be free' and the analysis of the value and costs of weather and climate information lead directly into a review of how weather events and weather information influence the agricultural and energy sectors of the world's commodity markets. The very important subject of marketing and communication is then considered, with particular reference to the opportunities and challenges afforded by the various sectors of the media, including the exciting developments of user-based videotex systems.

The interrelationship between weather and climate events and various activities may be described as 'The weather–economic mix', which is discussed in Chapter IV. It is recognized that not all activities are economic in nature, but many decisions have an economic context regardless of the nature of the activity. This chapter begins with an examination of the specific manner in which climate and various activities interact, at various levels ranging from local to international. This is followed by an examination of behavioural responses to particular weather and climate events such as drought and storms. The discussion here emphasizes the fact that a broader appreciation of these interrelationships would lead to more effective responses. The chapter then moves on to examine in depth the concept of 'climate sensitivity' as applied to both specific activities and national economies. The idea of climatic impacts is then introduced, particularly from the viewpoint – as expressed in the World Climatic Impact Programme – that nations cannot only maximize the benefits of correctly utilizing the atmospheric resource, but can also reduce the unavoidable losses associated with its changeable nature. 'Climate solutions to problems' are then examined as well as the problems associated with 'measuring' the costs of weather and climate variations. The chapter also includes a detailed analysis of the weather and climate sensitivity of the New Zealand economy, an examination of New Zealand income and product accounts from a weather and climate viewpoint, plus consideration of the political realities of climate sensitivity. An underlying factor discussed in the chapter is the conflicts of goals associated with those who must respond to weather and climate events. As an illustration, an operator of a ski resort may welcome additional snow (and a favourable forecast thereof), whereas the operator of a fishing lodge – downstream from the ski resort – would be anything but pleased with such a prospect. Similarly the announcement of the likelihood of thick fog at an airport may well be greeted with enthusiasm by hotels, restaurants, and rental car companies in the vicinity, but with dismay by airline operators and those needing to travel. That is, gains to some may well be losses to others.

Chapter V, 'Assessing the impact of weather and climate', focuses on how the impact of weather and climate may be assessed in *economic* terms. To do this the various data banks linking the parts of the weather-economic mix are examined. This is followed by an in-depth analysis and explanation of the need to assess weather and climate information in terms of its significance to various areas and populations. This type of assessment may be described as the 'weighting' of weather and climate information, in the same manner that a consumer price index 'weights' the various components of the 'consumer basket' according to their importance. Decision-makers such as national producer boards, or energy authorities, now believe that the weighting of weather and climate information is essential to their deliberations. This is especially true in Canada, the United States, and New Zealand. This new approach elevates applied climatology from the classical textbook age (still current in many national meteorological services and educational institutions), to the operational and planning age. Specific examples of the 'why, what, and how' of the weighting concept are given, together with the development and application of 'climatic confidence indices' on both a national and international scale. These latter indices may be considered to be the climatological equivalent of 'leading economic indicators' but *because of their availability in real-time*, they may in certain circumstances also be considered to be *predictors* of 'leading economic indicators'.

Chapter VI, 'The assessment process', develops ideas relating to the assessment process applied to real-world problems, and leads directly from the impact studies cited in Chapter V. Specific examples are first given, assessing the economic, social, political, and strategic costs associated with severe droughts and an intense cold wave. This is followed by activity assessments which specifically relate the weather and climate resource to the transport, retail trade, manufacturing, construction, road construction, electric power, and agricultural sectors. These assessments incorporate the use of suitably weighted weather and climate information, and show clearly that for any time period, what is 'favourable' for one sector is not necessarily 'favourable' for other sectors.

In Chapter VII 'Weather- and climate-based forecasts of economic activities' are discussed which take the assessment process one stage further, namely through the use of commodity-weighted weather and climate information in formulating weather/climate-based forecasting models of economic activities and production. The various types of prediction models are discussed, followed by specific examples relating to the development and use of weather/climate-based forecast models for the dairy and wool industries. In addition, the development and use of weather/climate-based economic forecasting models are considered, including their real-time application to national business activity indicators.

The issues and challenges likely to affect resource management in the next twenty-five years are highlighted in Chapter VIII, 'Forthcoming challenges and opportunities'. The 'information opportunity' provided by the real-time availability of a vast amount of weather and climate data is first discussed, particularly from the viewpoint of climate 'change(s)' and political realities. Following this, an examination is made of the political and strategic aspects of econoclimatic analyses, including their impact on national intelligence agencies. This leads to a discussion of the sensitivity of nations to a changing climate – both now and what is likely in the year 2000. The political realities of possible changes in the climate, and the even more important political realities of actual variations in the atmospheric resource, are then examined. The important – and as yet unresolved – question of the ownership of the atmosphere leads into a discussion of the issues relating to the deliberate and inadvertent modification of the atmospheric resource. The disposal of effluents into the atmosphere is considered, including the very important environmental issues concerned with the carbon-dioxide question, the acid rain problem, and a 'nuclear winter'. A futuristic agenda for 15 April 1994 is then presented in the context of a 'weather administrator's day'. It suggests that within a decade the meteorologist will be making significant economic decisions with considerable political overtones. Finally, the potential global and national issues which need to be considered before we reach the twenty-first century are assessed.

I The atmosphere as an élite resource

A. Understanding the Atmospheric Resource

1. The setting

The atmosphere is an élite* resource which may be tapped, modified, despoiled, or ignored. Most societies also forecast the availability of the various components of this resource, such predictions being primarily based on knowledge of the nature of the atmosphere through observations made and collected globally in real-time. But these forecasts not only require information on the past and present weather and climate, but also information concerning the interrelationships between the social, political, and economic actions of societies and individuals and the atmospheric events. Together, this information assists in developing the ability of individuals and societies to manage this élite resource, thereby improving the economic, political, and social outcome of the many activities that are sensitive to weather and climate.

The international community through United Nations Conferences on Food, Water, Population, Energy, and Climate, as well as many individual nations, have in recent years become aware of and accepted the view that there are certain physical limits to the availability of natural resources. Included among these resources is the atmosphere. One of the major problems in regard to this resource, however, is the recognition that the atmosphere is not only an élite resource, but *also* that it is a variable resource. This variability can be evaluated in ways which are useful to both political and economic planning and management, and of prime importance is the *impact* of the short-term variations in its various components – particularly those related to water and temperature. These impacts will clearly continue to increase in importance as the demand for food and water increases, and the relative cost of energy increases; hence it is very important to understand the increasing impact that variations in the weather and climate will have on economic, social and political activities (Maunder 1978a).

Of course it is often argued that information about the atmospheric

* The term 'élite resource' was first used by Dr W. R. D. Sewell in a paper presented at a conference in Brisbane, Australia in 1985. 'Élite' refers specifically to the unique and special aspects of the atmosphere which set it apart from other natural resources.

resource cannot be used in economic, agricultural, and political planning, until reasonably accurate climate forecasts are available. It must be appreciated, however, that most economic planning, and most political decisions involving climate-sensitive activities, are made on the basis of the information (including forecasts) *actually* available, and *not* on what 'could be' available. Clearly, decisions *have* to be made. Thus, qualified meteorologists and climatologists are needed to provide the best possible guidance to decision-makers. Obviously, the economies of many nations are dependent to a relatively large extent on next year's weather and climate, yet how many advisers of top-level decision-makers take the actual and forecast weather and climate into account? Further, how should meteorologists and climatologists (whether government or private) provide the appropriate guidance and expertise to these advisers?

One of the main functions of many national meteorological services is to apply meteorological science for the 'benefit of the community'. Included in this function are *forecasts* of the weather and the climate, but an equally important function is the provision of *information* about the weather and climate. Closely linked with the 'benefits to the community' is also the knowledge of the real relationship between economic productivity and variations in the atmospheric resource. Further, if the atmosphere *is* considered a resource, then the innovative words of A. A. Miller (1956) that 'we have to learn to live within our climatic income' have become very pertinent as we move towards the twenty-first century.

2. Information and decision-making

People, firms, and governments in their pursuit of useful and profitable enterprises must make many weather- and climate-related decisions. Such decisions are made for many purposes and are based in part, if not in whole, on the information available (Maunder 1970a). Accordingly, whenever information *is* used in decision-making in connection with some weather- or climate-sensitive activity such as forecasting Australia's wool production, or the world's coffee price, or Canada's heating energy requirements, it is essential that the information be in a form appropriate to the problem and the area concerned.

The cost of providing such information is of course important; however, such costs are minimal when it is considered that there is the capability of applying the flow of information about the weather and the climate in a much more deliberate manner both to monitor and eventually control significant weather- and climate-sensitive aspects of national, regional and local economies (Sewell *et al.* 1973). For instance, the analysis of the weather and the climate of one country by another, for purposes other than international aviation and shipping – such as to forecast agricultural production in a competitor country – is a reality. Monitoring of the atmospheric resource, as well as some limited control of the atmospheric resource, are therefore of fundamental importance, and are matters which will continue to have economic, social, and political consequences at the very top level well into the twenty-first century.

In discussing the variability of weather and climate, Mason (1976) and others have noted that an increasing world population and a desire to raise living standards have increased the pressure on the natural resources of food, water, and energy. Mason correctly pointed out that the balance between the supply of and demand for such resources could be seriously affected by marginal changes in climate. He further emphasized that it is important that planning studies should recognize the variable and unpredictable nature of climate. In addition, Mason predicted that any changes in the weather and the climate may have a greater economic and social impact in the future than in the past.

National meteorological services are of course very much aware of the increasing need for accurate and timely weather and climate information to be available to decision-makers. But information about the weather and the climate is much more than tomorrow's forecast or the average January temperature. Indeed, many decision-makers have become convinced that the real-time availability of appropriate weather and climate information is of considerable benefit in providing more relevant answers to the vagaries of the 'economic climate'.

The atmospheric resource is also fundamental to the agricultural competitiveness of many countries. A key question is: How should these nations plan for the most efficient and flexible marshalling of these resources to gain maximum benefit – both for each nation and for an increasingly food-deficient world? One answer to this question is that agricultural decision-makers must be provided with *real-time* weather and climate information, appropriately 'weighted' by economic activities and areas, which is available on a time-scale which provides sufficient lead-time for decision-making. A specific example of this type of information is its use in forecasting New Zealand's dairy production. Specifically, the New Zealand Meteorological Service provides weather-based (one-, two-, and three-monthly) predictions of dairy production to the New Zealand Dairy Board, these predictions being based on the known relationship between the differences in the soil moisture and the differences in the dairy production. An explanation of this econoclimatic model is given in Chapter VI, but an essential aspect of the model is that it provides a system which transforms point-based weather and climate data to area-based economic data.

B. CHANGING DIRECTIONS IN CLIMATOLOGY

1. An overview

Although it is an old field of scientific endeavour, climatology has until recent years usually been considered an adjunct of the central concerns of meteorology. In particular, within geography, climatology is often associated with conditions near or at the 'surface'. It is also usually related to *physical* geography, despite its strong linkages to people and economic activities. In addition, those who define (or *used* to define) climate as the '*average* state of the atmosphere', clearly set it apart from atmospheric dynamics, the interest of most meteorologists. Moreover, thinking of climate in terms of averages has tacitly led to the view that atmospheric variability can be described by quasi-stationary time series. The occurrence of substantial climatic changes over the millennia was, of course, acknowledged, but the climatologist and the meteorologist had neither the data nor the conceptual tools to deal with such changes.

Public attitude and policy also tended to consider conditions averaged only over the recent past, with the result that political behaviour displayed a general obliviousness to long-term changes or the occurrence of events such as droughts in Africa. Recently, an appreciation of the real atmospheric variability has been growing in the scientific community – but has yet to be recognized in the political and economic communities – and a major effort in what might be called 'meteorological climatology' has been initiated in several areas, at both the national and international level. What follows in this section is a personal view of the changing directions in climatology – past, present, and future. Hopefully it provides an insight for those who for various reasons still think of climatology as a second-class endeavour.

2. Climate data

It is relevant to first note the changes in the use of climate data over time. Initially, most if not all climate data were used to compile 'statistics' of the climate of places. Later, these 'statistics' became part of the climate archive, and more recently have formed the basis of climate data banks. This first use of climate data is fundamental to most studies of the climate, and with the current emphasis on global climate monitoring in a near real-time, these data are essential for placing the present climate in its correct 'historical' setting. Education was the second use of climate data, as is evident from looking at any of the 'classic' textbooks on climatology published in the first four decades of this century. The education element initially relied heavily on the availability of the climate archive – which in those days was mainly monthly averages of the various climate elements.

The next use of climate data developed during the 1930s when it was realized that the climate *did* vary, and that the climate archive could be used for planning in the medium and long term; for example, in the design of bridges to withstand extreme wind gusts, and in the optimal economic use of the land/climate resource in designing cropping systems. The fourth use (and *not* – it should be emphasized – the first, second, or third use) of climate data was for research, for until the late 1960s research on climate was mostly confined to a rather traditional and mainly descriptive track.* Indeed, it was not until the World Climate Conference in 1979 that attention was focused on a 'new look' *climate* research programme. Allied with the emphasis on climate research has been a vigorous approach – in a few quarters – in the use of climate data and information in operational decision-making. Many of these applications involve the real-time monitoring of the climate and they provide essential information especially for the up-market decision-maker.

Finally, the use of climate information in impact studies should be mentioned. This use has developed rapidly during the last decade and was recognized at an international level at the World Climate Conference. Subsequently, the World Meteorological Organization (WMO) approved the World Climatic Impact Studies Programme in which the United Nations Environment Programme (UNEP) is the lead agency.

3. The past (to the 1950s)

Climatology in its earliest meaning was the study of the world's 'normal' weather conditions as varying with latitude and season. However, over the centuries climate and climatology gradually achieved a new meaning in that people learned to understand the importance of other factors. Thus climate came to mean the *synthesis* of the atmospheric conditions over a region or at a place as influenced by all environmental circumstances. By the beginning of this century, however, the words 'climate' and 'climatology', at least among meteorologists, came to imply the study of the average, or rather the most probable, weather conditions. More specifically, the climate of places was obtained as a result of a statistical treatment of the meteorological observations taken at the available observation network. But an important distinction should be made between the viewpoint of the French (and possibly Russian) schools and the English-language school. For example, Wallén (1968) commented that in the French language, and in particular among biologists, geographers and agronomists of the French academic school, the definition of climate *is* essentially 'a synthesis of the atmospheric conditions as emanating from an interaction between the air on one side and

* There were notable exceptions. For example, C. W. Thornthwaite started the Laboratory of Climatology in 1948 (at Centerton, New Jersey), and he and J. R. Mather were closely involved in basic climatological research in the 1950s. Penman in England and Budyko in the Soviet Union were also very active in basic climatological research in the 1950s.

other environmental circumstances on the other'. Wallén further noted that the French and Russian schools consider climatology as an *overall* scientific subject in which meteorology forms that part which deals with the physical processes in the atmosphere. In contrast, climatology from the viewpoint of meteorologists of the English-language school became essentially *part* of the overall subject of meteorology, defined as an approach to describe weather phenomena occurring over a period of time.

Developments in climatology during the first half of this century were highlighted by the work of several well-known climatologists. Among these developments were those pioneered by Thornthwaite (1953) who used the term 'topoclimatology' in calling for what he saw as a *new* climatology – a climatology separate from the immediate concerns of the meteorologist and rid of the idea of *average* weather. To Thornthwaite, this was a climatology focused upon the Earth's surface and concerned with the exchanges of heat, water, and momentum at the 'bottom' of the meteorologist's atmosphere. Thus, he saw an active, positive role for geographically oriented climatologists quite different from their more accepted role as 'the keepers of the records'. Thornthwaite's view of the vital role of geographical climatology is now fairly well accepted, although it must necessarily be expanded to include pertinent findings from related disciplines, as well as the more spatially and temporally extensive climatologies that have been made possible by advances in computational methods.

Another well-known climatologist of this period (and later) was H. H. Lamb, who effectively summarized the pre-1950 situation in which the *static* nature of climate was to the fore. Lamb (1982) noted that the conclusion that 'climate is essentially constant' was at odds with the acquired wisdom and experience of previous generations. He further stated:

> The assumption of constancy was, however, a convenient one for those practical operations using climatic statistics for planning. And no questions needed to be asked about which years the observations covered. It even meant that valuable statistical techniques could be developed to derive estimates of the ultimate extremes of temperature etc. and the average recurrence intervals (known as the 'return periods') of rare events. Climatology became essentially the bookkeeping branch of meteorology – no more and no less.

4. 1960–70 changes

Many changes to the climatological scene took place in the 1960–70 period particularly in regard to the traditional viewpoint of climate constancy. Indeed Smith (1961), in considering the *significance* of climate variations in their association with agricultural production, said:

> if . . . we examine recent climatic variations through the eyes of a statistician, we may be forced to conclude that none of them are 'significant'. . . . This may

> be acceptable to the pure scientist, but it could tend to infuriate the man who has experienced what *appear to him* [italics mine] to be a radical change in the circumstances in which he has to live and work.

Thus, although some climatological relationships may not be statistically significant, they may nevertheless be very important to people, plants, and animals, and the economy of an area. Moreover, in the case of most farmers, the critical weather factor is two 'bad' seasons in succession, whatever their 'statistical' significance.

Another feature of the climatological scene in the 1960s was the beginning of studies linking weather and climate activities to socio-economic activities. Indeed, with few exceptions it was not until the early 1960s that serious consideration was given to the impact of climate on socio-economic activities. Even then the idea that an economist or a sociologist could be usefully employed in a national meteorological service was far from reality. However, it is now well recognized that climate is part of, and at times contributes a great deal to, the economic and social structure of national and international activities (see Maunder 1985a).

The 1960s further saw quite dramatic changes in the way the world viewed climatology, and not surprisingly in the way climatologists viewed the world. These views are admirably summarized by Lamb (1982):

> We live in a world that is increasingly vulnerable to climatic shocks. After some decades in which it seemed that technological advance had conferred on mankind a considerable degree of immunity to the harvest failures and famines that afflicted our forefathers, population pressure and some other features of the modern world have changed the situation. In the years since about 1960, moreover, the climate has behaved less obligingly than we had become used to earlier in the century.

5. Socio-economic developments

While the 1960s brought significant changes to the climatological scene, it was not until the 1970s that there was general recognition that climate is not only a resource but can also have a profound influence on the economic and political scene. The reasons for this recognition are noteworthy: first, J. D. McQuigg wrote a PhD dissertation on *The Economic Value of Weather Information* (McQuigg 1964); at the same time the United States Weather Bureau (1964) prepared a report on *The National Research Effort on Improved Weather Designation and Prediction for Social and Economic Purposes*, and the Australian Bureau of Meteorology (1965) were responsible for a conference on *What is Weather Worth?* A World Weather Watch Planning Report (Thompson 1967) on *The Potential Economic and Associated Values of the World Weather Watch* was also written in the mid-1960s, while the author of this book wrote a PhD dissertation on assessing the significance of climatic variations on New Zealand's agricultural incomes (Maunder 1965). A further significant development occurred in 1965 when a symposium was convened on 'The Economic and Social Aspects of Weather Modification', resulting in the publication *Human Dimensions of Weather Modification* (Sewell 1966). Participants in the symposium included economists, sociologists, biologists, geographers, political scientists, planners, lawyers, meteorologists and climatologists.

This symposium was organized primarily as a result of concern over the political, legal, social, and economic aspects of weather modification. Understandably, it soon became obvious that the so-called *human dimensions* of weather modification were of even greater significance when the *unmodified* atmosphere was concerned. As a result the United States National Science Foundation in 1968 convened a symposium on the 'Human Dimensions of the Atmosphere' (see Sewell *et al.* 1968), again with participants from both the social sciences and the atmospheric sciences. In the following year the Executive Committee of WMO established a panel of experts to study the present and future role of meteorology in economic and social development. The report of this panel was published in 1974 (see Schneider, McQuigg, Means, and Klyukin 1974).

During this period *The Value of the Weather* (Maunder 1970a) was published. The viewpoint expressed in the Preface that the book 'will focus attention on an aspect of our environment which for so long has been very much neglected' is still pertinent. Indeed, this new focus of attention is shown clearly in the following 'Statement on Climatic Change' adopted by the Executive Committee of the World Meteorological Organization in 1976:

> It is important to emphasize that information regarding the impact of climate variability on human activities is essential for application in the decision-making process. The methodology to be developed for this purpose therefore should aim at making it possible to present ultimately the impact of climate variability in terms of production figures, costs, or other similar measures which can be used directly by the economists, planners and politicians.

6. The 1970s

The 1970s continued to focus attention on the socio-economic implications of weather and climate, with the UN specialized agencies being well to the forefront. In particular the mid-1970s saw the holding of World Conferences on Food, Population, and Energy, and in 1979 WMO organized the World Climate Conference. During the 1970s there were also important contributions to the 'non-scientific' aspects of meteorology and climatology – notably by the author of *The Genesis Strategy*, a critical commentary on climate and global survival (Schneider 1976), and by Dr M. H. Glantz who is one of the few social scientists seriously involved in the social aspects of weather and climate (see, for example, Glantz 1977a,b). Mention should also be made of the work in climate and socio-economics of the Canadian Climate Centre (see Phillips and McKay 1980); the Oklahoma Climatological Survey (see Eddy *et al.* 1980); the Assessment and Information Services Center (formerly the Center for Environmental Services) of the US National Oceanic and Atmospheric Administration (see CEAS 1980, 1981, 1982); the Climatic Research Unit of the University of East Anglia (see Lamb 1979, 1982, and Wigley, Huckstep, Ogilvie, and Farmer 1985) which has promoted an awareness of the effects of climatic variations on man and society; and the SCOPE volume on *Climate Impact Assessment: Studies of the Interaction of Climate and Society* (Kates, Ausubel, and Berberian 1985).

The various parts of the World Climate Programme are also closely associated with economic and social matters. However, it is all too easy for *any* programme to lose sight of its goals, and White (1982) rightly and forcefully comments:

> The World Climate Program should support a component to analyse the economic and social consequences of climatic events. The problem is difficult. Understanding such consequences takes the program into non-scientific realms. But if the World Climate Program is to be of utility to Governments and others who must make decisions, then it seems to me it must provide information about social and economic consequences. It has to answer the question 'So what?'

7. The 1980s

The increased activities and concerns of the 1970s continued unabated into the 1980s. A typical national response was in Canada where the Canadian Climate Centre of the Atmospheric Environment Service was designated as the lead agency for the Canadian Climate Program. One of the prime aims of the Canadian Centre is to allow for a much closer association between traditional climatology (e.g. data management, quality control, archiving services), application climatology (e.g. applications, impact studies, hydrometeorology), weather and climate monitoring, and weather and climate prediction.

At the Ninth Session of the WMO Commission for Climatology Mr D. W. Phillips presented a lecture on 'Developing a National Climate Program – a decade of Canadian experience' (see Phillips 1985). In his lecture, Mr Phillips noted that Canada has developed a National Climate Program (CCP) to meet the needs of Canada for information and advice on climate and its variations. When fully developed, the programme will allow for a greatly enhanced climate data and prediction service and the development of better methods of applying climate data and information to activities in the economic/social and resource sectors. The programme will also allow for a better understanding of climate and its interaction with the oceans and the possibility of climate change caused by increasing amounts of CO_2 and other atmospheric contaminants. It will also allow studies to be made of the impact such change might have on Canada's social and economic activities. Mr Phillips said that although the programme is being co-ordinated by the Atmospheric Environment Service several other Services and Departments are actively involved. Many of the provinces are also participating since many areas of provincial responsibility are markedly dependent on climate. In addition, many non-government groups in universities, private industry, professional associations and individuals with common interests in climate have expressed interest in the programme. A key feature of the Canadian Climate Program is the active participation of a Canadian Climate Board which serves as a steering and overall policy and priorities body.

There was also considerable concern over the future impact of the increasing amount of carbon dioxide (CO_2) in the atmosphere (see also Chapter VIII), and in discussing this problem, Hare (1980) noted – in regard to Canada – that the CO_2 changes do not imply disaster or great disruption for the Canadian economy, but they do argue for readiness, and for a determination to make good use of them. Hare noted that, if this is done, Canada may on balance profit from the predicted increase in CO_2 in the atmosphere.

Other developments in the early 1980s were the much greater appreciation of the importance of weather and climate information in decision-making, the 'operational' activities associated with the World Climate Programme, and the greater awareness of the need to monitor the weather and the climate on both a daily and historical time-frame basis. These developments are discussed in Chapter II. In addition, there has been considerable activity (and concern) over the climatic and environmental consequences of acid rain, 'greenhouse' gases, and nuclear war. These important subjects are examined in Chapter VIII.

8. Directions for the future

One of the goals for climatology, identified by several national planning authorities, is the development of a climate forecasting capability. Indeed, although 'useful' weather or climate forecasts months or seasons in advance may not be feasible in the next decade, it is clear that any work leading to an improved understanding of the physical processes governing climate, or the role of surface or human influences in climatic variability, will contribute significantly to improved climate forecasts. To achieve these desired advances, it is clear that an improved data collection and analysis system is needed, together with better quantitative models, and a wide range of applied climate studies. All three developments are necessary and must proceed hand in hand, for models are of little use without adequate input data, and the applied studies are needed to clarify situation-specific cause-and-effect and feedback relations that must be included in models.

With respect to the economic, social, political and strategic aspects of climatology, there has been considerable progress during the last two decades. However, there is still a reluctance in many national meteorological services to recognize fully the importance of these aspects. There is also a reluctance by many scientists in the meteorological profession to appreciate that most, if not all, of our 'political masters' need to be convinced of the benefits to society of weather and climate information, if governments are to continue to finance most observational and monitoring programmes, as well as many research and forecasting activities.

Until very recently most national meteorological services were staffed at the professional level with physicists and mathematicians, and most of the activities of such services were concerned with short-range weather forecasting – particularly for aviation – and the compilation of traditional climatological summaries. During the 1970s some changes to this traditional viewpoint were made – and it is appropriate to note the imaginative thinking of many people in the WMO Secretariat since the early 1970s, such thinking being clearly reflected in many of the programmes of WMO.

The close connection between the difficulties involved in making weather forecasts and the difficulties involved in making economic, social, or political forecasts is also worth emphasizing. Indeed, such connections point clearly to the even more difficult problem when one tries to link the meteorological and climatological system *with* the economic, political, and social system. It can of course be done and, in my belief, it *must* be done if meteorology and climatology are to have any real political and economic impact in the future. It is therefore evident that climate studies of the future must not be confined to the analysis of atmospheric data. Rather, such studies must consider the human problems of food, health, energy, and well-being. The economic, social, human, and political implications of the climatic understanding developed, must of course be applied at all levels – local, national, regional, and international. To the degree that this is done, meteorologists and climatologists will make a positive contribution to decision-making in both developing and developed countries, whatever their political systems.

C. WEATHER AND CLIMATE: IS THERE A DIFFERENCE?

1. Changing viewpoints

Television, radio, newspaper, and business journal reports during the 1970s made many people – including politicians – aware of a succession of climatic vicissitudes that had worldwide economic, political, and social repercussions. For example, the Sahel drought of 1970–5 brought widespread famine, and the freezing weather and drought in the Soviet Union in 1972 led to large grain purchases abroad. The 1970s was also a period in which climatology moved to the forefront as a significant scientific discipline.

These developments point to the changing viewpoint of what in the English language is meant by 'weather' and what is meant by 'climate'. Traditionally, in terms of time, 'weather' has been regarded as the events that happen over a short period – usually a few hours to a few days, whereas 'climate' has been regarded as the events that happen (or happened) over long periods – usually years to decades to centuries. Unfortunately, the English language does not have a word which describes the important middle time period – that is weeks to months, and perhaps the English language should adopt the German word *Witterung*, or the Japanese word *tenkoo*, which are basically concerned with the weather *and* the climate of the recent past. Politically of course, it is the recent past (and the near future) that are most important, hence it is not surprising that the Assessment and Information Services Center of the US National Oceanic and Atmospheric Administration (NOAA) issues 'climate' impact assessments on a *weekly* basis. Some would argue that they are really 'weather' impact assessments, but the important feature of these assessments is that they discuss an important period of time as far as the political, economic, and social life of the United States is concerned. The word 'weather' should still of course be used to describe what happened over short periods of time – but the time periods generally accepted in this connection are now much more of the order of days, rather than weeks or months. Consequently, the word 'climate' is now often used to describe atmospheric conditions over quite short periods, and much shorter periods than that which was considered appropriate ten or twenty years ago. The multi-use of both 'weather' and 'climate' is also more common and this is perhaps the most appropriate solution, in the English language, to the 'atmospheric time-language problem'.

There is of course still an 'average climate' misconception. Indeed, although many informed decision-makers are very much aware of the significance of weather and climate variations on production, marketing, and consumption, others go from one period to the next with an almost complete disregard of the effects of both weather and climate variations on their activities. Clearly this disregard is to the detriment of both communities and nations.

2. Information for decision-makers

The *real* economic and political value of weather and climate information is of course only realized when it affects the way decisions are made. A continuing need exists, therefore, to assess and present the impact of weather and climate to decision-makers in terms which *they* can understand.

To assist such decision-makers, national meteorological services and several private meteorological companies prepare and disseminate a wide variety of weather and climate forecasts – ranging in time from a few hours to a few months.* In addition, a few national meteorological services and a few meteorological consultants prepare and disseminate specially evaluated commodity-weighted weather and climate information in real-time.

Irrespective of how people define weather and how people define climate, it is evident that the overall impact of *short-term* weather and climate variations will continue to increase in importance. Therefore, paramount to any discussion of weather and climate variations is an understanding of the increasing impact such variations will have on communities and nations. Despite this, most weather- and climate-related activities have until recently been planned as if the climate was stable. But during the 1972–5 period, the world was made very much aware of the catastrophic drought in the Sahel area** of Africa, with all its political and social implications. A decade later, similar weather and climate impacts are still occurring, as is evident by the disastrous drought in Ethiopia in 1983–5, as well as the impact of the 'El Niño/Southern Oscillation' in many parts of the world in 1982–3 (see Hare 1985b).

Clearly, to provide a better service for decision-makers it is necessary to identify those activities directly or indirectly affected by the weather and the climate. National and consulting meteorological services must also provide decision-makers with 'real-time' weather and climate information, appropriately weighted by economic activities and areas, and this information needs to be available on a time-scale which provides sufficient lead-time for decision-making.

Is there, therefore, a difference between weather and climate? In a real-time sense the answer must be no; it is also evident that the time-frame is becoming less important as the ability to monitor *all* aspects of the atmospheric environment in real-time becomes a viable and economic

* For example, the Swedish Meteorological and Hydrological Institute (SMHI) is developing a new meteorological information system called PROMIS-90, which covers the whole spectrum of predictions from very short-range to medium-range. In a paper discussing some aspects of PROMIS-90, Liljas (1984) notes that in order to establish a cost-effective and developing weather service the Swedish weather service has to work, both operationally and with developmental tasks, after the principle that 'weather information has no value until it is used with success in weather sensitive plannings and decisions'. Liljas (1984) further comments that the requirements for very short-range forecasts are increasing, but that high costs are involved in the development of a meteorological information system for the whole spectrum of predictions from the very short-range to the medium range. However, Liljas also correctly emphasizes the gap that exists between the large technical and economic efforts that are devoted to the collection and processing of data, and our lack of knowledge of what and how the weather-dependent society would really like to know.

** The drought in the Sahel area was particularly severe in 1972–3, and also in 1983–4, but the rainfall was at least 25 per cent below average for the whole 1972–84 period.

reality. In this regard it is relevant to note two of the decisions of the Ninth Session of the WMO Commission for Climatology held in Geneva in December 1985. First, in terms of monitoring the weather and the climate in real-time, the Commission for Climatology endorsed the re-affirmation by the WMO Commission for Basic Systems (at their extraordinary session held in Hamburg in October 1985) of the principle of the 'free' exchange of World Weather Watch (WWW) data; second, the Commission for Climatology also noted that many data requirements for climate application and impact studies are met by the operational exchange of data through the WWW system.

D. THE REAL ECONOMIC CLIMATE

1. Area, time, and business aspects

The phrase 'economic climate' has recently become very fashionable amongst politicians, economists, company directors and the news media. But, what is the true 'economic climate' of a nation? Webster's dictionary gives one definition of climate as: 'the prevailing temper or environmental conditions characterising a group or period'. This definition suggests that the 'true' economic climate not only involves consideration of the whole political, social, and economic life of a nation, but also environmental factors including the weather and the climate.

However, although a glance at almost any newspaper indicates the importance of weather conditions to the world of commerce,* few substantial studies have been completed on the specific effects of weather on the business community. Indeed it seems almost paradoxical that the value of information about the weather and the climate to business and commercial activities appears to be ignored not only by the majority of meteorologists and climatologists, but also, rather surprisingly, by the businessmen themselves. There are exceptions, of course, and the relatively flourishing consulting activities performed by meteorologists in the United States and a

* For example, the following quotation from the *Financial Times* of 14 December 1985 highlights the commercial impact of weather and climate:

> Coffee prices have been rising virtually without interruption all week on the London and New York futures markets in anticipation of severe drought damage next year to the Brazilian crop, which normally accounts for about 30 per cent of world exports. The London robusta futures market closed last night at £2,148 per tonne, £99 up on the week and its highest level for six months. . . . Although consumers are well stocked at present and there is no immediate shortage of coffee, traders are clearly expecting considerable tightness in supplies next summer, as the Brazilian crop is harvested. . . . The crop is expected to be about half of this year's 30m bags. Yesterday, the Brazilian authorities underlined the seriousness of the situation by declaring a state of emergency in the two worst-hit states, Rio Grande do Sul and Santa Catarina.

few other countries are evidence that some business enterprises are willing to pay good money to know more about a particular aspect of the economy that influences their business, namely the weather and the climate and its variations.

The value of weather and climate to any area – small or large – requires the identification of activities in that area that are affected directly or indirectly by changes in the weather and the climate, and an analysis of the manner in which a given change results in gains or losses to such activities. The problems of evaluating weather in an area have been considered by a number of investigators, including, for example, in the 'pioneering' 1966–73 period, McQuigg and Thompson (1966) on natural gas consumption, Maunder (1968a) on agricultural incomes, and Maunder, Johnson, and McQuigg (1971a and b) on road construction. In most cases, however, these and most later investigations have either been spatially restricted or they have been related to 'long periods' of at least a month. Such studies are not therefore designed to contribute to the more general question: 'What is the value of the weather to a large area (such as the United States, France, or Argentina) over a short period of time (such as a week)?'

However, in an analysis by the author on retail trade (see Maunder 1973a), a large area and a short time period were used in a study of United States weekly retail trade. These factors of a 'large area' and a 'short time period' were chosen deliberately for two reasons: first, because variations in nationwide economic activities are important to various high-level decision-makers, and, second, because only weather over a short period such as a week has any real practical meaning to the millions of low-level decision-makers who go about their day-to-day and week-to-week activities. As previously noted, it may be argued that there is little point in such an analysis, since any relationships that might be derived cannot be used because accurate weekly weather forecasts are not available. However, it must be assumed that such forecasts *will* become available, and that decisions based on such forecasts will be made. For example, five-day forecasts are issued on a regular basis in the United States – among other countries – and are widely distributed by the news media. It is reasonable to assume therefore that in the United States both the news media and the National Weather Service are serious about the news-worthiness and value of such forecasts, including the use of such forecasts in decision-making. Of course, it does not necessarily follow that the decisions in the world of commerce, are, in fact, based on the effect that the present and future weather has and will have on business. However, if business (including business on a national scale) chooses to ignore the weather and the climate, as well as the forecasts of the weather *and* the climate that are available, should it not be part of the meteorologist's responsibility to educate the business community to better utilize such forecasts? Furthermore, if meteorologists are *really* serious about predicting the weather and the climate, then one justification for even more effort in this direction is the potential *use* of this information by a whole host of producers, and consumers, from all areas of the community.

Traditionally, national meteorological services have observed, collected, and processed synoptic weather information in real-time. Recently, in a few countries including the United States and New Zealand, such information has also been analysed in an *applied* sense in real-time, so that key econoclimatic indices are now available – not only regionally and nationally, but also, if required, on an international basis. Thus 'climate productivity indices' and 'international weather indices' for key commodities are a reality. Furthermore, the availability of such information through videotex type systems requires an awareness by both meteorologists and climatologists of the marketing potential of their services, and also by the decision-makers who can use such information.

The need for both accurate and timely weather and climate information to be available to decision-makers is well established. But information about the weather and the climate is much more than tomorrow's forecast; indeed, many decision-makers now consider that appropriate weather and climate information is of considerable benefit to them in providing more relevant answers to questions relating to the vagaries of the 'economic climate'. But not all decision-makers are convinced, and in this regard the following observation is relevant: 'our challenge is to produce practical information that can be readily understood and integrated in a smooth and timely fashion into the planning process'.

These observations by McKay (1979) are very pertinent, and underline the difficulty facing the applied climatologist in convincing decision-makers of the *real* worth of their 'product'. McKay noted further that closer involvement with the user is essential to ensure viability, relevance, and real benefits from new information. Further, he emphasized that users want improved consultation on how and what information is provided, and they want the institutional barriers removed that tend to block communication and the use of climatic information in planning and investment.

These comments *are* important, and meteorologists and climatologists in a number of countries have in fact been to the forefront in providing this kind of useful information. However, the atmospheric scientist still faces a challenge both in educating users in the potential value of weather and climate information, and in the marketing of such information.

2. Information, economics, and politics

It is important that the meteorological profession breach the dams of ignorance, indifference, and inefficiency which unfortunately prevail in analyses of many of the applied aspects of the subject. They must also assist the decision-maker to come to terms with both the political *and* the atmospheric economic climate. In this connection it is relevant to highlight the development during the last 10 to 15 years of a new field of meteorology/climatology, namely *socio-economic climatology*. Many regard it with more than a degree of scepticism and this is ably demonstrated in the following comment by Sir John Mason, who at the time was Director-General of the British Meteorological Office.

> Meteorologists and climatologists need, and should welcome, the help of economists and others in assessing the likely economic and social impact of climatic changes but this report of a task force convened by the International Institute of Applied Systems Analysis (IIASA) in Vienna does not inspire confidence.

These remarks (see Mason 1981) relate to a book, *Climatic Constraints and Human Activities* (Ausubel and Biswas 1980), and both the book and Mason's critical analysis (see also Chapter IV) provide much food for thought. But the *real* economic climate is important; indeed it is pertinent to ask why the comments of an eminent meteorologist like Sir John Mason on economic aspects of meteorology are often met with considerable scepticism – interestingly for a variety of reasons – by some sections of the meteorological community, yet the same eminent meteorologist's writing on cloud physics, numerical modelling, etc. are highly respected. Perhaps the answer lies in the fact, demonstrated frequently in 'letters to the editor', that we all think it is easy to do a better job than a social and political scientist, or an economist, but much more difficult to do a better job than a physical scientist.

The 'other side of the meteorological fence' can in fact be quite different. Nevertheless, the relatively few exponents of socio-economic climatology can take heart in that their influence is far greater than their numbers would indicate. For example, it is very evident from the reading of any submission for funds (to governments, granting agencies, etc.), or any annual report of activities, that the economic, social, strategic, and political justification of meteorology and climatology is today always given prominence. The reasons for this emphasis are many, including the influence of socio-economic climatologists, but in addition the 1970s and the 1980s have clearly brought a variety of economic and social trends which increasingly demand more rational and effective responses to atmospheric events. The emergence of more sophisticated and often more weather-sensitive technologies was another major contributing factor in that they all called for a greater understanding of the atmosphere and of the interrelationships between the land, climate, water and human resources. For example, *Business Week* (13 October 1980) showed clearly that the weather and the climate cannot be taken for granted. Under the headline 'Food Prices May Actually Damp September's Consumer Price Index', it stated:

> The Bureau of Labor Statistics weekly spot commodity index for nine foodstuffs fell 9.4% between mid-August and mid-September. A big factor was a drop in meat and livestock prices, as the summer drought and soaring feed prices induced U.S. farmers to slaughter their livestock sooner than usual.

It is not the task of this book to belittle economists but it is essential that applied climatologists provide guidance to economic forecasting groups. Much of the necessary thinking for this guidance involves disciplinary areas peripheral to the normal activities of climatologists as well as economists, but the problems are real, and in many cases it is the 'atmospheric' component of an economy that is most often ignored. A key factor in providing this guidance is the availability of weather and climate data in real-time or near real-time. Indeed the meteorological global telecommunication system not only allows climate data to be collected in real-time, but also permits a conversion of these data – also in real-time – into at least provisional estimates of economic and social impacts.

One specific aspect of these new products – which is discussed in detail in Chapter VII – is their use in weather-adjusting national economic indicators, so that the true economic climate of a nation may be ascertained (Maunder 1982a). It is shown, for example, that it is possible to 'adjust' the *Business Week* 'Index of United States Weekly Economic Activity' (which uses electricity consumption as one of several economic indicators) so that it more truly reflects economic strength rather than environmental 'strength'. Similarly, it should be possible to adjust consumer price indexes so that they more clearly reflect the influence of weather on prices.

The challenge facing the weather forecaster, the weather services expert, and the climatologist, is to produce practical information that can be readily understood and integrated in a smooth and timely fashion into the operation and planning process. The chances of success in this regard are improved when the operation and planning process is understood – they are much improved when the user is convinced and involved, and also *pays* for the information.

3. Weather and climate and human well-being

The journal *Newsweek*, in a major story in its 9 May 1983 issue on 'Africa: The Drought this Time', quotes a district administrator in Ethiopia: 'We have no seeds because we have already eaten them. . . . If the rain falls, people will need seed and oxen, but they don't have either. . . . Worst of all, of course, they don't have rain.' Then, in an FAO Special Report on *Foodcrops and Shortages* issued on 15 April 1983 it is stated: 'in Asia, wheat crop prospects in Bangladesh, India and Pakistan remain good. . . . In six West African countries crop failures towards the end of 1982 due to drought have resulted in a tight domestic food supply situation. . . . Recent rains in

Australia may have broken the drought, this improving prospects for planting the 1983 wheat and coarse grain crops.'

Such quotations clearly point to *one* significant aspect of climate and socio-economics, namely the dependence of *people* on favourable weather and climate conditions. But there is of course much more to the subject than these happenings, however distressful they are to the people and governments concerned. For example, although last week's weather is history, the measurement of its economic and social impacts – using conventional economic techniques – is not known until several weeks or months after the event. But, as all meteorologists are aware, the meteorological global telecommunication system not only allows weather and climate data to be collected in real-time, but also permits a conversion of this data – also in real time – into at least provisional estimates of economic and social impacts. Indeed it is quite feasible using appropriate weather and climate data to make a forecast of economic and social *data*, which at the time of making the forecast do *not* exist. This is one reason why the real-time climate and socio-economic mix is so important, a reason which has significant political overtones.

In contrast, such real-time aspects of the weather and climate were probably not considered at all when Huntington many decades ago wrote on climatic determinism. At least in the English language Huntington was the best-known advocate of a causal relationship between climate and human inventiveness and his views are set forth most completely in his last book *Mainsprings of Civilization* (Huntington 1945). Very much earlier – indeed as early as three or four thousand years ago – the various cultures of the day were aware of the significance of weather and climate events. However, with few exceptions it was not until well into this century – notably the early 1960s – that serious consideration was given to the impact of climate on socio-economic activities. However, it is now well recognized that climate *is* part of, and at times contributes a great deal to, the economic and social structure of national and international activities. Accordingly, it is no longer a surprise to find economists, sociologists, and marketing experts not only employed by national meteorological services, but also highly respected for adding that 'extra dimension' to the meteorological and climatological scene.

E. Weather and climate: a serious game

A survey of the meteorological and climatological literature over the past few decades shows that many facets of the atmospheric environment have been examined. These cover a wide range, but only a fraction of these publications is concerned with what the public considers to be weather. Within each facet there is a variety of specialists and generalists, and although there is some cross-fertilization, most meteorologists and climatologists – if they do not become administrators – tend to become specialists.

The various aspects of what the author called 'The Weather Game' are described in a Presidential Address to the Meteorological Society of New Zealand (see Maunder 1983).

This address looked at many different aspects of meteorology and climatology (most of which are discussed in other sections of this book) and each aspect was considered in terms such as its overall importance, application, impact, political significance, and possible role in the international scene. It was noted that each of the topics discussed could be viewed in the form of a game – with its specific rules, protocol, history, adherents, and sceptics. But like all real games, one must play the game to achieve satisfaction, and one must also be part of the 'administration' of the game (or otherwise be very influential) if one hopes to change any of the rules by which the game is governed.

One specific aspect of the 'weather/climate' game is the influence of the specialists who in most cases do not cross 'boundary fences'. For example, in October 1985 a conference to consider the CO_2/climate issue was held in Villach, Austria. (See also Chapter VIII, p. 312.) This joint UNEP/WMO/ICSU assessment considered five aspects of the CO_2/climate issue, namely (1) the rate of release of CO_2 into the atmosphere as a function of future energy developments, (2) the carbon cycle and future concentration of atmospheric CO_2, (3) the climate response to increase atmospheric CO_2 concentrations, (4) the impact of climate change on terrestrial ecosystems, and (5) the impact of climate change on society and the management of social and political responses to climate change. To assist the conference, experts on each of these five aspects of the CO_2/climate issue were asked to comment on the draft reports. In this it is interesting to note the widely varying number of people involved in each of the advisory groups which reflects the specialist interests of meteorologists and climatologists. For example, there were forty-one experts involved in aspect (3) on the climatic response to increased atmospheric CO_2 concentrations, but only twelve people involved in aspect (5) on the impact of climate change on society.

Much earlier than the 1985 Villach conference, there was also concern about other 'pollution' issues. For example, Professor Fred Hoyle, the internationally acknowledged astronomer and cosmologist, in discussing the 'way ahead' at a lecture in 1970 at Victoria University of Wellington, New Zealand, said that it 'must lie in adapting ourselves to the environment, not in trying to adapt the environment to whatever fancy we have at the moment'. Clearly, the atmosphere is part of this environment, and pollution of the environment has caused and is causing serious concern in most parts of the world. The pollution problem has, however, been brought about largely because of man's ignorance – notably of the fact that the atmosphere is *not* an infinite reservoir but a finite resource which, like all resources, can be misused.

The value of meteorological and climatological information has been recognized for some time as an essential ingredient to the weather/climate game, as has the need to consider the urgent problems concerned with considering the atmosphere as a resource. But, much more could and should be done in this regard, and this book endeavours to provide the vision and lateral thinking that is (and will be) necessary if meteorology and climatology are to become really involved in the ultimate 'uncertainty business'.

II The institutional framework

A. HUMAN RESPONSES TO THE ATMOSPHERIC RESOURCE

1. The media speak

There is an increasing awareness that information about the weather and the climate can play an important role in the decision-making processes associated with management of weather-sensitive activities and enterprises. This includes national and international planning. For example, the US Conference Board magazine *Across the Board* (June 1977), in a cover page article 'The uncertain earth', stated: 'Official concern is understandable. As many of us had to relearn last winter, our way of life is predicted on the favors of the weather. . . . By one estimate, there is about $270 billion's worth of weather-sensitive industry in this country alone.' Similarly, *Business Week* (2 August 1976), in a report under 'Economics', stated that 'while there is not enough evidence to conclude that the world's climate is deteriorating, it is clear that climate has become more variable'.

Similar media comments could be provided from almost all business journals and for almost every week since the early 1970s, clear evidence that the world is *not* becoming less dependent on the weather and climate as people and nations become more 'developed'. Indeed the comment in *Time* (31 August 1981) on 'trouble down on the farm' although referring specifically to the Soviet Union, could very easily have been applied to other nations and for other periods. *Time* in fact – in their comment – neatly summarizes the often precarious position that people find themselves in when trying to live within their climatic income.

> On the gently rolling plains of southern Russia and the Ukraine, stunted stalks of wheat and corn lay flat on the rich black earth, blighted by drought and wind. In the lower Volga region, rain mercilessly pelted burgeoning grain; harvesting combines stood idle as farmers watched the crop sink into the mud. The forecast is bleak this summer in the *kolkhozy* (collective farms) and *sovkhozy* (state farms) of the Soviet grain belt, where capricious weather has caused a third consecutive bad harvest – with an anticipated shortfall of 51 million metric tons in Soviet grain production.

As is evident from reading 'between the lines' of such statements, decision-makers using weather and climate information are now found at the highest levels in companies and government departments, as well as at the ministerial level in governments. It is also evident that several specialized intelligence agencies are also involved in decision-making which involves the use of weather and climate information. But, while the state of all national economies is dependent to some extent on next year's weather and climate, the question remains as to how many top-level decision-makers take the weather and climate and the expected weather and climate into account? Further, would more top-level decision-makers take the weather and climate and expected weather and climate into account if there were more direct advisory communication between their advisers and the meteorological system?

2. Atmospheric interactions and decision-making

Until recently, much of applied meteorology and applied climatology was concerned solely with the interactions between the atmosphere and physical or biological processes. For example, there has been considerable work on such processes as the flux of water through the layers of the soil within which crops are grown, the effect of temperature and humidity on animals, and the forces exerted by the wind on buildings or aircraft. But it is clear that a much more comprehensive concept of applied meteorology and climatology must be used, including the interactions between the atmosphere and people as decision-makers, since there are clearly situations in which the impact of weather and climate on physical or biological processes is influenced *both* by the weather and climate events that occur, and by the choice of alternatives made by people.

In many areas, the emergence of more sophisticated and often more weather- and climate-sensitive systems has created a need for more rational and effective responses to atmospheric events. This need for better and more useful information about the atmosphere has come about as a result of two important concepts developed from the ideas of McQuigg (1964). They are: (1) that *information* concerning the atmosphere, such as the past weather and climate, the present weather, and the future weather and climate, is an important resource, and (2) that given an understanding of physical-biological-sociological interactions with the atmosphere, and given sufficient *information* about atmospheric events, people can at times use their management ability to improve the economic and social outcome of many weather-sensitive activities.

There is an increasing awareness that *information* about the weather and the climate can play a very important role in the decision-making processes associated with the management of weather- and climate-sensitive enterprises. Included in these enterprises are aspects of national economies concerned with productivity, such as the *total* retail trade, *total* rice production, and *total* electric power consumption, which are the concern of high-level decision-makers such as ministers of national governments, or directors of national companies. The question of whether a useful relationship can be obtained between weather and climate indicators and national economic activities over national space scales is valid; nevertheless, the continued publication of weekly and monthly indicators of economic activity in business journals such as *Business Week* and *The Economist* is an indication that high-level decision-makers are interested in this type of information. Further since some decisions are undoubtedly based on this information, it appears useful to incorporate into the decision-making process the factors which are associated with the weather and the climate.

B. SPECIFIC NATIONAL RESPONSES: THE ROLE OF THE PUBLIC, EDUCATION, AND PRIVATE SECTORS

1. National business and government

When the economic planning processes in many countries are examined, it is evident that many top-level decision-makers (in both the government and private sectors) appear to be unaware of the potential value of weather information (past, present and future). Important questions arise from these matters, notably: (1) what is the role of the meteorologist in national economic planning and in advising top-level decision-makers? (2) what is the function of a government meteorologist or national meteorological service? and (3) what is the function of a consulting (or private) meteorologist or company in these processes?

It is obvious that the answers to these questions will vary from country to country, particularly where both government and private meteorologists are active. For example, in the United States, more than 400 professional meteorologists and climatologists are employed in over eighty consulting meteorological companies. The role of the 'public service' meteorologist/climatologist in the United States may therefore be quite different from that in most countries where 'private' meteorologists/climatologists are few or non-existent. The situation in the United States is discussed by Epstein (1976) who, speaking for the National Oceanographic and Atmospheric Administration (NOAA), stated: 'The industrial meteorologist represents an indispensable link in our ability to bring the best meteorological services to the nation. We have important jobs to do, and we do them best when we work together.'

2. The New Zealand approach

The functions of the New Zealand Meteorological Service are set out in the *Civil Aviation Act, 1964*, which states that the Service is: (1) to provide a meteorological service for the benefit of all sections of the community; (2) to promote the advancement of the science of meteorology; and (3) to advise the Minister and government departments on all matters relating to meteorology.

The question of the role of a government (or public) servant in fulfilling the functions stated in such an Act is a little less clear, but to at least one permanent head of a government department in New Zealand (the then Secretary of Defence, J. F. Robertson, as quoted in the *Evening Post*, 25 September 1976) the answer is in no doubt. He stated:

> The public servant sees his role clearly as part of a department which is the instrument by which the Minister gives effect to his Government's policy. A department or a Ministry has no other form than as an instrument of Government to implement Government policies.

There is a further important question which asks what are the obligations of the meteorologist in such matters. This is also not entirely evident from the Act, but at least New Zealand's Minister of Science in 1976 (the Hon. L. W. Gandar) was quite clear regarding the obligations of scientists, when in opening the National (New Zealand) Physics Conference he stated:

> Scientists have an obligation to keep the public informed of new developments in research and their possible impact on daily lives. . . . We, and I mean here the public, scientists, and politicians can choose well only if we are well informed about scientific research and we are made aware of the implications in its applications.

It may be inferred, therefore, from the statement by the Hon. L. W. Gandar and the implications of the official role of the New Zealand Meteorological Service, as stated in the *Civil Aviation Act*, that meteorologists (whether government or private) should become more involved in advising top-level decision-makers and assisting in national economic planning. Clearly, the opinion in many countries today is that this should happen, for the alternative is that weather and climate information will be considered only in an 'ivory tower' atmosphere, or left to be dispensed by weather 'entertainers' on television, by 'disc jockeys' on commercial radio, or by journalists. Clearly, weather and climate information is far too valuable to be treated in *only* those ways.

At the party political level in New Zealand, there has, as far as I can ascertain, never been any official policy on meteorology and/or climatology. For example, little if any reference was made to the 'atmospheric economic climate' in the 1981 election manifestos of either the National, Labour, or Social Credit parties, or the manifestos of any of the four major parties in the snap election in July 1984. However, despite these shortcomings (or blessings!) the Director of the New Zealand Meteorological Service (J. S. Hickman) made some very pertinent remarks in an editorial in the August 1982 issue of *Weather and Climate* which in the absence of any official government policy or party political policy are clearly worth emphasizing. He stated:

> We have a previously undreamt of opportunity to develop the . . . interface between the Meteorological Service and the user of meteorological information. . . . Whoever develops this interface may well control the market place for weather information.

3. The United States approach

In contrast to the situation in New Zealand, the *official* concern of the United States in regard to climate matters is noteworthy. For example, during 1977, the US Interdepartment Committee for Atmospheric Sciences issued a comprehensive *United States Climate Program Plan* to guide the response of the United States to the growing number of climate-induced problems and to anticipate the national and international effects of fluctuations in climate. The resulting National Climate Program Act was signed by the President of the United States in 1978.

Regrettably, the National Climate Program that came into being as a result of this Act has to many people been much less impressive than the Act may have led people to believe. A serious lack of funds is one reason for the less than modest programme that has been carried out so far. Other reasons include the very tardy approach to *state* climatological programmes by a majority of the fifty-one states. In addition, the emphasis of the United States Government on activities such as the World Climate Programme of WMO has tended to focus attention on the larger global, strategic and economic issues which result from the 'nation-atmosphere' interface, rather than local issues.

Nevertheless, in some specific areas progress and developments have been impressive. For example, at the top level in business and government in the United States there is increasing awareness of the value of weather and climate information. Specifically, the US Department of Commerce, through its National Oceanic and Atmospheric Administration (NOAA), established in 1978 a Center for Climatic and Environmental Assessment (CCEA). This Center was an organizational response to two questions: first, how does the United States extend her abilities for observing and predicting short-term weather events to predicting longer-term fluctuations; second, how does the United States put the knowledge so gained to work in ways that are useful in terms of crop production estimates, resources management, energy utilization, etc. The establishment of this new Center permitted the availability of a family of products not previously obtainable, notably, early warnings and assessments of crop yields, global weather briefings linking climate and socio-economic systems, and assessments of the risk of damage to national resources by climate variations.

4. The universities

The following comment appears in a brochure outlining the aims of the United Nations University:

> It is often assumed that once information is available it naturally flows to useful outlets: in fact the reverse is much more frequently true – much of the world's knowledge lies unused behind dams of ignorance, indifference, and inefficiency.

As is stated, the United Nations University seeks to breach these dams and channel knowledge where it is needed. Such aims are laudable, and clearly the universities have a very important function in the higher education aspects of meteorology and climatology. Regrettably, however, the status of meteorology and climatology in many universities is rather low, and in many cases non-existent, despite the excellent work done by many meteorologists, climatologists and associated academics in non-atmospheric science departments, in the field of meteorology and climatology during the last thirty to forty years.

For example, New Zealand – along with many other countries – is still awaiting a university department of meteorology and/or climatology, and of even greater concern in New Zealand is the reluctance of the two agricultural universities to give any *real* recognition to the role that the meteorological and climatological discipline could play in their educational environment. Indeed the disparity between the university situation in some areas of the United States where the importance of meteorology and climatology is clearly recognized and appreciated, and the New Zealand situation – where it is not appreciated – can only be deplored. Such a disparity, of course, is not restricted to a United States/New Zealand comparison; but it would be fair comment that the United States is one of very few countries in which the atmospheric sciences *and* their social, economic, and environmental connections have been *fully* recognized by university administrators.

A further matter of concern is that many national meteorological services continue to appoint university-trained staff who are exclusively qualified in mathematics and/or physics, but give few opportunities to those equally highly qualified applicants in engineering, statistics, economics, agriculture, resource management, or geography. Some pertinent comments relating to this situation were made by Bernard (1976) who recommended the appointment to meteorological services of personnel who have achieved an advanced level of expertise specially orientated towards the applications of meteorology to development, such as agronomical engineers, hydrologists, or geographers, and who have undergone additional meteorological training following their basic university education. Indeed, as Bernard rightly commented: 'The purely physical and mathematical approach of conventional meteorologists results in their being too impervious to the scientific and technical applications of meteorology for socio-economic progress.'

Fortunately, in some countries the suggestions of Bernard and others are being acted upon and, with the co-operation of the universities and the more enlightened outlook of many national meteorological services, the situation is now much improved.

C. THE INTERNATIONAL SCENE

1. International responses and concerns

During 1974 three important world conferences took place: the World Population Conference, the World Energy Conference, and the World Food Conference. In these a start was made to study realistically the fundamental changes that are needed to ensure the optimum use of the world's resources to meet the demands of the future. These conferences were brought about because of a number of factors, chief of which was the growing concern that there are physical limits to the availability of resources. Indeed it was during 1973 – one year before the UN Conferences were held – that the world first began to receive sharp warnings of some of these limits, particularly concerning the supply of energy and of food, and it became apparent that our ability to make efficient and humane use of the world's resources through existing institutions was more limited than we had thought.

Included among these resources are those of the atmosphere and those of food; indeed the production and availability of food is perhaps man's most pressing weather-sensitive activity. Further, with the available world reserves of some crops at all-time lows – when related to the *need* for food and the price many people can afford to pay – those weather and climate variations associated with food production are becoming much more important. This variability can be evaluated in ways which are useful to planning and management and, in regard to food production, this knowledge must be used for planning intelligently national and world food security programmes if these are to be successful.

In considering the atmosphere as a resource, and in particular as an élite resource, there are three important factors to be considered: first, monitoring and understanding its variability; secondly, predicting its variability; and thirdly, assessing the impact on consumption and production of this variability. Furthermore, because the *impact* of short-term variations in the available atmospheric resources can expect to continue to increase in importance because of the growing demands for food and energy, any change in the climate for the worse could well result in food shortages and energy supply problems that will be far more critical in some areas than will occur if the climate remains *normal*.

Appropriate international meteorological planning must therefore be evolved if we are to live within the limit of our atmospheric resources. But, if this is to be accomplished, the international politician and the planner must become more weather orientated, for only then will optimum use be made of the climate resources of the 1990s and beyond. Central to this meteorological planning is the need for a much more comprehensive monitoring and analysis of the world's weather and climate, both to detect and predict changes, and to understand the consequences of such changes.

2. The World Meteorological Organization

The World Meteorological Organization (WMO) is a unique international organization which, among other things, provides an open forum for discussions on a wide range of meteorological and climatological matters. Most importantly it allows the small nations of the world (both developing and developed) to have a considerable influence on how this United Nations specialized agency is organized. Indeed from an organizational point of view – particularly in regard to standards of observations, exchange of information, formats, etc., WMO is very much respected among the various specialized agencies of the United Nations. But perhaps the greatest contribution of WMO is through its Secretariat and Officers, and through its various technical committees and commissions, who collectively are very much aware of the economic, social, and political implications of international meteorology and climatology.

The major work of the WMO is carried out by its eight Technical Commissions, each headed by an elected President and Vice-President. One of these Technical Commissions is the Commission for Climatology and the history of this commission provides an indication of WMO's response to the changing nature of what is meant by climate and climatology. Indeed the difference between the English and French academic schools in their 'definition' of climate and climatology – as discussed in Chapter I – is perhaps best illustrated by the difficulties that the Congress of the World Meteorological Organization has had in agreeing to a name for this Commission. The first session of the Technical Commission held in 1953 was called the 'Commission for Climatology', and this name continued for the second, third, fourth, and fifth Commission meetings held in 1957, 1960, 1965 and 1969. By the sixth meeting however, held in 1973, the name had been changed to the 'Commission for Special Applications of Meteorology and Climatology', and this name was retained at the seventh meeting held in 1978. A further name change occurred at the WMO congress in 1979 when the Commission was renamed the 'Commission for Climatology and Applications of Meteorology'. This name was subsequently used at the eighth meeting of the Commission held in 1982, but the WMO Congress in 1983 made the climatological circle complete by changing the name once again to the 'Commission for Climatology'.

The new 'Commission for Climatology' is of course significantly different from the earlier Commissions of the same name. This is primarily for two reasons: first, many climatologists and meteorologists now have a much more enlightened view of climatology and the climate system; second, the needs of the various peoples of the world for weather and climate services and information are now significantly different from those of the past. The *new* 'Commission for Climatology' is also very much aware of its 'lead role' in matters pertaining to the important World Climate Programme, and it will clearly play an important role in shaping both climatological *and* meteorological decision-making during the coming decades.

3. Weather and food: A case study response of international organizations

In 1971 and 1972 world agricultural production was depressed as a result of unfavourable weather in several countries;this in turn led to a serious decrease in world food reserves and resulted in high food prices. The situation was further exacerbated by a boom in economic activity in the developed countries leading to an increased demand for commodities, and by worldwide inflationary forces accentuated by monetary instability and speculative activities. This was later aggravated by sharp increases in energy and fertilizer prices in 1973 and 1974.

An important aspect of the national and global food supply situations is the role of the meteorological community in supplying weather and climatic information which may or may not affect the supply and marketing of food. Indeed, on an international level, the implications of country A being able to monitor and predict – and use to *their* advantage – country B's weather, and therefore country B's potential crop production, are wide-ranging and involve high-level political and even military implications at both national and international levels. The international implications of providing more weather information are therefore considerable, and in this regard the far-sighted Agrometeorological Programme in Aid of Food Production, submitted by the Secretary-General of the World Meteorological Organization (WMO) to the Seventh Congress of WMO in May 1975, is pertinent.

WMO has always given considerable attention to agricultural meteorology and, mainly through the activities of the Commission for Agricultural Meteorology, has achieved substantial progress in this field. Each Congress of WMO has also adopted a resolution on this subject which has served as the basis of WMO activities in agricultural meteorology during the subsequent four-year period. In particular the Sixth Congress (1971) adopted Resolution 15 entitled 'Agrometeorological Services in Aid of Food Production' which requested the Executive Committee and the Secretary-General (of WMO) to seek means to improve collaboration with other international organizations in the field of agricultural meteorology, to participate as appropriate in international programmes in agrometeorology, and to encourage training of agrometeorological specialists particularly in developing countries.

In the first part of the WMO financial period 1972–5, this resolution formed the basis of WMO's activities in agricultural meteorology and useful progress was made. In 1973, however, as a result of adverse weather conditions in some key agricultural areas, the world faced a serious shortage of food. Greatly increased attention therefore began to be given by the United Nations and other bodies to all factors relating to increasing world food production and hence to agrometeorology. Thus in October 1974 when the sixth session of the WMO Commission for Agricultural Meteorology was held, it was clearly recognized that WMO would be called upon to increase substantially its activities in this field. The basis of a WMO Agrometeorological Programme in Aid of Food Production was accordingly drawn up in order to enable the benefits of these proposals to be used without delay in negotiations with the United Nations and other organizations.

In November 1974 the United Nations World Food Conference was held and WMO gave substantial support in the preparation of a number of documents. The Conference adopted a number of resolutions, several of which called upon WMO, by name and as a member of the UN family of organizations, to take certain action. The General Assembly of the United Nations subsequently endorsed these resolutions by adopting Resolution 3348 (XXIX). Included in the World Food Conference resolutions was Resolution XVI, which, among other things, requested WMO, in co-operation with FAO, (1) to provide, as part of the Early Warning System, regular assessments of current and recent weather on the basis of the information presently assembled through the World Weather Watch, so as to identify agriculturally significant changes in weather patterns; (2) to strengthen the present global weather monitoring systems in order to make them directly relevant to agricultural needs; and (3) to encourage investigations on the assessment of the probability of adverse weather conditions occurring in various agricultural areas, as well as investigations leading to a better understanding of the causes of climate variations.

These resolutions were studied by the Secretary-General of WMO who drew up a new and greatly expanded WMO Agrometeorological Programme in Aid of Food Production. This programme responded to the new responsibilities which WMO was called upon to face and, as was pointed out by the Secretary-General at the time, was fully compatible with the purposes and procedures of WMO. The new Programme was submitted to the Seventh Congress of WMO, held in Geneva in May 1975, and the following examines some of the factors which were discussed at that Congress.

The proposed WMO Agrometeorological Programme in Aid of Food Production was divided into six components, concerned respectively with a proposed World Agrometeorological Operational System (or Watch), agricultural development and planning, agrometeorological data, research and investigation, technical assistance, and training.

In the case of agricultural operations, the provision and utilization of appropriate short-term and medium-term weather forecasts were considered essential. For example, the document considered that the ideal forecast for agriculture should be meteorologically accurate in both space and time, as well as *useful*. Viewed in this context, it was considered that general-purpose forecasts fall short of the more specific needs of agriculture. Accordingly, the provision, wherever technically and scientifically possible, of short-range and medium-range forecasts for *specific* use by the agricultural sectors of nations should receive priority consideration, with the understanding that such forecasts must be expressed in terms *which are meaningful to farmers*.

The important questions of food management and food reserves were also addressed, and it was recognized that all nations, developed or developing, would benefit from the free exchange, on a voluntary basis, of objective weather- and climate-based estimates of expected agricultural production on a regional and global scale. It was considered that at the *national* level, such estimates of crop yields would be useful in connection with the management of National Food Security Systems, the arrangements for the purchase of food in case of a shortage, the early recognition of areas with a developing food crisis, and the modification of national food policies in the light of surpluses or shortfalls in national and international food supply and demand. Similarly, at the *international* level, weather- and climate-based estimates of agricultural production were considered to be useful in connection with the management of a World Food Security Programme, the determination of donor and recipient areas in case of surpluses and shortfalls in food production, the early recognition and continuous monitoring of potential areas of crop failure, and the planning of food relief programmes for areas of anticipated crop failure and food shortages.

Part of the WMO Agrometeorological Programme in Aid of World Food Production submitted to the Seventh WMO Congress therefore involved weather-based crop-yield assessments, under which a number of meteorological centres were to be assigned international responsibility for preparing and disseminating such assessments (based *exclusively* on meteorological information) on a global basis. It was proposed that the dissemination of these assessments would be made on fixed dates and in real-time, using the Global Telecommunications System of the World Weather Watch. To safeguard the voluntary nature of the scheme, it was further proposed that predicted crop-yield assessments in respect of a country should only be disseminated if that country did not explicitly oppose such dissemination.

Many of the proposals submitted in 1975 by the then Secretary-General of WMO (Dr D. A. Davies) in the proposed Agrometeorological Programme in Aid of Food Production were supported by the Seventh Congress, but in the opinion of many Members (i.e. the nations), the provision of weather-based crop-yield assessments was considered to be 'premature'. Such opinions in the main were based on the expressed view that agrometeorological expertise was – at that stage – not sufficiently well developed to provide crop-yield assessments on a regional and global basis with any real degree of precision. However, it is pertinent to suggest that perhaps a more fundamental reason for the non-acceptance of the plan at the Seventh Congress was (and still is) the far-reaching economic, social, political, and even strategic aspects of such information if it is used in the *wrong* way. In other words, in certain circumstances weather information can become almost a too powerful tool, which if *incorrectly* used can in the minds of some governments or political leaders create rather than solve problems.

Considered from an international standpoint, it was clear – even at that time – that there was a need for such information, and the author (see Maunder 1975) noted that it would be unwise to speculate as to how long a system such as that described for predicting agricultural production from the already available weather data of the World Weather Watch system could remain 'suppressed'. As it happened the whole question of climatic monitoring became much more acceptable in the post-1983 period, mainly because most countries accepted the fact that they could not remain *climatic islands*. The question in 1974 was however quite different; nevertheless, several groups within WMO were at that time already promoting practical programmes in these areas. For example, one of the Working Groups established at the Sixth Session of the Commission for Agricultural Meteorology, meeting in Washington in October 1974, was concerned with weather and climate as related to world food production. This Group met in Geneva in June 1975, and considered in some detail plans for implementing the decisions of Congress relating to the agrometeorological programme (or agrometeorological 'activities' as was decided by Congress) in aid of food production. In particular, consideration was given to: (1) how and in what way the 'summarised past and present meteorological and climatological information' could be made available and communicated to the FAO Secretariat; (2) how and in what way 'special agrometeorological data' could be obtained and exchanged at the national level (and ultimately at the regional and international level) for the assessment of crop prospects as required by agricultural authorities; and (3) the examination of available and currently used crop-weather models, and recommendations for keeping abreast of and encouraging the development of suitable techniques for the weather-based assessment of area crop yields under various climatic conditions.

At the meeting of the Working Group it was pointed out that the Seventh Congress of WMO had placed great importance on the development of agrometeorological networks and services, research on crop/weather relationships, and on the application of meteorology to agricultural planning at the national level. A substantial amount of the budget had also been allocated for consultants, technical missions, agroclimatological surveys, etc. to the developing countries, with the aim of meeting requirements at the national level. With this in mind the Working Group proposed a number of courses of action for consideration by the Secretary-General of WMO, in consultation with the President of the Commission for Agricultural Meteorology.

Included among the proposals were: (1) that WMO should arrange for, and urge all Members to transmit CLIMAT data in real-time and that a good way of making such data available to the FAO Global Information and Early Warning System would be to route it through GTS (the WMO Global Telecommunication System) facilities in Rome; (2) in furthering the provision of summarized meteorological and climatological information, representative regional information summaries, including maps using CLIMAT and other data as appropriate, should be prepared by WMO Regional Meteorological Centres on a monthly basis, and provided on a voluntary basis to the WMO and FAO Secretariats and interested Members; (3) that such information should be exchanged between regions, and there should be a feedback of such information to National Meteorological Services for agricultural planning and decision-making; and (4) that the President of the Commission for Agricultural Meteorology should initiate the appropriate action in order that the reporting and transmission of the (then) synoptic rainfall observations be made *mandatory* as soon as possible, and that a proposal for a uniform method and schedule of reporting these observations be initiated.*

The Working Group also discussed the usefulness of an improved regular exchange of meteorological data over the GTS, and emphasized that such an exchange was extremely important for agriculture, because of its usefulness in the assessment of agricultural prospects and in the monitoring of climatic situations detrimental to agriculture, as well as the usefulness of such data in the real-time application of models for crop-yield assessments. However, although it was considered that a great potential already existed (in 1975) for an improved real-time exchange of CLIMAT and SYNOP messages, it was recognized that these messages do not contain all of the parameters which are of special agrometeorological importance. The Working Group felt that a solution might be the development of a new 'AGMET' message to be issued every ten days, and containing those elements which do not appear at present in the SYNOP or CLIMAT messages, such as total radiation, evaporation, extreme events harmful to agriculture, as well as an appropriate presentation of other elements of direct interest to agriculture in particular areas. It was considered that this new message would allow the

* At the session of the WMO Technical Commission on Basic Systems (CBS) held in Hamburg in November 1985, it was finally agreed that the reporting and transmission of the rainfall observations should be mandatory.

possibility of receiving information from meteorological stations which are not on the synoptic network, and that this advantage would be particularly useful in those areas where the present SYNOP network of stations is related mainly to the needs of aviation.

While some aspects of these questions have been acted upon and are now operational, many have not; nevertheless, considerable overall progress has been made, particularly in the Climate Applications Referral System (CARS) of the World Climate Applications Programme, which was developed following the World Climate Conference in 1979. Another positive development resulting from the World Food Conference and the whole 'agrometeorological programme/activities in aid of food production' has been the development of an operational 'Global Information and Early Warning System on Food and Agriculture'. This system is managed by the FAO and the following extracts from their publications in 1977 and 1985 illustrate the type of weather/agricultural information that is monitored and made available for 'restricted official use'.

The first two extracts relate to 1977 and were published in *Food Information*, a monthly supplement to the FAO *Food Outlook Quarterly*.

> The outlook for 1977 world wheat and coarse grain crops is mostly favourable, but a reduction from the 1976 record is anticipated, mainly because of lower wheat plantings and low soil moisture in North America, and drought in China, North Africa and Latin America. Based on early indications, FAO forecasts 1977 world wheat production at 5% below last year, and world coarse grain production at about 2%.

These highlights relate to crops; pastoral products are also mentioned in the 1977 reports, and the following comment relates to meat products:

> First quarter estimates confirm the expected downturn during 1977 in beef and veal production in North America and Western Europe. . . . Lower cattle slaughter in the Southern Hemisphere exporting countries, combined with the downturn in production in North America and Western Europe, has sent up beef prices in international trade in recent months.

Such reports have been published monthly and quarterly since 1977, and the following extracts are taken from the January 1985 issue of the FAO publication *Foodcrops and Shortages*. The report first notes that in January 1985 there were 'abnormal food shortages' reported in:

AFRICA:	Angola, Botswana, Burkina Faso, Burundi, Cape Verde, Chad, Ethiopia, Kenya, Lesotho, Mali, Mauritania, Morocco, Mozambique, Niger, Rwanda, Senegal, Somalia, Sudan, Tanzania, Zambia, Zimbabwe.
ASIA:	Bangladesh, Jordan, Kampuchea, Laos, Lebanon, Viet Nam.
CENTRAL AMERICA:	El Salvador, Nicaragua.
SOUTH AMERICA:	Bolivia.

In this context 'abnormal food shortages' indicates 'an already existing shortfall in basic food supplies in a country below usual consumption requirements, caused by crop failures, interruption in imports or disruption of internal distribution, which may require outside assistance'.

Similarly, 'unfavourable crop conditions', defined as 'a condition likely to result in a shortfall of production as a result of a reduction of the area planted or adverse weather conditions, plant pests, diseases and other calamities' were reported in:

AFRICA:	Botswana, Ethiopia, Lesotho, Mozambique, Sudan, South Africa.
ASIA:	Bangladesh, Kampuchea, Viet Nam.
CENTRAL AMERICA:	Colombia.

Foodcrops and Shortages then highlights the world food/weather situation as at January 1985 (see Table II.1), and the following are extracts from the FAO publication:

> Food shortages are already being reported in Sahelian countries. The position is expected to deteriorate further in the coming months. Substantial migration of people and livestock has been observed, particularly in Chad, Mali and Niger. Seven Sahelian countries face exceptional food emergencies in 1984/85.
>
> Good harvests are anticipated in most countries of Central and South America, reflecting favourable crop conditions. The wheat crop now being harvested in Argentina is estimated at 12.5 million tons, the second highest harvest on record reflecting exceptionally high yields in the main producing provinces.
>
> Good snowcover has so far prevented widespread frost damage to winter grains in western and eastern Europe and the U.S.S.R., despite exceptionally cold weather during January.

These highlights are then followed by the country-by-country reports. Examples for four countries are now given; these reflect both the positive and negative aspects of the weather and the climate which is so often a 'two-edged sword'.

> *Burundi* (as of 20 December, 1984)
> The latest estimates of the 1985 first season crops (maize and beans) are less optimistic than earlier forecasts. The crop is now expected to be below normal, due to excessive rains, subsequent attack of plant diseases and lack of pesticides.
>
> *Ethiopia* (as of 3 January, 1985)
> Following the drought-reduced harvests of 1982 and 1983, particularly in the north, the complete failure of the secondary crop of May 1984, and the extremely poor 1984 main crop, the critical food supply situation will continue throughout 1985. An FAO crop assessment mission which recently returned

Table II.1 Selected national crop production prospects – as at January 1985

Country	*Estimated crop index for latest completed crop year**	*Crop*	*Crop production prospects for crop year in progress*				
			*Plantings***	*Weather conditions*		*Overall crop conditions***	*Harvest period*
				Season to date	*Latest month (rainfall)*		
Cyprus	52	Wheat	Increased	Normal	Below normal	Normal/ Below normal	Jun.–Aug.
	96	Barley	Increased	Normal	Below normal	Normal/ Below normal	May–Jul.
India	123	Wheat	Average	Normal	Below normal	Normal	Apr.
	117	Rice	Average	Normal	Normal	Normal	Dec.
		– Rabi	Average	Normal	Normal	Normal	Apr.
	111	Coarse grains	May–Jul.	–	–	–	Sept.–Oct.
Indonesia	119	Rice – main	Increased	Favourable	Normal	Good	Apr.
Iraq	27	Wheat	Nov.–Dec.	–	–	–	May–Jun.

Source: *FAO Foodcrops and Shortages*, 21 January 1985.
Notes: * Estimated as a percentage of the five-year moving average.
** Compared with a normal crop year. In cases where the crop has not yet been planted the planting period is indicated under 'Plantings'.

from Ethiopia has estimated the 1984 production of main season cereals crop and pulses for both peasant and state farms at 4.4 million tons which is around 32% below the 1980–82 average.

Argentina (as of 15 January, 1985)
During December, seasonably dry weather allowed the continuation of the harvest throughout the wheat belt and the completion of the sowing of sorghum and soya crops. . . . Following the exceptionally high yields achieved in most of the area harvested so far, the estimate of the 1984 wheat output has again been increased to around 12.5 million tons. This would be the second highest harvest on record.

Canada (as of 12 January, 1985)
Statistics Canada, puts the estimate of Canadian wheat production in 1984 at 21.2 million tons, 20% lower than the 1983 output; the decline is due mainly to a 4% reduction in the area planted to wheat and an 18% drop in the average yield for the Prairie Provinces, where crops were affected by drought. However, the quality of the crop is generally better than last year.

4. The World Climate Programme

The World Climate Programme (WCP) arose out of the World Climate Conference held in Geneva in 1979 in response to the increasing recognition of the need for a much more comprehensive monitoring and analysis of the world's climate. It was also – at least in part – a reaction to the relatively poor track record of meteorologists engaged in forecasting the weather at various time-scales. Indeed, although the aims of the Global Atmospheric Research Programme (GARP) and the various components of the World Weather Watch (WWW) were and still are laudable (see Ashford 1982), the benefits have not always been as great as many would have liked. A further fact of considerable importance is that GARP, WWW, and similar *weather* programmes were basically (or at least initially) responses to the concerns of meteorologists and not necessarily responses to the concerns of society.

Central to the aims of the World Climate Programme is the need for a much more comprehensive monitoring and analysis of the world's climate, both to detect and to predict changes, and also to assess its impact. But while the aims of the World Climate Programme are laudable, the comments by R. M. White (an astute commentator and one of the motivating forces behind the 1979 World Climate Conference) are worthy of close examination (see White 1982):

> The World Weather Program was just that – a 'weather' program. It was directed at improving our ability to forecast the weather. . . . The objective of the program was to improve weather services with very little attention as to how these weather services would be applied to serve society. The World Climate Program however is not just a scientific effort, but also an effort to understand the social and economic implications of climatic variability and change.

A recent publication of the World Meteorological Organization (1983) provides a useful summary of the various parts of the World Climate Programme. The overall programme is 'officially' concerned with (1) data, (2) applications, (3) research, and (4) impact studies; but a more useful breakdown of the programme (as suggested by Dr J. L. Rasmussen, President of the WMO Commission for Climatology from 1982) is: (1) climate system monitoring, (2) research, (3) services – that is, applications to agriculture, energy, transport, general public, etc., (4) climate prediction: 30-day, 90-day, interannual, longer-time, and (5) impact studies. The various World Climate Programme activities are also concerned with assisting governments to introduce climatic considerations into the formulation of national policies. In all instances the key consideration is relevance to the society we live in and the benefits of the programme to mankind.

D. MONITORING AND UNDERSTANDING THE WEATHER/CLIMATE RESOURCE

1. Observing the weather

A vast and complex weather reporting network (from land, sea, and the 'upper' atmosphere) is an essential component of many national meteorological services. An excellent example of how vast a network is required is given by New Zealand, which although small in area, gathers weather data from almost half of the Southern Hemisphere. Moreover, because New Zealand is geographically isolated its responsibility for monitoring weather conditions is very large. For example, as well as New Zealand land-based weather stations, the New Zealand Meteorological Service collates daily weather information from stations as far north as the Equator and as far south as the South Pole, and in an area extending to the west of Australia and east to Easter Island.

Some stations outside New Zealand are fully manned by New Zealand Meteorological Service personnel (Raoul and Campbell Islands for example), while many others throughout the Southwest Pacific have New Zealanders in charge and are administered from New Zealand. Reports from this network of stations are part of the World Weather Watch Programme, organized by the WMO. Many national meteorological services have similar observing networks, and all rely to a considerable extent on weather observations from the global weather community of nations.

The kind of weather observations made vary a little from country to country in terms of their number, spacing, elements observed, and their frequency of observation, but the situation in New Zealand where there are basically three kinds of observation stations (rainfall, climatological, and synoptic) is fairly typical of most countries. For example, in New Zealand the 1700 rainfall stations send in monthly reports, as do the 360 climatological stations which provide more detailed information including sunshine duration, wind, cloud cover, pressures, temperatures, and so on. This information is not usually used in a real-time sense for forecasting but rather for compiling background reference-type climate information so as to give an overall picture of present, recent past, and past weather and climate patterns.

As in all countries, however, the synoptic weather stations along with observations from ships, floating buoys, aircraft and other 'remote sensors' are the key to day-to-day forecasting and to providing real-time weather information to assist decision-makers. Reports every three to six hours are available in New Zealand from up to 180 synoptic stations (some of these are automatic stations), and hourly reports from up to 35 of these stations. In addition, most New Zealand synoptic stations provide with their 9 a.m. report information on the rainfall, sunshine, radiation, soil temperature, wind run, and maximum and minimum temperatures, during the previous twenty-four hours, and this information forms the basis for a real-time weather information service.

In New Zealand, the Meteorological Service's National Weather Forecasting Centre in Wellington is the receiving centre for a massive amount of weather information. For example, approximately a third of a million figures are received and processed every day. Each figure tells something of the weather conditions in one particular place at a particular time. These reports come in a variety of ways: radiotelephone, telephone, an aeronautical fixed telecommunication network (AFTN), post office telex, and telegram. An efficient communications network is therefore essential to the smooth running of any national meteorological service.

2. Climatic data sources

In order to research many aspects of climatology there is a need to extend the existing data-collection networks in order that the available observations may provide the most representative data. New sources of data (not necessarily instrumental) also need to be investigated. Observations by satellites or by other means also provide information on the Earth's radiation budget, cloud distribution, atmospheric composition, and the significant topoclimatic effects that influence the exchanges of heat, water, and momentum must be included in the catalogue of data sources. There is also a need to develop *climate indices* that can be used to describe the climate of fairly large areas for use in modelling work, and in particular the need for *commodity-weighted climatic indices* to be available for political or 'administrative' areas, and relatively short-time periods, for use in real-time decision-making (see Chapter V).

As an example of recent more specialized data-collection work, it may be noted that using oxygen-isotope ratios in calcareous shells, and the interpretation of plantonic biota in deep-sea sediments, members of the CLIMAP project (1976) constructed sea surface summer temperatures for the last 18,000 years. However, while such techniques permit resolving time on the scale of centuries or millennia, a finer temporal and spatial resolution is necessary for building predictive atmospheric models of decadal- and annual-scale variability. Improved resolution may be obtained for more recent conditions using radio-carbon dating combined with dendrochronological and pollen evidence of environmental conditions. Further detail may also be obtained for the period of the historical record, prior to instrumental observations, through the careful analysis of written historical evidence.

In the past climatologists have also examined long records of instrumental observations of temperature and precipitation (where they could be found) and evaluated them for consistency and accuracy. However, spatial coverage of such data is usually very poor, and only a very careful use of historical documents provides a means for obtaining data beyond the most populated and longest-settled regions. A recent example of such work is by Moodie and Catchpole (1976), who applied the technique of 'content analysis' to journals of the Hudson's Bay Company to determine the dates of annual freeze and breakup of estuaries along Hudson Bay between 1714 and 1871.

Many national meteorological services are also making a significant effort to improve the completeness of operational weather and climate data. Efforts to obtain such data from ocean and remote land areas, to improve the use of data from satellites as well as to organize the management of the data collected are also necessary to provide worldwide, consistent, compatible data in a timely manner. At the international level the Global Atmospheric Research Programme (GARP) of WMO represented the first real international effort to collect a worldwide data set defining the state of the global atmosphere. The data collected were and still are being used to test and improve our understanding of the dynamics of the global circulation, and our ability to forecast the weather/climate resource in the medium range (a few days to a few weeks) and the longer range (a few weeks to a few months). The GARP observations have also helped to determine the necessary operational observing network for making such forecasts.

The increased efficiency in collecting and disseminating meteorological data has contributed to a resurgence in recent years in the development of synoptic climatological models. For example, new techniques to evaluate pressure system trajectories, and quantitative approaches to determine air-mass characteristics and frequencies, have been applied to a variety of environmental problems (see, for example, Kalkstein 1979). One of the advantages of such an approach is its inherently holistic quality, for – at least as far as the Northern Hemisphere is concerned – many meteorological elements may be assembled and analysed within one cohesive framework – the air mass.

The collection, processing, verification, and organizing of climate data from so many sources present a major challenge. Further, because it is apparent that a single year or even several years of observations will not provide sufficient information to study either regional or planetary climates, it is clear that an improved climate monitoring system must be developed. At the same time it is important to understand the economic attainable levels of observational accuracy, the appropriate parameters which should be monitored continuously for climatic research, and the special attention that needs to be given to satellite systems for monitoring topoclimatic effects, including sea surface temperature, soil moisture, and aspects of the radiation budget.

3. Climate change and climate variations

'Climate change' is paramount in any consideration of weather and climate as a resource. However, several aspects of the term have to be studied before any worthwhile analysis of the problem is considered. They include: what is a climate change; is there evidence for climate change in terms of specific *changes* in temperature, rainfall, etc.; if so, how is the climate changing; what are the possible causal mechanisms; can these mechanisms be modified by man; are the effects of a climate change predictable – and at what degree of accuracy; and, finally, what are the economic, social and political implications of climate change?

Although several terms are in common use when describing climate change, such as 'variations', 'trend', 'oscillation', 'periodicity', as well as 'change' itself, there appears to be little uniformity in their use and acceptance. For this reason, it is perhaps desirable to suggest that climate or *climate state* is 'the totality of weather conditions existing in a given area over a specific period of time'. If this is accepted, it *may* be considered that the concept of climate change becomes (or should become) important only when a relatively long time period is considered. But such a viewpoint is far too restrictive. Indeed, because of the finite resources available in the world, the importance of monitoring, analysing, understanding, and forecasting variations in the available atmospheric resources over *all* time periods (days, weeks, months, seasons, years and decades) is of far greater importance, whatever the 'true' meaning of climate change may be. In particular, the impact of short-term variations in the available atmospheric resources is expected to continue to increase in importance mainly because of the increasing climate sensitivity of most national economies. In any discussion of climate change, it is important therefore to understand the increasing impact climate change will have on economic, social and political activities.

4. Monitoring climate variability

Central to many aspects of regional and national planning is the need for a much more comprehensive monitoring and analysis of the world's climate, both to detect and to predict changes. Sewell and MacDonald-McGee (1983) in assessing the climate scene in Canada noted that the matter of climate variability is rapidly taking on the aspect of an important policy issue. They comment:

> At present, however, understanding of the manner in which climate affects human activities in Canada, and the ways in which the latter may cause changes in climate are not well understood. Despite a recognition of this deficiency by various researchers and by officials in agencies concerned with environmental management, progress in mounting research has been very slow. Problems associated with the management of a hitherto unrecognised resource, climate, seem certain to become a major focus of policy formulation before the end of the century.

However, regardless of the political, economic, and social impact of any long-term trend, it is evident that much more significant year-to-year variations occur; indeed it is of considerable economic significance to note that in most cases trends in a climate series over time – to which it *is possible to adjust* – are usually very small by comparison with short-term variations. Further, since many political regimes have a 'thinking lead-time' of only a few years, any real economic and political significance of a longer-term trend in the climate is very difficult to infiltrate into the political decision-making process. But long-term climate variations, and their predictability or otherwise, *do* infiltrate the political decision-making process, as is evident from this comment which was published on the 'commodity page' of the *Financial Times* (3 November 1983):

> An air of certainty has crept into recent projections of how the global climate is going to change in the future. The majority view is that the general trend is firmly in the direction of a warming of the climate. Only a decade ago, in the wake of the disastrous harvests and droughts of 1972 in the Soviet Union and sub-Saharan Africa, the consensus was equally determined that we were heading for a new ice age. . . . So why the sudden change, and can we place any more faith in the latest set of forecasts?

While these comments may be typical of how the press (and the influential press at that) see the work of meteorologists and climatologists, the implications are very clear that it is also time for climatologists to put their house in order. As part of the World Climate Programme, WMO is indeed doing just this, and Unninayar (1983) has highlighted some of the concerns:

> Measurements taken over a period of time, on a global basis, enable the construction of the history of climate, and studies on climate trends, climate variability and the interactive processes which are involved in causing such changes. The understanding of these processes is vital to formulating the physical-mathematical description of the climate system in numerical models. As a priority requirement it is considered necessary that a mechanism be formulated, relying on existing research/analysis/operational centres, to

> compile and disseminate summarized information on the present status of the climate system, changes from past years, indication of trends, and significant events of regional and global consequence.

The politician, the planner and the decision-maker must also become more weather and climate orientated if optimum use is to be made of the weather and climate resources of the 1990s and the twenty-first century. Central to this 'planning' – as already noted – is the need for a much more comprehensive monitoring and analysis of the world's climate, both to detect and predict changes, and to understand the consequences of such changes. A necessary first step in this regard is to understand the specific application of weather and climate information to economic, social, and political problems.

III Events and information

A. Data and information

In the preface to *Quarterly Predictions*, published by the New Zealand Institute of Economic Research, it is stated that

> economic forecasting is a chancy business for in addition to our own imperfect understanding of the way the economy works, there are also likely to be errors from chance factors such as changes in the weather at home and overseas, from inadequacies in our basic information, and from the unknown effects of political events.

In the light of these pertinent comments, the responsibilities of meteorologists and climatologists to provide decision-makers with the most appropriate weather and climate information are very clear.

It is, of course, often very difficult to assess the increased economic, political, and social benefit that may arise from an improvement in the use of weather and climate information. Indeed in a simple manner one may consider that the economic outcome of a climate-sensitive process under management is influenced by four factors (Fig. III.1). First, weather events (that is, what actually occurs or occurred at a specified place); second, weather information (that is, what is reported to have occurred, or what is forecast to occur, or what an analysis of climate information tells us about what has occurred in the past); third, non-weather events (such as the actual price structure and current government policy); and fourth, non-weather information (such as the expected price structure, or the expected government policy). In all cases, the economic and, in some cases, the social and political outcome of a weather- or climate-sensitive process under management, is subject to some degree of uncertainty, principally because at the time the most appropriate alternative is chosen, the decision-maker does not know the actual value of either the 'weather information box' or the 'non-weather information box'.

Weather and climate information may of course constitute a number of separate items. It may, for example, consist of (1) the weather existing at the present moment; (2) the weather and the climate that is expected to exist at a specified time in the future; and/or (3) analysis and interpretation of the records of the weather and the climate that has existed in the past, including the recent past. All three types are of value (see Maunder 1970a, 1981a) but the specific type of weather or climate information required is (or should be) determined by the kinds of problems which are confronted and the

FIGURE III.1 The relationship between weather events and information, non-weather events and information, the choice of alternatives, and the economic outcome of a weather-sensitive process under management

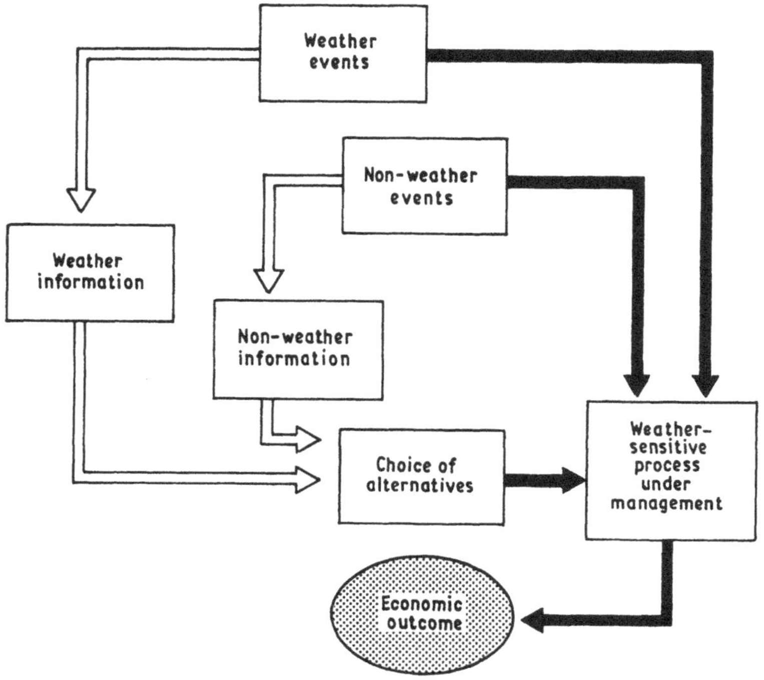

Source: After Maunder (1985a), from Maunder (1970a) and McQuigg and Thompson (1966).

associated decision-making involved. For example, an analysis of the past weather and climate is useful for assisting in *planning* the location and design of many types of facilities, such as irrigation schemes, water storage systems, and agri-business enterprises. On the other hand, relevant real-time or near real-time weather and climate information is essential for any *operational* type decisions such as marketing, agricultural production forecasts, and analysis of agri-business trends. Further, present weather and short-period forecasts of future weather are most useful in making *operational* decisions such as the scheduling of irrigation, estimating energy demand, or the forecasting of agricultural production and the 'futures' prices of agricultural products.

Applications of meteorological and climatological information have been considered by a number of investigators, including those at a special symposium on 'The Business of Weather' held over twenty years ago in Chicago. Hallanger (1963) made some very pertinent remarks to businessmen at that time which are possibly even more relevant today than they were in 1963. He stated:

> the key point in the realisation of the potential value of weather information to your activity [is that] . . . the meteorologist, working as a team with your people, must become familiar with your operation. Only then is he in a position to identify the true weather problems. Only then can he provide the appropriate weather information in the most useful form. Only then will the return exceed the cost by the greatest amount.

Thus, the potential value of weather information to an economic activity and to a particular operation will be realized only when the qualified meteorologist and/or climatologist, working with the co-operation and backing of management, develops the information most suited to the specific needs of the activity.

Such observations are not new, indeed several investigators including Demsetz (1962), McQuigg (1964), Rapp and Huschke (1964), and White (1964) had during the mid-1960s effectively demonstrated the 'value' of weather information, and had advanced the seemingly reasonable thesis that weather information has (or accrues) economic value because it makes possible better management decisions during the operation of a weather-sensitive process.

Nevertheless, it is true to say – even twenty years after these pioneering studies were made – that it is difficult to compute, or even make reasonable estimates of, the increased economic value that might result from improvement in the accruing of weather and climate information. Two of the reasons for this state of affairs are – as first suggested by McQuigg and Thompson (1966) – the dearth of real economic data from weather- and climate-sensitive processes, and the expense, undesirability, or even danger of conducting experiments in which the accuracy of weather and climate information used by a decision-maker is deliberately varied. Some techniques and models are, however, now available for evaluating both the weather and climate information, and a selection are discussed in this book.

B. WEATHER AND CLIMATE INFORMATION

1. The value of weather and climate information

To determine the value of weather to any area or any country, one must first identify activities in that area or nation that are affected directly or indirectly by the weather and the climate, and, second, analyse the manner in which a given change in the weather or climate results in gains or losses to such activities. The problems of assessing the value of the weather and the climate for various areas (usually commodity-type sectors of a nation) have been considered by a number of investigators. In most cases, however, these investigations have either been spatially restricted or based on relatively long periods. Such studies are not, therefore, designed to contribute to the question many decision-makers ask: 'What is the value (both positive and negative) of weather to a large area (e.g. the United States) over a short period of time (e.g. a week or a month)?'

In 1973, the author suggested that there *may* be little point in such analyses, since any relationships that might be so derived could not be used for the planning of stocks or the allocation of specific resources until reasonably accurate weekly and monthly weather forecasts were available (Maunder 1973a). At the time it was assumed that such forecasts would soon become available – at least within a decade – and that 'high-level' decisions based on such forecasts would or could be made. However, in the light of the now 'recognized' value of *weather information* (such as soil moisture, heating degree-days, etc.) in *real-time*, as distinct from the 'still to be recognized' value of medium- and long-range *forecasts*, the emphasis which was placed on weather forecasts in 1973 may not now be justified. A distinction should however be made between the rapidly improving track record of 'atmospheric circulation' forecasts, compared with the poorer track record of the *word* weather forecasts. Indeed, the 'correct' translation from the forecast 'atmospheric circulation' to the *weather* resulting from this circulation has yet to be achieved to the satisfaction of many users. But, irrespective of the 'real' value of weather *forecasts* to users, it is evident that the value of weather and climate *information* has increased, particularly in regard to the key sectors of food production and energy consumption, and in explaining economic and political vagaries both regionally, nationally, and internationally.

As noted earlier, the potential value of weather and climate information can be realized only as the flow of this information is made to affect decision-making. The decision-maker in national economic planning must therefore be convinced that there *is* value in weather and climate information (Maunder 1977a). However, even with perfect weather-economic production models, major problems still exist in the acceptance by decision-makers – both within and outside meteorological circles – of this new information. In particular, the acceptance and use of appropriate commodity-weighted weather and climate information (see Chapter V), and forecasts of production resulting from this information (see Chapter VII) offer a continuing challenge to the meteorologist and the climatologist in the fields of marketing and education. An early and positive contribution in this emerging area was the establishment, by the National Weather Service of the United States, of a Center for Climatic and Environmental Assessment in the United States in 1974. The development of this Center (under the leadership of the late Dr James D. McQuigg at Columbia, Missouri), and a similar Special Services unit in the New Zealand Meteorological Service in 1978 (under the leadership of the author), permitted the availability of a family of products not previously available, notably early weather-based assessments of agricultural production and the real-time monitoring of weather variations on a regional, national, and commodity level.

The traditional role of the atmospheric scientist in this area was, until the late 1970s, one of caution (Maunder 1972e, 1977c), but it is now recognized that the meteorological community (government and non-government) must not only give a strong lead in this area, but must also make sure that decision-makers at the top-level know how to use such information. Indeed, the need for the meteorological and climatological community to give a strong and positive lead in these areas cannot be overemphasized, for regrettably most top-level decision-makers only think of meteorology and climatology in terms of the weather forecast – and usually only in respect of the weather forecasts that were wrong.

2. Presentation of weather and climate information

The presentation of weather and climate information is a highly specialized and an important form of communication; indeed, since weather and climate have a significant effect on many activities and businesses, a large number of people are called upon to make weather- and climate-sensitive decisions. Meteorological and climatological organizations aim at satisfying the requirements of large numbers of customers who can be divided into 'consumers' or 'households', and 'producers' or 'firms'. Consumers are generally those who *use* weather and climate services to maintain or increase health, safety, and general welfare, whereas producers are basically concerned in accruing benefits to their 'firms' through the use of weather and climate services in their professional, commercial, or industrial activities. For example, 'producer' type weather and climate information is used by the agricultural, energy, retail, and transport industries, whereas 'consumer' type weather and climate information is basically used by the general public for their day-to-day activities, as well as by civil defence, health, and recreation organizations.

During the past twenty years many advances have been made in the atmospheric sciences; these involve not only improvement in the science and techniques of weather and climate forecasting, but also the much greater awareness of the real need for weather and climate information (including but not restricted to forecasts) by society. For example, many national meteorological services provide on a routine basis weather services relevant to a wide variety of users, such as temperature/humidity probabilities, snow warnings, fog forecasts, heavy rainfall forecasts, lightning suppression probabilities, seasonal forecasts, weather-based agricultural production forecasts, frost warnings, rainfall probabilities, wind forecasts, and 'week-away' weather forecasts. It is, nevertheless, fair to say that in general, weather infomation to the public has not improved as much as it should have, in either its presentation or its impact. In addition, there is an inevitable change through time in the reaction of users (including the public) to weather and climate information. It is most important, therefore, that atmospheric scientists are aware of society's changing desire for, and reaction to, weather and climate information.

3. Weather and climate information and decision-making

The need for a more rational and effective response to atmospheric events, and the associated requirement for better and more useful information about the atmosphere, has resulted from an increasing awareness that information about the weather and climate can play a very important role in the decision-making processes associated with the management of weather- and climate-sensitive enterprises. Included in these enterprises are aspects of regional, national, and international economies concerned with agriculture, energy, trade, tourism, insurance, construction, and prices.

At the international level, the problem is considerably more complex, but it is evident that all economies – whether these are monitored by international organizations, national governments, or private companies – are sensitive to weather and climate variations, the most obvious examples being food and energy aspects of economies. In addition, it is important to note that since it is the relatively 'short-term' variations from the 'near average' weather and climate that most governments and international agencies have to contend with on a day-to-day and week-to-week basis, the recognition that 'short-term' variations are important – economically, socially, and politically – could well help to avoid problems such as those which first arose in 1973, when grains were 'traded' between a number of countries to an unnecessary degree, described by one writer as 'the great grain robbery' (see Trager 1975). Of course the 'long-term' variations are also important, but to most if not all governments it is *this* season's 'crop' which is of paramount importance and not what *may* happen to the crops next decade.

Several investigations have effectively advanced the seemingly reasonable thesis that weather and climate information has or accrues economic value because it makes possible better management decisions during the operation of a weather- and climate-sensitive process. Nevertheless, it must be stressed that although it is often suggested that the provision of better information (both weather and non-weather) will lead to better decisions, a great deal depends on the way in which the information is presented to, and used by, the consumer. Clearly, it is the presentation of the 'package' to the consumer that is critical to the success of the marketing strategy of both government-funded and 'private' meteorological services, and there is strong evidence that this aspect of the 'weather business' will become much more important in the future.

In this regard, a recent review of assessing user requirements and the associated economical values for both short-range weather forecasts and current weather information made by Murphy and Brown (1984b) is noteworthy. In commenting on the nature of user requirements they emphasize three factors:

> First, it should be recognized at the outset that requirements will vary from user to user, even for individuals involved in essentially identical activities. . . . Second, the degree of sophistication of users – or, more accurately, the degree of sophistication of their information-processing and decision-making procedures – has important implications for the manner in which user requirements can and should be met. . . . Third, studies of user requirements should be sufficiently detailed to ensure that it is possible to differentiate between activities that are weather sensitive and activities that are weather information sensitive.

In discussing the third factor Murphy and Brown correctly point out that many operations that are affected by weather conditions may not be sensitive to weather information, either because of the presence of constraints on the options available to the user, or because the state of the art of observing and forecasting the weather is such that information with the required accuracy or with a sufficient lead-time cannot be provided. They further note that many studies of user requirements and economic values have not been sufficiently detailed to distinguish between these two conditions and, as a result, have led some investigators to misjudge the requirements for and overestimate the values of meteorological information.

The number of decision-makers interested in weather and climate information varies considerably in time and space. For example, in the United States, the number of 'general public' decision-makers using weather forecasts might vary from a few million on a relatively unimportant weather day to over 100,000,000 on a day with considerable weather activity throughout the nation. By contrast, during quiet winter days when field work is at a minimum, the number of farmers in the United States interested in weather information may be only a few thousand, this number increasing in the spring planting season to several million farmers. Similarly, in the case of aviation and marine interests, the number of decision-makers in the United States may vary from day to day from a few thousand to over 100,000. But, all decision-makers are *not* of equal importance. For example, the economic value in the United States of weather-related decisions ranges from the 100 million or more who use weather information for convenience only, which may have a value of a dollar or so, to a few very large decision-making organizations where weather and climate information will have a value in millions of dollars.

The *real* value that can be placed on short-term variations in the weather and the climate therefore requires the identification of those specific activities directly or indirectly affected by such variations, and a quantitative analysis of the consequent effects that they have on such activities. But assessing the *real* value of a 10 per cent reduction in, say, summer precipitation in Australia during three consecutive summers, or the impact of a 2°C decrease in spring, summer, and autumn temperatures in Europe during the same season – both of which can result from relatively small changes in the 'climate', are questions which, until very recently, have received little attention.

4. Realizing the value of weather and climate information

In most countries, changing social patterns, trends in population growth from rural to predominantly urban areas, the accelerating demand for energy, the changing impact of markets, and the emergence of more sophisticated and often more weather- and climate-sensitive systems, have created a need for more rational and effective responses to atmospheric events. In 1968, the expenditure by the governments of the world on meteorology was estimated to be in excess of $US1000 million (Thompson, J. C. 1969), a comparable figure for 1985 probably exceeding $US12,000 million. It is relevant, therefore, to ask three questions as they apply to any country or region: first, what is the current and potential value of weather and climate services; second, how will prospective advances in the atmospheric sciences improve the usefulness of such services; third, what are the potential benefits or losses associated with the deliberate or inadvertent modification of both the weather and the climate?

Until the early 1960s, questions like these, let alone attempts to answer them, were of little concern to governments or to the meteorological community. In the last decade however, several factors have contributed to a need for much more serious 'meteorological bookkeeping'. But can these questions be answered? In some cases the answer is no, while in others the answers can only be given imprecisely. Nevertheless, attempts to answer such questions must be made for two main reasons: first, both scientific and technical advances have brought about increasing understanding of the atmosphere; second, complex sensing, communications, and data processing systems are capable of producing a virtually unlimited flow of data. These two factors have involved large costs, but as McQuigg (1972a) pointed out over a decade ago, while budgetary considerations are important, there is a more fundamental reason for being interested in the economic impact of weather and climate information. The reason is that people can effectively apply the rapidly improving flow of weather and climate information in their deliberate attempts to improve the significant weather- and climate-sensitive aspects of an economy.

In assessing the value of weather and climate information, the concept of the 'weather and climate package' is useful. What, for example, is the value of weather forecasts published in newspapers, the value of aviation weather forecasts to commercial airlines, the value of forecasts of extreme weather to catchment and energy authorities, or the value of published climate data? Moreover, how does the value of *this* weather and climate information compare with the value of *other* weather and climate information which could be provided with present technology and finance, and the potential information which could be provided if technology and finance were improved?

Regrettably, answers to most of these questions are not known. In particular, very little is known about how efficient the present 'weather and climate package' is, or whether it is economically desirable to provide more weather- and climate-related information. Further, if a better 'weather and climate package' was made available, would the users of meteorological and climatological products know how to use it, and how would these customers use the 'package' to their advantage?

It is evident from the above considerations that much of the potential value of weather and climate information to an economic activity or to a particular operation can be realized only when the meteorologist and the climatologist working with the co-operation and backing of management (including, where appropriate, government) develop information about the weather and the climate most suited to the specific needs of the activity or operation. This observation is not new; nevertheless, it is true to say that little real effort has been made, since the pioneering 'weather/economics' studies were published in the mid-1960s, actively to promote in-depth studies of the value of weather and climate information to regions or nations or even individual companies or user groups. Accordingly it is still difficult to compute, or even make reasonable estimates of, the increased economic value that might result from improvement in the accruing of weather and climate information.

It is clear that the decision-making involved in managing a weather- and climate-sensitive aspect of an economy or a business must take many factors into account. The decisions involved are made by many people with a wide spectrum of responsibility, experience, and training (both in the meteorological and non-meteorological sense). But it must be appreciated that the real potential value of the flow of weather and climate information in this context can only be realized when it affects decisions. Accordingly, the suggestions first put forward by McQuigg (1972a) are to the point. He stated:

> In some important instances, the greatest contribution a meteorologist can make is to find a way to establish real communication between a certain group of decision-makers and the meteorological system. This is not to say that all a National Weather Service has to do is to provide information to management (or to the general public), and then wait for the economic benefits to become obvious to all concerned. But it surely is true that no economic benefits will accrue to the meteorological information system from a weather-sensitive segment of the economy in which the persons making management decisions are (1) not aware of the availability of meteorological information; (2) aware of the potential value of meteorological information, but do not have any channels through which such information may be received.

C. WEATHER FORECASTS

1. Balancing scientific enquiry with the value of the product

Much work has been accomplished on modelling the physical atmosphere. However, much less research has been done on the *non-scientific* aspects of atmospheric modelling. Thompson (1966) and Maunder and Sewell (1968) pointed out this deficiency many years ago, and an appraisal of the relevant literature at that time – and since – shows that little thought is usually given to the *end product* of the meteorological chain. This chain comprises many parts, from the basic observations, through the 'world of mathematics, physics and computers', and then through a *translation programme* to the weather forecast and finally to the user. A basic question at the end of this sequence is the value of the product. Simply, what is 20 mm of rain or, for that matter, any rain worth to an individual, an area, or a nation; moreover, can any figure (in dollars or an equivalent measure) really be placed on the overall *benefits* or *costs* of a correctly forecast dry period of two weeks to the farmers of Indonesia or Kenya, or the value of a correctly forecast cold snowy day to the people of New York, Moscow or London?

Suppose for example that a major snowstorm is forecast to move over London tomorrow. Do the people of London accept this, and adjust accordingly, or convince their politicians and their scientists that they should try to prevent the snowstorm occurring (perhaps by altering its path towards Paris instead)? In either case, if we know something about the impacts of a snowstorm on London, then we would be in a better position to decide which of the above actions we would take. But *do* we know? In looking at this question of adjustments to tomorrow's or next week's weather, four items stand out as being most important. These are: the use and value of weather information; the economic and social aspects of weather forecasting; weather modification and tomorrow's weather; and the weather sensitivity of various activities.

These items are important because of a number of factors. First, if the currently available use of present weather forecasts is *not* known, then any improvement in the weather forecasts – from the user's viewpoint – may not only be a waste of time, but also a waste of public funds devoted to them; second, it is probable that *more effective use* of weather forecasts would achieve better 'results' than an improvement in the accuracy of the weather forecast; third, if we know what tomorrow's weather is going to be, *and* what the *effect* of that weather will be, then we are in a position to alter our production schedule, decide whether to go ahead with tomorrow's activities, or possibly call upon a weather modification agency or company. However, it is clear that if *we do not know* the impact of tomorrow's weather on an activity, or an area, then we are hardly in a position to know how much to modify the weather if we should want to do so (and have the capacity to do so), and we surely should not modify the weather if we do not know what the results of such modification will be.

Clearly, then, unless we know the *real* impact of tomorrow's forecast weather, should we not question the wisdom of trying to forecast the weather, unless it is purely for entertainment or convenience purposes? In this regard one may note that in the fiscal year 1965 the United States Weather Bureau spent $94 million on weather services, and $8 million on research and development, and of the latter less than $250,000 (or 0.3 per cent) was devoted to social science research. Comparable up-to-date information is not readily available, but it would be surprising if more than 1 per cent of the budget of any national meteorological service was used to assess the end products of the weather forecasting chain. For example, the budget for the 1984/85 financial year of the New Zealand Meteorological Service of $NZ16 million, showed that much less than $160,000 (or 1.0 per cent) was devoted to the *non-scientific* aspects of weather forecasting. However, with the growing recognition of the 'user-pay' principle by many governments and of course all business enterprises, it is evident that much more of the budgets of national meteorological services must be devoted to these aspects of weather forecasting.

2. The weather forecast: Its use and value

Despite the availability of global computer analyses of the current atmospheric system, and equally powerful computer-based forecasts of the atmospheric system likely tomorrow, recent improvements in forecast accuracy (especially in the medium-term range) have not been as rapid as a lot of people would have liked. Of course, many users appreciate very much the accuracy and reliability of the official forecast, and this is clearly indicated in a recent letter to the New Zealand Meteorological Service from a Federated Farmers' organization. In this letter they state that 'each successive forecast allows ready confirmation (or otherwise) of the likelihood and intensity of any imminent weather patterns over the next 12 hours'. Such monitoring of the weather forecasts can only be commended, and is really the only sensible approach to coping with the vagaries of the weather in many areas of the world.

For many years, the larger meteorological agencies have put substantial research efforts into long-range weather forecasting. The limited success they have had so far shows that there is no quick or easy solution to the problem. Nevertheless, always at the forefront of the desire for better weather forecasts is the *apparent* need by the public and specialist users for *longer-range* forecasts. This is despite the fact that many decision-makers in weather-sensitive industries are not in a position to use such forecasts, because they do not really know how the *present* and *past* weather affects or has affected their operations. For example, Glantz (1977a), in assessing the value of a long-range weather forecast for the West African Sahel, asked respondents in agriculture, meteorology, political science, economics, and medicine *what they would have done* in the Sahel if they had had accurate rainfall and temperature forecasts six months in advance of the beginning of the 1973 growing season. Their reply in essence was not encouraging and Glantz in summarizing their responses stated:

> Although a reliable long-range weather forecast is not yet possible, it may not even be desirable for many parts of the world until that time when some essential adjustments to existing social, political, and economic practices have been undertaken.

The key question is, of course, 'where does each nation fit into this?' Have, in fact, the essential adjustments to existing social, political, and economic practices been undertaken? It is suggested that in many cases there is considerable room for improvement. Indeed, it seems strange that many businesses seem to be quite content to ignore not only the effect of the present and past weather and climate on their operations, but also the impact of the future weather and climate (both in their own country and their competitor countries) on their activities, with a consequential loss of profits to their 'shareholders', which in many cases are the taxpayers of their country.

3. Weather forecasting: Analysis into words

Weather forecasting is the activity associated most readily by the general public with meteorological services and although far from being the only service provided, it is, on a day-to-day basis, clearly the most important. But, what is involved in making a weather forecast? Let us consider – by way of an example – the first national forecast of the day, issued by the New Zealand Meteorological Service at 5.30 a.m. This forecast is based on surface and atmospheric observations from a network of reporting stations, as well as on satellite pictures received during the night.

The satellite photographs used by the forecaster are taken over New Zealand, Australia, and the Pacific, and provide the forecaster with information on the cloud structure, movement, development, and significance in terms of weather. In addition, infrared satellite photographs give valuable information on the heights of clouds and the temperatures in the lower atmosphere. The atmosphere is, of course, three-dimensional, and the forecaster must analyse and use maps depicting the upper level airflows as well as those near the surface.

The forecaster begins by analysing the most current weather map and then compares this map with previous maps to make a prognosis, which is a projected weather map for some specific time in the future (usually 24, 36 or 48 hours). In addition, the forecaster is assisted by several computer-based predictions of the state of the atmosphere at specific times in the future. Such numerical weather predictions are particularly important when considering the large oceanic regions in which there are very few land or ship observations.

The New Zealand Meteorological Service's computer also allows high-speed processing of all incoming data, including satellite signals, and handles communications including the distribution of forecasts and real-time weather and climate information. In this regard it is significant to note that in an average year the Meteorological Service prepares for New Zealand's 3 million inhabitants, 66,000 forecasts for radio, 20,000 forecasts for newspapers, 78,000 forecasts for marine interests, 56,000 forecasts for international aviation users, 72,000 forecasts for domestic aviation users, and 2000 forecasts for television. In addition, over 200,000 individual telephone calls are received from customers requiring specific weather and climate information, and in the three main centres over five million telephone calls a year are received on the automatic 'ansatel' weather forecasting service.

On the basis of the various prognostic maps, one final version is drawn, and the forecaster will then start preparing a *word weather forecast* (rather than lines on a map!). Indeed it is important to emphasize that although weather forecasting is based on scientific principles, the 'human touch' is very important. A computer, for example, may quite accurately predict the general weather situation, such as the location of anticyclones, fronts or cyclones, but the skill of the (human) forecaster is needed to interpret this in terms of the *weather* to be expected. In particular, local variations in weather patterns are a very real problem, and only experience, research, and a good local knowledge will help overcome this. The intangible skills, such as the gift of communicating with the many customers for weather services, are also extremely important, and these are discussed in detail below.

4. Weather forecasts and the user

Although the improvement of weather forecasts has for a very long time been one of the prime objects of meteorologists, and rightly or wrongly it is the weather forecast that gives meteorology the image which it has acquired, the *translation* from the weather forecaster to the weather consumer is not always done with the same care and attention as the *translation* from the weather map to the weather forecaster. The translation procedure is in fact an important aspect – too often ignored – in the weather forecast matrix. Some attempts have of course been made to convey to the consumer some of the *inner* thoughts of the weather forecaster, and the probability forecast used extensively in the United States for precipitation is one of the more useful applications of this concept.

But there are still many aspects of weather forecasting of which we are in many cases totally ignorant. For example, consider the following:

(a) How many people read the weather forecast in their local newspaper or listen to a weather forecast on the radio, or look at the television weathercast? More important, how many remember what they read, hear, or see?
(b) How often do the 'thoughts' of the weather forecaster get through to the user of the forecast? (This depends in part on how the thoughts of the weather forecaster are *translated* into the weather forecast, and how the weather forecast is *translated* through the medium of radio, television, newspaper, or videotex to the user.)
(c) What effect does the weather forecast have on the average user? How many users reschedule any of their activities because of the forecast and does it make any *real* difference whether the forecast is accurate or not?

Despite these 'unanswered' questions, it is evident that the mounting losses of property and income which result from extreme weather events, the ability of people to modify their atmospheric environment either deliberately or inadvertently, and the ever-present 'control' which the atmosphere (the rain, the snow, the wind, the humidity, and the sunshine) exerts over each of us and our economic and social life, are all significant. The predictions of the weather forecaster are, therefore, of considerable economic and social value. These predictions include weather forecasts issued several times a day to the general public and the following comments provide some understanding of the general public's desire for, and appreciation of, weather forecasts.

During the past three decades many advances have been made in the atmospheric sciences. These include a significant improvement in the 'correctness' and usefulness of weather forecasts, and a greater awareness of the demand for weather forecasts by a wide variety of users. For example, the Atmospheric Environment Service in Canada provides weather forecasts relevant for road maintenance, water control, forestry, agriculture, scheduling of power production, oil deliveries, retail sales campaigns, as well as for specialized marine and aviation activities. It is, nevertheless, fair to say that to the *public*, weather forecasts provided by most national meteorological services have *not* improved in presentation or impact. Moreover there is an inevitable change (through time) in the reaction of consumers to weather and weather forecasts. It is, therefore, important that weather forecasters are aware of the public's reaction to forecasts. Several surveys of the public's reaction to weather forecasts have been made and the following are a sample.

Over thirty years ago a survey of University of Toronto students on weather technology was published by the Controller, Canadian Meteorological Service (1949). In 1967 a similar study was conducted at the (Canadian) University of Victoria (Maunder 1969). In each case the sample was small and not without bias. For example, both groups were relatively young and it is possible that younger people are more willing to accept the discomforts resulting from the weather than older people. Second, the normal weather patterns of Toronto and Victoria are not directly comparable. Despite these reservations and possible bias, it is believed that the two surveys provided useful information. Among other things the students surveyed were asked to indicate the *importance to them* of the various items in a weather forecast. In the 1967 survey 48 per cent indicated that the precipitation expected was the most important item (with an average ranking of 1.4 on a 1 to 4 scale), compared with only 9 per cent who considered that the winds were of most interest (average ranking 3.5). The 1967 survey, in fact, indicated quite clearly that in Victoria a weather forecast should emphasize the precipitation expected, the temperature, the state of the sky, and the winds, *in that order*. By comparison the 1948 University of Toronto survey indicated that the temperatures were the most important (average ranking 1.78), closely followed by precipitation (average ranking 1.84). These results show that the question of just how important are temperatures, snow, rain, wind, cloud, sunshine, etc. to the user of the local weather forecasts must be taken into account, and that an ideal forecast (from a terminology-wise point of view) in one area – or for that matter during a particular time of the year – may not be ideal in another area or another time.

In the 1967 survey (Maunder 1969) the opportunity was taken of asking some additional questions related to the use of the newspaper, the radio, and television as a means of obtaining weather forecasts. The overall impression from the survey was that the great majority (of students) neither read, listened to, nor watched the weather forecast. If this survey is at all representative, it may therefore be much more appropriate for weather forecasters to concentrate on the *minority* who *are* interested in their product, than on the majority who *are not* interested. The basic difficulty, however, is to find out the type of person who *is* interested in the forecaster's product, and then provide that customer with the information required.

A more recent survey was made in New Zealand in 1984 when a 'face-to-face' interview of 1000 farmers was conducted for the New Zealand Meteorological Service by AGB: McNair Surveys Ltd, a professional client and consumer organization specializing in social and marketing research. The survey was conducted – as part of a larger 'farm survey' – in August 1984. A sample of 1000 farms was selected from the 61,800 farmers in New Zealand. Technically, the farm sample was a stratified random probability sample. All interviews were personal face-to-face interviews, enabling the use of showcards and display material where necessary, as well as helping to retain the respondent's interest and ensuring a high response.

Two principal questions were asked in the survey: the first related to the preference for one or other of the (then) two weather presentations on New Zealand television, and the second question asked about the sources of weather forecast information (radio, television, etc.), and their relative importance. Answers to the first question showed that overall there was almost equal preference for the weathercast presented by forecasters from the Meteorological Service and the weather forecasts presented by contract staff (non-forecasters) of Television New Zealand. There were, however, marked regional variations, with 66 per cent of the farmers interviewed in one area (Canterbury) preferring the presentation of the professional forecasters, compared with only 31 per cent of the farmers interviewed in another area (Auckland). The regional differences relate in part to the type of farming carried out, but they are also influenced by the fact that the professional forecaster presentation is shown late in the evening, compared with the early evening presentation of the other weather forecast presentation (along with the 'main' television news programme).

The second question provided answers which may well have application outside of New Zealand. Essentially, the survey of the 1000 farmers showed that the 'specialized' weather forecasts – such as the 4–5 day forecasts broadcast over a nationwide *but* non-commercial radio network – attracted only about 20 per cent of the possible audience, a percentage which is similar to the radio listening preferences of all New Zealanders. In contrast 72 per cent of the farmers said that (on an average day) they viewed the 'popular' television presentation of the weather, and 46 per cent said that they listened to the weather forecasts from the local or community 'commercial' radio stations which in New Zealand are supported by advertising. The survey also showed that only 22 per cent of the farmers obtained weather information on a daily basis from the morning newspaper.

A number of factors emerged from the survey, the key item being that even 'specialized' users such as farmers are fairly traditional in their source of weather information. In particular, the 1000 farmers surveyed indicated that despite the 'best' intentions of the New Zealand Meteorological Service, most farmers do *not* listen to, watch, or read the specialized and/or professional presentation of the weather, but prefer the more 'popular' presentations on commercial radio. Nevertheless, there is – in New Zealand – still a sizeable number of people (including farmers) who *do* want the more specialized and professional presentation of the weather, and it is important that both national meteorological services and private meteorological companies continue to recognize the 'market' importance of their more informed customers.

All of the surveys cited clearly indicate that the attitudes of the customer towards forecasting terms is important and changes from time to time, and from place to place. In particular, the meteorological profession must not only appreciate that time may alter the consumer's *understanding* of weather forecasts and weather forecasting terminology, but also that time may alter the consumer's *requirements* for weather forecasts. Indeed, more study along these lines may in fact be just as worthwhile in the long run as more accurate forecasting, however desirable the latter may be.

5. The value of weather forecasting: Future aspects

In two issues of the journal *Weather* in 1967, J. S. Sawyer, a well-known British meteorologist, discussed weather forecasting and its future (Sawyer 1967). Among his observations at that time were that the economic value of 10- to 20-day forecasts would be considerable, and that the possibility of producing such forecasts was one of the main incentives of the World Weather Watch. Nevertheless, it was disappointing to note at that time (see Maunder 1968b) – and also today – that except for the above observation, Sawyer considered weather forecasting and its future only as a physical science. For example, very little if any mention was made on the likely effect of weather modification, consumers' changing attitudes to the information contained in a weather forecast, methods of presenting weather forecasts, and whether in fact people really want to know what next week's weather is going to be.

Some of the problems in looking at tomorrow's weather have already been examined. But if tomorrow's weather, and what we do with tomorrow's weather, is to be more a matter of choice than chance, what can and should be done? First, more research relating to the *process* of decision-making in the management, use, and possible modification of the atmospheric resource is needed. That is, studies are required to identify what decisions are made, who makes them, and what factors appear to influence the outcomes. What decisions, for example, are made about adjustment to a snowstorm by individuals, private industry, and various levels of government? Second, more research is required on the role of weather information in the decisions relating to adjustments to the weather. For example, does an increase in the amount of information, or an improvement in the accuracy of weather forecasts, necessarily lead to changes in production schedules or alterations in human activities? To what extent is the meteorologist's view of the value of increased information borne out by the manner in which people *actually* use the information which he or she provides? It is evident that knowledge of these matters would be of considerable value to those involved in designing weather forecast 'programmes', as well as to those who are concerned with developing policies to encourage more efficient adjustment to weather variations, and more efficient use of weather forecasts.

D. THE BENEFITS AND COSTS OF WEATHER AND CLIMATE INFORMATION

1. Actual and potential aspects

An important aspect of the weather and climate sensitivity of various sectors of an economy is the actual, realized, or potential value of information about the weather and the climate. The relationship is of course not simple, and in many cases it is easier to assess the value of weather and climate information to a highly sensitive but very small economic sector than to a slightly sensitive but relatively large sector of an economy. For example, the raisin industry of the United States, although relatively small economically, is highly sensitive to some kinds of weather and climate information (see Kolb and Rapp 1962, and Lave 1963) which therefore have a potentially high value. In contrast, the manufacturing industries of most countries, although relatively high in overall economic importance, usually have a relatively low sensitivity to all but extreme weather and climate events as well as weather and climate information (see Palutikof 1983). Accordingly such information, even potentially, has a relatively low value.

Various studies have shown that the benefits of proper weather and climate information are high (see, for example: Gibbs (1968), Thompson (1967), and Berggren (1971)), and in a significant study by Hipp (1972), on the then proposed European Centre for Medium-term Weather Forecasting (ECMWF), the estimated benefit/cost ratio for the centre was assessed to be at least 25:1 by 1980. The reason for the high benefit/cost ratio, according to Hipp, was associated with the anticipated co-operation that this project would give to the participating countries, notably the opportunity to reap a substantial *national* benefit at a fraction of the normal cost if the country had approached the problem itself.

Several national attempts have been made to place a dollar value on weather and climate information including weather forecasts. Some of the first such analyses showed a benefit/cost ratio varying from about 5:1 (e.g. Tolstikov (1968), for the Soviet Union), to about 20:1 (as estimated by Mason (1966) for the United Kingdom). A similar early survey made for New Zealand (Maunder 1972c) showed that the then *potential* value of meteorological services in 1968/69 was about $32 million, compared with the then costs of providing these services of about $1.8 million. This analysis therefore showed a potential benefit/cost ratio of about 17:1. It is of course extremely difficult to calculate in any detail the various benefits – particularly those associated with *public* weather forecasts and the provision of *public* weather and climate information, but it is now evident that the costs of providing weather and climate information in most countries is relatively small compared with the potential benefits. A few examples of the difficulties involved in assessing the benefits are now given. It should be emphasized, however, that very few in-depth studies of the *actual* benefits to various sectors of an economy have been made. Indeed, the required reference to studies made more than twenty years ago points clearly to the dearth of 'hard' information on this subject.

2. Agricultural benefits

In the United Kingdom, Mason (1966) estimated the annual benefits of weather forecasts to agriculture to be at least 0.5 per cent of the gross farm income. Translating this percentage to a country like New Zealand for the year 1984/85 would give a benefit of $NZ22 million (0.5 per cent of the 1984/85 gross farm income of $NZ4,500 million), equivalent to about $NZ1/farm holding per day. Whether this value per farm holding per day is too high or too low is debatable, but it does give a first estimate of the probable value. Such a value will of course vary from area to area and country to country; it will also vary according to what weather and climate forecasts are available, how the weather and climate forecasts are used, and how much of their potential value is being realized. There is also the difficult question of assessing the value of climate forecasts (as distinct from weather forecasts), but in some circumstances climate forecasts (one, two or three months ahead) would clearly have a potential value considerably in excess of the 0.5 per cent of gross farm income suggested for weather forecasts.

The value of real-time and historical weather and climate information (as distinct from forecasts) must also be assessed, and with the advent of electronic instant-recall information systems, the potential is considerable. No firm data are available on the value to agriculture of these *videotex information systems*, but it is reasonable to suggest that these new systems – together with the traditional availability of real-time and past weather and climate information – have a potential, if not actual, value to agriculture of at least an additional 0.3 per cent of gross farm income.

3. Building and construction benefits

In the United States, Russo (1966) estimated that the total weather losses in the building and construction industry at 2 per cent of the value of production, whereas in the United Kingdom, Mason (1966) estimated the total weather losses as 3.5 per cent of the value of production. Mason further assessed that 10 per cent of these losses could have been *avoided* through the appropriate use of the weather forecasts *available at that time*. Again, applying the United States survey to a country like New Zealand (with 3 million people) would mean that, of the $NZ2700 million of building and construction expenditure reported in the 1978/79 census of building and construction in New Zealand, the actual losses due to weather would have been about $NZ50 million, of which 10 per cent (or about $NZ5 million), using Mason's estimate, could have been avoided through the optimum use of weather forecasts.

In terms of the value of weather and climate *information* to the building and construction industry – as distinct from weather forecasts – it is reasonable to assume that these would be of lesser value, but would probably be at least 0.1 per cent of the annual value of the building and construction industry, or about $NZ3 million ($NZ1 per person) in terms of the 1978/79 building and construction industry in New Zealand. Thus, the total value of weather forecasts and weather and climate information to the building and construction industry in New Zealand in 1978/79 could have been of the order of $NZ8 million. In New Zealand about 30,000 *house units* or equivalent are constructed in an average year; the total value of weather and climate information (including forecasts) to the industry would therefore be the equivalent of about $NZ250 per *house unit*. As with agriculture, these estimates will vary considerably from area to area, and country to country, but the data do give an indication of their probable value.

In a much more recent paper the use of weather forecasts at construction sites in Sweden is reviewed by Wahlibin (1984). He notes that as is to be expected at a modern construction site, the current day's activities are very much predetermined, but that plans for the coming day such as decisions to cast concrete (or not) rank high and are clearly dependent on the weather that is forecast. Wahlibin also discusses the economic value of weather forecasts to construction in Sweden. Users in the market were asked to give their estimate of the value of weather forecasts, the values ranging from 5000 to 10,000 Swedish kronor, with an average of 6900. Wahlibin further comments:

> In relation to an estimated mean turnover per month . . . , the user estimates (of the value) are 1–2 per mille of the costs during the test period. This is the same estimate as that made by the investigators in the 1980–81 user test. Aggregating estimated values and comparing to the costs of the forecasts, the value is 7 times the price paid by the association and the cooperative company buying the forecast service.

4. Manufacturing benefits

The value of appropriate weather and climate information to the manufacturing sector of an economy is probably less (on a percentage basis) than to either agriculture or construction. However, in several food exporting countries, such as the United States, Argentina, and New Zealand, some aspects of the manufacturing industry are particularly weather sensitive. For example, in New Zealand one significant weather-sensitive manufacturing sector is the large food processing industry which in the mid-1980s had a turnover exceeding $NZ3000 million (equivalent to $NZ1000 per person). If only this industry is considered and the value of weather and climate information is assessed at a nominal 0.25 per cent of the total turnover, then the value of weather and climate information to the food manufacturing industry in New Zealand could be expected to be at least $NZ7 million (equivalent to more than $NZ2 per person). The *total* value of weather and climate information (that is past, current, and forecast weather) to the total manufacturing industry in New Zealand would of course be considerably in excess of these values.

5. National benefits

The total benefits of weather and climate information to any nation obviously include items in addition to those just discussed. In particular, the energy sectors in most countries are not only large but usually very weather sensitive – in terms of requirements for heating, cooling, transportation, and agriculture. In addition, there is the value of *public* weather forecasts as well as the value of more specialized weather and climate information used for such things as ship routing, advertising, and highway maintenance.

In the New Zealand situation – as already noted – it was originally estimated by Maunder (1972e) that the potential value of weather and climate information in 1968/69 was $32 million compared with the *then* costs of providing these services of $1.8 million. A more recent analysis by the author indicated that the benefits to agriculture, building and construction, and manufacturing total at least $80 million, which compared with the current annual expenditure (in 1985) of $16 million gives a benefit/cost ratio of 5:1. It is evident that if all sectors of the New Zealand economy were considered the benefits would be considerably higher; it is therefore reasonable to suggest that in most countries an analysis of the benefit of weather and climate information to all weather- and climate-sensitive sectors would provide a benefit/cost ratio of at least 5:1, and more likely a ratio exceeding 10:1.

With these benefit/cost ratios in mind, it is relevant to note that a common theme through most of the analyses of the benefits of weather and climate information is the difficulty of obtaining information as to the *real* value of weather information. Over twenty years ago Mason (1966) had some pertinent comments to make on this subject which are clearly worth repeating:

> The benefits of the weather to the community, being partly social, are difficult to evaluate in purely financial terms. One cannot apply commercial criteria based on the price the customer is willing to pay because the basic service is free. The economic value of a special service for a particular customer can usually best be judged by the customer himself, but he may be reluctant to divulge this for fear that the charges will be increased. In some cases it is sensible to ask to what extent would a particular industry suffer in the absence of a weather service, but others, such as aviation, could hardly have become established in these circumstances. Despite the difficulties, I think it important to attempt these cost–benefits studies even though, as in most economic problems, neither the data nor the assumptions are very firm.

E. FOOD COMMODITIES, WEATHER INFORMATION, AND POLITICS

The growth in the world population requires, according to Baier (1977), production increases of almost 30 million tonnes in food and feed grains *per year*. This means that the 1980 production should have been some 100 million tonnes above the record output of 1976 – this increase being equal to three times Canada's normal production – and by 1985 world production should have exceeded the 1976 record by an amount equal to the current production of the United States. Baier further noted that the achievement of this goal requires a concerted global approach to monitor the tightening food supply/demand situation (see also Baier 1974). Information provided by agricultural meteorologists is clearly essential in providing more lead-time for planning and decision-making at the national and international level.

The use and value of crop-weather models in this context are discussed in detail by Baier (1977), and, using such models as examples, he rightly pointed out that the exploitation of both the vast amount of climate data in the archives of national weather services, and the real-time weather information available over the WMO Global Telecommunication System (see Robertson 1974), presents a real challenge to agrometeorologists at three levels: (1) at the international level, the input of summarized past and present weather and climate data to the FAO Global Information and Early Warning System on Food and Agriculture (see Chapter II for details), as well as the monitoring activities of the World Meteorological Organization in association with developments following the establishment of the World Climate Programme; (2) at the national level, agrometeorological techniques permit the monitoring of agricultural production prospects as a basis for planning agri-business operations, export/import policies, price supports and other government controls of the national economy; (3) at the local level, a meaningful interpretation of weather and climate in terms useful to researchers, extension officers, and farmers in improving crop production and the economics of agriculture can provide substantial economic and social benefits.

The international implications of providing *more* weather and climate information are considerable, and as discussed in Chapter II a proposal for an Agrometeorological Programme in Aid of Food Production was submitted by the Secretary-General of WMO to the Seventh Congress of WMO in May 1975. This programme included the assessment of regional and global crop conditions using current meteorological data, and a number of meteorological centres were to be assigned international responsibility for preparing and disseminating predicted crop-yield assessments on a global basis. It was considered, specifically from the viewpoint of food management and reserves, that all nations would benefit from the free exchange of an objective, weather-based advance estimate of expected crop yields on a regional and global scale, since such predictions could be used to provide more lead-time for planning and decision-making. However, as pointed out in Chapter II, this assessment and prediction scheme was considered to be 'premature' by the Seventh WMO Congress.

There will, of course, always be competitive advantages to those who know how to use information, and especially to those who first use such information. In certain circumstances, therefore, weather information can become almost too powerful a tool. It is, none the less, clear that there is a need for such information, and one wonders how long a system such as that described for predicting agricultural production from the *already* available weather and climate data of the World Weather Watch system can remain 'suppressed'. Conversely, and regrettably, perhaps the time is also not too far distant when the present 'free' exchange of the world's weather and climate information will be openly questioned, for there is little doubt that the use of weather and climate information, particularly in relation to its use as an early predictor of production, has many political and strategic overtones.

F. THE COMMODITIES MARKET

1. Weather and the futures market: A United States example

Of all the topics discussed in this book perhaps the most obvious money-related weather subject is the use of weather and climate information in the 'futures market'. Indeed, as will be well known to those familiar with the futures market, large financial gains (and losses) are almost everyday occurrences. In contrast, in considering the *stock* market, few investors would consider that there is any proven link between weather variations and the prices on the stock exchange (although this could be challenged); nevertheless, the more astute investors appreciate only too well that droughts and floods often have subsequent effects on the prices one has to pay for shares, or the prices one is able to realize when shares are sold. However, in the commodities or futures market, prices clearly vary – on the basis of weather information – daily and even hourly.

The commodities market is the exchange where commodities such as wheat, corn, soybeans, sugar, wool, cocoa, and frozen orange juice are bought and sold for delivery at a stated time in the future. The market exists for a number of reasons, but primarily the market enables people who *actually* trade in commodities such as farmers, chocolate firms, and wholesalers to reduce the risk they face of losing money because of changes in the prices of specific commodities. The futures market also allow producers to get a guaranteed price for raw materials they haven't yet produced. Further, they allow manufacturers to know exactly how much they will have to pay for raw materials they will need at some time in the future. Obviously, the prices for commodity 'futures' are related to a number of things, but of prime importance is the anticipated demand for, and supply of, a particular commodity at some future date. Consequently, there is a close relationship between the actual, reported, and expected weather conditions in the various producing areas, and the price at which 'futures' in a commodity will be bought and sold. Reports of the weather that has occurred, and forecasts of the expected weather, therefore have a real dollar value as far as the commodity markets are concerned, and consequently influence the prices that the consumer will eventually pay for these commodities.

The commodities market is perhaps the best example of weather and climate sensitivity in the business world, particularly as the market reflects not only the sensitivity of the *actual* weather and climate on commodity prices, but also – and in particular – the influence of *information* about the weather and climate. For example, it is well known that *reports* of frosts in Brazil, or freeze conditions in Florida, can affect commodity prices before confirmation of the *actual* weather conditions are known. But it needs to be emphasized that although weather and climate data are collected, analysed, and assessed in real-time, the data are at best *sample* surveys. Thus, *unofficial* reports of weather conditions – from secondary networks – although not part of the 'official' record, are clearly part of the information package available to commodity market dealers and their customers, and are naturally acted upon according to the 'sensitivity' of the markets. Of course, in addition to the weather that has occurred, or is reported to have occurred, there is the weather and the climate that is expected to occur in the future: tomorrow, next week, next month, or next season. The commodities market is therefore a 'melting pot' of past, present, and future weather and climate information, and it is in this melting pot that perhaps the *real* sensitivity of weather and climate is measured.

The role of the meteorological community in supplying weather and climate information which may or may not affect the market is a very important aspect of the commodities market. In a few countries, including the United States, the supply of such information is done by both the government-funded national weather service and private meteorological companies, and the commodities pages of the *Wall Street Journal* almost daily have a meteorologist (usually from the private sector) commenting on the future price of corn, wheat, or orange juice, or a weather-related explanation of what has happened or is likely to happen on the commodities market. Two typical examples of such comments from the *Wall Street Journal** during the January–March period of 1981 are as follows:

> Spring is warming the Florida orange groves, but juice processors are increasingly worried about the aftermath of January's freeze. Rumors of frost damage to the new Valencia crop propelled last week's futures prices to a peak of $1.41 a pound, up from as little as $1.27 the prior week and 78 cents before the frost.

Soybeans rose more than 21 cents a bushel amid reports of dry weather in South America and U.S. growing areas. Hot, dry weather in Brazil and Argentina during the last month is leading traders to reassess the outlook for the soybean crops there.

News of weekend rains in the Grain Belt pushed prices lower, with May-delivery wheat dropping 7.50 cents, to $4.33 a bushel. Generally improved weather is prompting some analysts to look for larger crops.

Although to many people not familiar with the United States commodity market, the weather-related comments quoted from the *Wall Street Journal* may appear to be well beyond the 'back of the envelope' explanation, or the more usual use of weather and climate information in medium- and long-term planning; such comments are very real, and in the United States – and some other countries – are acted upon daily by consumers and producers as well as by investors and speculators. However, despite the obvious sensitivity of the commodity market to weather information, even in the sophisticated United States commodities market the overall emphasis is on providing and assessing fairly *basic* weather information, and not specific *commodity*-weighted weather information. In the United States and presumably other futures markets, there is therefore still a large untapped market for the more imaginative meteorological consultant, to assist those people who really do want to know whether to buy or sell their wheat, wool, coffee, or soybeans. Further, in terms of weather and climate sensitivity, perhaps the commodities markets provide the real key to real-world weather and climate sensitivity, rather than the more academic approach which is so often removed from real-world problems.

2. Commodity prices: A specific example

The impact of a specific climate event on commodity prices can be quite considerable. Naturally, such price changes affect export markets, the ability to import, as well as consumer price indices or their equivalent. By way of example, the dramatic price change of one key commodity, soybeans, which occurred in the United States between the comparatively normal 1979 season and the very hot and dry 1980 season (Fig. III.2) is now considered.*

The price of soybeans in the United States varied during the 1979 season from a little over $6.50/bushel in January, to a peak of about $7.40 in midsummer, before decreasing to about $6.30 in December. In marked contrast, the price of soybeans in 1980 reached a low value of under $5.90 in May, and then – mainly due to the anticipated weather-related shortfall

* Clearly, similar examples could be cited for other commodities. For current examples, see the commodity page of the *Financial Times* or the *Wall Street Journal*.

FIGURE III.2 Changes in the US price of soybeans: 1979 and 1980

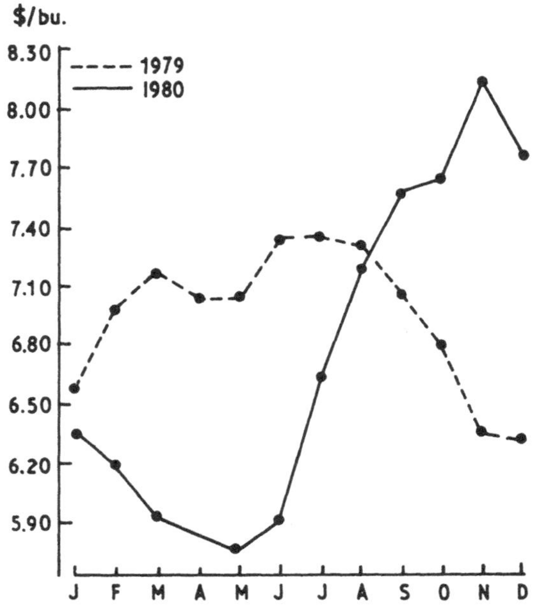

Source: After Maunder (1985a), from Liebhardt (1981). Data from Center for Environmental Assessment Services (1981).

in production – increased very rapidly to reach a peak of over $8.00 in November. The economic and social implications of such weather-induced prices are many and, coupled with similar price increases in other key food commodities, it is evident that significant weather and climate variations can play a very major role in overall price structures, both nationally and internationally.

G. COMMUNICATING WEATHER AND CLIMATE INFORMATION

1. Communications

Information about the weather and the climate is a key factor in many socio-economic aspects of weather and climate enterprises. Indeed, in several countries specialist information companies can provide users with any kind of weather and climate information they require – at a price. Most of the information marketed by such companies is highly time-dependent but whether the question relates to, say, the optimal routing of a ship from Calcutta to New York, or weather and soybean production in Brazil, information about the past, present and future weather and climate conditions is available to assist the decision-maker.

The presentation of such weather and climate information is, of course, a highly specialized and important form of communication. Specifically, the weather and climate package can be viewed in three phases: first, the *preparation* of the package – the technical task of the meteorologist or climatologist; second, the *presentation* of the package through radio, television, newspapers, journals, or videotex systems; and third, the *use* or *application* of the weather and climate package by the consumer. In all cases education and marketing is the key: first, education of the meteorologist and the climatologist in the techniques of the media; and second, marketing of the weather and climate package to the consumer in the best possible way. To do these things, meteorology and climatology must broaden its vision by actively encouraging research in the social, economic, and marketing aspects of the profession. This could be done by encouraging meteorologists to look anew at some of these problems, and weather forecasters, because of their particular interests, may be the best people to do this. In addition, employment of social scientists, economists, and marketing experts in meteorological services should be given particular consideration, and sponsoring of research by the specialists should be undertaken. But of prime consideration should be the realization that meteorology, and particularly the communication of weather forecasts and weather information, embraces many things besides the physical sciences.

2. The popular and the business press

The *Press* (Christchurch, New Zealand), in an editorial on 10 October 1981, stated the following:

> The time is long past when weather forecasting was something of a national joke and when the predictions of meteorological services were regarded – no doubt unfairly – more as an expression of opinion than as scientifically based forecasts. Greater knowledge of weather patterns, more sophisticated equipment and detailed information gained from satellites have enabled forecasting that is a reasonably sure guide to the weather.
>
> Specialised weather forecasts are supplied for people who have to make decisions of much greater consequence than domestic plans. When forecasts are reliable, decisions that may have important economic consequences can be made with greater confidence. Builders, farmers, foresters, fishermen, all need sound forecasts: serious errors may be disastrous for them.

Unfortunately the media are not always so constructive or astute – indeed the *Evening Post* (Wellington, New Zealand) in an editorial 'Weather watch' on 22 August 1981, referring to the new computer at the British Meteorological Office, stated:

> Hope springs eternal! The new computer . . . will at least be able to tell you that it's useless to waste 6 days at Lord's . . . for the one hour play possible each day . . . come to think of it, this could have an electrifying economic effect. Company directors . . . instead of lounging around the Long Room . . . may actually be doing some work. Perhaps there really is method in the British Met. Office's madness!

The *weather game* the media often play is, of course, quite fascinating but despite the comments expressed above there *are* many influential newspapers and business publications which clearly have a much greater awareness of what it is all about. Indeed, in recent years, editors of many influential economic, news, and planning publications have become very much aware of the importance of weather information in national and international business activities, and its growing importance in national planning. For example, *Newsweek* (2 January 1984) stated:

> As ever, human error and aggression, flukes of nature and plain bad luck left ordinary people suffering. . . . Even the weather seemed especially bad around the globe, apparently because of the shifting of the Pacific current called El Niño. Drought and famine afflicted large areas of Africa, Australia and Brazil, and floods washed out homes and farms in Europe. And in the United States, a heat wave withered some of the world's best grain fields, sending clouds of priceless topsoil blowing in the wind.

Similarly, *Time* (9 August 1976) under the heading 'The World's Climate: Unpredictable' pointed out that 'Record rains and floods soaked some areas, while droughts parched others, with potentially serious social, economic and political effects.' Then the *United States News and World Report* (5 July 1976), in a special report on the Third Century (of the United States), quoted from a study of the Central Intelligence Agency that:

> In the cooler and therefore hungrier world the U.S.'s near monopoly position as a food exporter . . . could give the U.S. a measure of power it had never had before – possibly an economic and political dominance greater than that of the immediate post World War II years.

The human aspects of climatic uncertainties are usually well reported in the media. *Newsweek* (29 August 1983), for example, vividly describes the 'long' drought in Brazil in the following words:

> For five years a relentless drought has parched an area of northeastern Brazil five times as large as Italy, ruining crops, property and countless lives. Just a few miles away, the towns and cities along the north-east coast remained relatively unaffected by the bad weather. But last week the drought's devastating consequences finally reached the picturesque coastal cities. For three days, thousands of peasants – most of them women and children – invaded the town of Crato in Ceará state, demanding food, water and jobs.

The 'purely' economic aspects of drought are also well appreciated by business writers such as those of *The Economist* (7 August 1976) in a cover-page article on 'Europe's Scorched Earth', which noted that it is the 'extension of drought from farming to industry which must really scare Europe's planners if the rains do not come'.

Media comments such as these could be provided for almost any period. A key question here in understanding weather and climate sensitivity is why did the editors of such business journals consider that there was indeed something newsworthy in the events that they reported. The answer probably relates to no more than a 'feel for the situation', but it could reflect a real understanding of how environmental factors (including the weather and climate package) really do influence the minds of people and nations. Whatever the reason, it is clear that climate sensitivity – including its social and political overtones – is recognized by at least some of the media, and is not restricted to either the data-rich industrialized nations or the data-poor developing nations.

One important aspect of the daily newspaper in communicating weather and climate information which should not be overlooked is, of course, the daily weather column or, in some of the more enlightened newspapers, the daily weather *page*. Such columns or pages provide readers with 'all the editors believe they need to know' about the local, regional, and global weather scene, and there appears to be a wide difference of opinion as to what should be published. Clearly it is not possible to comment critically on what – in the context of weather information – is (or should be) published in the world's daily newspapers, but certainly mention should be made of the enlightened viewpoint of the management of *US Today*, which publishes a full colour page of weather information in both its European and Asian editions. In addition, the 'domestic' edition of *US Today* not only gives readers a full page (in colour) of the United States weather scene, but also provides nearly every day an informative comment, description, and/or diagram of a meteorological event of social, economic, or political importance.

3. Television weathercasting

Most meteorological services are very much aware of the increasing need for accurate and timely weather and climate information to be available to both the public and the more specialized user. But information about the weather and the climate is much more than tomorrow's forecast or the noon temperature in Tokyo, Tahiti, or Toronto. Irrespective of the value of this type of information, however, the public quite rightly still wants to know what the weather will be like tomorrow, and what better way is there to present this information than through television.

The 'Weather' is in fact a prime television programme, and in many countries it is deliberately put at the end of a news programme to hold an audience. In some countries the role of the weathercaster on television is really in its infancy, especially when compared with the 'all weather' television channel in the United States which uses twenty-seven meteorologists and operates twenty-four hours a day through a cable television network. However, whatever the type of weathercast available the viewer is always keenly interested in how it is presented, and what is presented. An article in *Time* (17 March 1980) on 'The Wonderful Art of Weathercasting' is full of gems, particularly the following:

> It is an odd and specialised calling: not exactly journalism, not exactly meteorology, not exactly soothsaying, not exactly show business, but parts of all four.

But weathercasting is also more than this; indeed most people have definite views of what they like and what they don't like. In many countries the 'art' of weathercasting has had a long period of development, and today in the United States a television company will spend many hundreds of thousands of dollars to attract the 'right' weathercasters, and provide the 'right image' for the viewer. Nevertheless, there can be too much of a good thing, and the following extract from the *Evening Post* (Wellington, New Zealand) of 24 August 1985, written by Francie Brentzel, highlights the fact that 'getting the forecast right' is what it is really all about.

> So I said to him, 'New Zealanders just don't know how to "mediatise" the weather like in the States'. 'What do you mean?' he said. 'Well, last night this very distinguished-looking television news reporter said, "We are forecasting fine weather in Wellington tomorrow" '. 'What's wrong with that?' he said.
>
> 'What's wrong! That was it. That's all he said. I want more, I tell you. I want to know the average temperature for this date and the all-time high and low temperatures for this date since biblical times. I want to know high tide and low tide, the exact moment of sunrise and sunset, and the phase of the Moon. I want to know the barometric pressure, if it's rising or falling, and if there are any small craft warnings.'
>
> 'Why do you want to know that?' he said. 'We don't have a boat.' 'Don't interrupt. I want to see the weatherman point to cities on an electronic map; then I want the satellite still-photos of New Zealand speeded up to simulate weather systems' movements across the country. I also want to know the humidity, the pollution index and whether or not I should stay indoors if I have a heart condition.'
>
> 'There is no pollution in New Zealand.' 'Well, don't you think the people of New Zealand would like to be told! But let me finish. I want all this information superimposed over scenes of people living the weather – you know people sunbathing, people struggling against the wind, people with their umbrellas turned inside out. And, if the weather is really bad, I want to see a reporter out there on the street live asking folks how they're coping with the elements.'
>
> 'Look,' he said, 'the man said it was going to be fine today in Wellington and it certainly is fine, isn't it?' 'It's just that I'm used to useless facts.'

4. Videotex systems

The 1980s have brought significant changes in the way weather and climate information is communicated, and the development of videotex systems in many countries offer a considerable marketing challenge to meteorologists and climatologists in the provision of past, present, and future weather and climate information. Indeed, it is evident that as we move into the late 1980s communication of information will be paramount in many weather- and climate-dependent decision-making activities. This is particularly so, as the various videotex systems become more common.

Videotex systems are specially designed to provide information to specific groups. For example, systems capable of providing information to groups such as wheat growers in Australia, grain storage companies on the Canadian Prairies, kiwifruit growers in New Zealand, urban planners in Scotland, international soybean buyers in St Louis, hydro-electric authorities in Switzerland, railway operators in Japan, television companies in the United States, or wool marketing companies in the Southern Hemisphere, are expected to be the norm well before the end of the present decade. The type of information that can be provided (and already is in some cases) to these specialized customers includes the whole range of weather and climate products, including – at least in theory – every weather and climate forecast, and every piece of weather and climate data available in every national meteorological service. That such a range of information will be available to any customer at the touch of a button creates a formidable marketing and educational challenge to meteorologists and climatologists, and this will only be successfully met if a bold and imaginative approach is used. But clearly, there are already developments in a number of countries which will mean that *communicating* weather and climate information will finally break through the communications barrier.

IV The weather–economic mix

A. THE SPECTRUM OF WEATHER AND CLIMATE VARIATIONS

1. An overview

Weather and climate fluctuations are producing and will continue to produce major economic, social, and political consequences. Some of these are relatively short-lived (although very dramatic at the time), such as the record low temperatures which affected several parts of the United States during January 1982, while other weather/climate events are much more significant such as those reported by the official Soviet news agency Tass on 28 January 1982. Stating that '1981 was one of the most difficult years for the economic development of the USSR', Tass commented that extremely unfavourable weather – a drought of rare intensity and duration – had adversely affected not only agricultural output but also a number of industries.

One application of this weather–economic mix has been the increasing demand for improved techniques for measuring the sensitivity of national and international economic activities to weather and climate variations. This has been due in part to the increasing impact that food surpluses and deficiencies are having on world political and economic events, to the increasing cost of energy and its implications for heating and cooling, and to the use of agricultural fertilizers. In addition, there is increasing national and international concern expressed in several areas that people – because they are not judiciously utilizing their climatic resources – are living beyond their climatic income.

The traditional response of the climatologist to interpreting regional and national economic indicators in terms of weather and climate is usually extremely conservative. However, as several national meteorological services now not only observe, collect, and process weather and climate data in real-time, but also analyse it in real-time on a commodity basis, it is evident that key econoclimatic indices can now be computed and made available not only regionally and nationally, but also, if required, on an international basis. The use and value of such indices are discussed in detail in Chapter V.

An important refinement of the weather–economic mix is the ability to use weather/climate information to adjust economic indicators. For example, the 'weather-adjustment' of the components of the *Business Week Index* of United States economic activity has been examined by Maunder (1982a) to demonstrate among other things that some economic activities such as electricity consumption reflect – at least in part – environmental conditions rather than pure economic 'strength'. Results from this survey of the weather adjustments that *could* have been made to the *Business Week Index* for the summer weeks in 1980 are discussed in detail in Chapter VII. It must be noted, however, that the concept of 'weather adjusting' economic indicators is usually considered to be beyond the scope of both the meteorologist and the economist, and many consider that such adjusting is meaningless. Nevertheless, weather information which *is* suitably weighted and specifically adjusted to take into account various economic distributions such as population, energy consumption, corn production, electricity production, etc., is available for such adjustments, which are not only desirable but essential if a proper interpretation of weather/climate sensitivity is to be assessed.

A final refinement of such weather/climate adjusting is the potential use of commodity-weighted weather information in forecasting the trend of national economic indicators. Further, since weather information is available in real-time, whereas weekly national economic indicators are subject to a publication delay of at least two or three weeks (or months, depending on the economic parameter), weather-based forecasts of economic activity can be made available one, two, or three weeks (or months, depending on the economic parameter) *before* the actual production/consumption information is available. Although attempts to weather/climate adjust national economic indicators and to make weather-based forecasts of such indicators are difficult and controversial, it is considered that such attempts not only provide economists, decision-makers and publishers of economic indicators with realistic adjustment factors, but they also provide needed assistance for the *economic* forecaster whose track record in many countries leaves a lot to be desired.

Indeed, it is considered essential for the applied climatologist to assert a more positive role in assisting economists, agriculturalists, politicians, and other decision-makers in a better appreciation of the weather/climate–economic mix. Essentially this involves a much better understanding of the weather/climate sensitivity of the various aspects of an economy. But in order to achieve this new level of understanding, a much more concerted effort will be required by a number of people in areas peripheral to the normal activities of those concerned with weather/climatic impacts. It is believed that such people can contribute in a very positive and useful way in assisting those decision-makers concerned with some of the world's important problems associated with the supply and utilization of food, fibre, and energy.

2. The atmospheric component

Extreme weather events, such as hurricanes, tornadoes, lightning, major floods, droughts, and hail, often result in catastrophic losses of property and income, and on many occasions serious losses of life. Estimates for the current extremes of hurricane costs in the United States, for example, indicate that the extreme *cost* in a year is likely to exceed $5000 million and may approach $10,000 million. Tornadoes, lightning, floods, droughts, snow, hail, and fog can also cause significant losses. It is nevertheless important to point out that near 'above average' and near 'below average' weather and climate conditions are also economically significant, and it is surely reasonable to suggest that it is the weather and climate events which do *not* make the newspaper headlines which are more important in the long run than those that do.

An important question to be asked, therefore, is whether it *is* the relatively small changes in the overall climate rather than the extreme events (or alternatively, whether it is the 'small' long-term events, or the 'larger' short-term events), which have the more important overall social, economic, and political impact in the long run. For example, a cooling in the Soviet Union in the summer by 3°C, a decrease in cloud amount over North America in the winter by 10 per cent, or an increase in the temperature of the South Pacific Ocean in the summer by 1°C, particularly if sustained for more than one season could well have social, economic, and political consequences which in total could be much greater than the destructiveness of hurricanes, severe rain storms, or tornadoes (see e.g. Alexander 1974; Maunder 1977a; Robertson 1974; Taylor 1974).

The long-term climate variations are also important. During the first half of the twentieth century, for example, much of the Northern Hemisphere experienced a period of unusually warm weather. However, the 1950s and, in particular, the 1960s showed significant signs of a change in some areas to a wetter and colder climate, while the 1970s and the 1980s have been a mixture of very mild and extremely cold seasons. By contrast, in New Zealand, and probably other Southern Hemisphere countries as well, the long-term weather variation is in many ways out of phase with that which has occurred in parts of the Northern Hemisphere, particularly during the last thirty years. For example, the mean annual temperature at Auckland (based on observations at the various sites being adjusted to the current observing site) showed an increase from about 14.5°C in the 1900–20 period to about 15.5°C in the 1950–80 period. In addition, the *nationwide* average temperatures for New Zealand in both July 1984 and July 1985 were among the highest average July temperatures for New Zealand since observations were first taken in the 1860s.

3. Political and marketing realities

Linked with both short- and long-term climate variations is the knowledge that the world's population is growing exponentially, and that there are now no longer exploitable virgin territories just 'over the mountains' because the world is, at least in a practical sense, finite. Of major concern is the fact that the crop-production areas of the earth are limited in extent and yield. McQuigg (1974) pointed out that this situation has led to climate becoming more and more a significant factor in grain production. That is, as grain reserves continue to fall, climate variability becomes more important as a sensitive controlling parameter in determining future supplies. Thus climate plays an ever more important role in international politics. The United States, for example, provides about one half of all the grain that moves across international boundaries, and so by withholding grain could exert considerable political pressure. Several news magazines have also commented on the power that the agricultural capacity of the United States could wield. For example, *Business Week* (15 December 1975) suggested that 'there is a growing consensus that the US should be as tough in using food power to achieve national objectives as it is in employing other economic capabilities'.

In the past, there have been many meteorologists and climatologists who were quite content to remain in ignorance of the value of their work. However, it is now very evident that the considerable expenditure on meteorology and climatology must provide appropriate returns – social, economic, and political. However, while budgetary considerations are important, a more fundamental reason exists for being interested in the impact of weather and climate information, notably that people and nations will undoubtedly try in certain circumstances to apply the rapidly improving flow of weather information in a much more deliberate manner, in order to monitor and control significant weather-sensitive aspects of national economies. The deliberate control or modification of the weather and climate of one country by another country may *also* become a reality. In this latter case, it is clear that such control or modification will – at least in the foreseeable future – be limited in area, but its effect on certain aspects of agriculture could be very considerable. Monitoring and control of the atmospheric environment are therefore of fundamental importance, and are matters which will continue to have increasing economic, social, and political consequences at the very top level.

As a postscript to the above discussion, it is perhaps relevant to note that few if any of the meteorologist's models of the physical atmosphere *go on* to consider the economic and social aspects of the variability of the atmosphere; in fact, until the 1970s generally only token consideration was given to the end products of the meteorological chain. But since it is now well recognized that weather and climate information has social, economic, strategic, and political value it is most important that such information be presented to the decision-maker in much the same way as any consumer goods are packaged for efficient and effective marketing, and further that such information is correctly used. But if the improved weather and climate package is to achieve its purpose, the appropriate national or international meteorological and climatological authorities and their controlling governments must be convinced that there are important customers who need the information that could be provided. The meteorological and climatological community therefore has a continuing task of providing ways to establish direct communication between top-level decision-makers and the meteorological system.

B. THE IMPORTANCE OF WEATHER AND CLIMATE VARIATIONS

It is evident that the world is becoming much more sensitive in the economic, social, and political sense to the resources of the atmosphere. Indeed, any variations in these resources, which in the past were easily accommodated, have now become very significant factors in both short- and long-term planning. For example, *Time* (27 June 1977) neatly summarized the impact of weather and weather information on coffee prices. It stated:

> For about two years coffee drinkers have bitterly watched prices jump from $1.45 a lb to more than $4. A crop-killing frost in Brazil in 1975 touched off frantic bidding by buyers who feared a shortage. . . . Now the U.S. Department of Agriculture forecasts that Brazil . . . will harvest about 17 million bags of beans in the crop year that begins October 1 – not far from double the 1976–77 crop of 9.5 million bags. In all, world production this year should increase about 14%. . . . At the same time . . . slack demand and the prospect of heavy harvests have driven down future prices on the New York Coffee and Sugar Exchange. Contracts for coffee beans to be delivered in July fell to $1.95 last week, a drop of $1.45 a lb since April 14.

Similarly, *Newsweek* (11 April 1983) noted the importance of weather on produce prices in the United States as follows:

> Produce markets in the East and Mid-west are already suffering, too. 'I can't remember a year with weather like this,' says A. H. Nagelberg, president of the New York Produce Trade Association. 'Wholesale prices are ridiculous.' The cost of West Coast-grown cucumbers has increased by 50 percent, and the price of a case of broccoli has risen from about $10 to $20. Asparagus, which normally sells in early April in the East for about $1.50 a pound, is going for as much as $3 – twice the price.

It is of course not surprising that in recent years the editors of influential economic, news, and planning publications have become very much aware of the importance of weather information in national and international business activities, and its growing importance in national planning. For example, the record cold 1976–77 winter in the eastern United States gave editors plenty to talk about, such as *Time* (21 February 1977) which in a report on 'Assessing the cold's damage' stated: 'Now that a thaw is in sight, economists can begin to measure the impact of the unforeseen cold on the economic recovery.' The verdict of *Time*'s Board of Economists: 'A sharp, but strictly temporary, setback.' In like manner *Business Week* (31 January 1977), in an article 'An old-style winter disrupts the economy' said that 'the most critical impact of the frigid weather is sharp curtailments of natural gas, which have resulted in widespread shutdowns of industrial plants'.

A similar situation has arisen in the reporting of events which are part related to weather and weather information. For example Schramm (1975), writing in *Challenge* under the title 'The perils of wheat trading without a grain policy', stated:

> In 1972, a momentous event made painfully clear a number of contradictions between America's domestic agricultural policies and its diplomatic goals. That event was the Russian wheat deal. Over one-fifth of the 1972 American crop was sold to the Soviets at bargain prices, resulting in such shortages in the domestic wheat and flour market that the price of a loaf of bread nearly doubled.

In contrast, news of a different nature in relation to food production in both the United States and the Soviet Union occurred in early 1977. For example, *The Economist* (15 January 1977) reported that the Soviet Union has just harvested a record grain crop of 224 million tonnes – more than enough to feed her 255 million people and 320 million livestock, while *Newsweek* (14 March 1977) quoted the (then) US Secretary of Agriculture Bergland that if there were a new wheat glut the United States might be forced back into the 'soil bank' programme, paying farmers not to plant all of their acreage.

The potential economic, social, and political value of past, present, and future weather is also made very clear in an article on 'Drought in Africa' in *The Economist* (11 June 1977). Under the title 'Acts of Man', the report stated:

> The Sahelian disaster prompted a guilty realisation that a new assessment of the underlying causes of drought was necessary to prevent it ever having such consequences again. . . . A succession of good years disguised the incipient calamity. Only a temporary deterioration in the climate was required to bring the whole system to its knees.

The experience gained, according to this article in *The Economist*, was that no grand design was going to transform the marginal savannah lands of the sub-Sahara, and that 'modifying the climate, growing a green wall of trees across the desert or piping desalinated water from the sea all belong to the realms of science fiction'.

More than two decades ago, when food surpluses were probably at their highest, Watson (1963) stated:

> Climate determines what the farmer can grow; weather influences the annual yield, and hence the farmer's profit, and more important especially in under-developed and over-populated countries, how much food there is to eat.

In the light of recent events, the potential value of information about the past, present, and future weather and climate is obvious.

C. Climate effect studies

Studies of the influence of the climate and of variations in the climate upon the human use of the Earth's surface contribute in a special way to what may be called geographical climatology (see Mather *et al.* 1981). These studies not only provide information for answering practical questions, but they also provide information that can be used to evaluate data network design, and clarify cause–effect and feedback relationships so necessary in developing useful climate models for predictive purposes.

Of particular importance is the need for better and more useful climate-effect studies in the areas of energy and energy production, agriculture and food production, water resources, and health. For example, problems of energy supply (and costs) have generated considerable interest in solar radiation as an alternative energy to fossil fuels. Further, studies relating to the measurement, calculation, and estimation of solar radiation income have escalated along with feasibility studies relating to solar-house construction. However, it is evident that the climatologist must also become more involved in engineering applications, since construction and design are highly dependent upon the effects of site, orientation, and microclimatic fluctuations.

In agroclimatology, much of the research emphasizes the relationship between climate variability and either the productivity of 'human-altered' environments such as the Iowa corn field, or the productivity of 'natural' environments such as the grasslands of the merino sheep or the range cattle. The economic, and in many cases the political, importance of large-scale weather fluctuations must also be emphasized, since they contribute greatly to variability in global food production. A key research area is determining the impact of weather and climate on food and fibre yields; this has led to a variety of attempts to model the physiological responses of plants and animals, and to develop yield equations. In addition, climatologists now have available many sophisticated devices to measure the internal stresses in plants and animals that have been induced by climate.

There is also considerable research activity in certain aspects of forest climatology which are directed toward the determination of forest-fire hazards and the accurate specification of the moisture content of forest fuels. In addition, the rapid transition to forest monoculture indicates an urgent need for additional studies on the desirability of this silvicultural practice as it relates to the effect of climate variability upon unstable human-produced ecosystems. New predictive models also need to be developed to estimate the probability of organism outbreaks as determined by the weather, and climatologists should continue to benefit from the resurgence of international interest in this issue, encouraged in part by developments through the International Biological Programme. Moreover, the application of pesticides and the initiation of biological control techniques to retard the spread of unwanted organisms (particularly insects) can only be effective if the relationships between weather and the outbreaks are better understood.

Hydroclimatic studies designed to measure and compute streamflow, to determine lake levels, and to quantify the effect of increasing urbanization upon water quality and quantity represent additional directions in applied climatological research. Existing water-budget models are available to evaluate some of these problems, but further research is needed on infiltration rates, precipitation intensity, snow-melt run-off, deforestation, and groundwater movement. The effect of acid rain is a particularly important new research direction, especially in parts of the United States, Canada, and Europe. The causal mechanisms involved are of concern to the meteorological community, but the effects of such precipitation on weathering rates and the dynamics of plant, animal, and fish populations represent important research directions for climatologists. Other hydrological priorities include monitoring of the world's ice masses, and continued research on evapotranspiration rates as they are affected by changes in soil moisture, groundwater, and vegetation.

In medical climatology, studies have usually been carried out by the medical specialist rather than the climatologist. However, increased knowledge of air pollution and better health statistics, along with improved climate models of possible climate–health relationships, now make it possible for climatologists and the medical profession *jointly* to study in some detail such things as fluctuations in respiratory diseases, epidemics, or the pattern of hospital admissions, and so learn how climate stresses affect bodily functions.

In addition, a number of investigators have emphasized the importance of climate on 'human efficiency'. For example, Bandyopadhyaya (1983) in his book *Climate and World Order: An Inquiry into the Natural Cause of Undervelopment* emphasizes the impact of climate on 'underdevelopment'* as follows:

> In India and other tropical countries I have noticed farmers, industrial labourers, and in fact all kinds of manual and office workers working in slow rhythm with long and frequent rest pauses. But in the temperate zone I have noticed the same classes of people working in quick rhythm with great vigour and energy, and with very few rest pauses. . . . I had no doubt at all in my mind that the principal explanation lay in the differences in temperature and humidity between the two climatic zones.

---

* A more extensive discussion on this subject is given in Chapter 2 of Bandyopadhyaya's book.

D. The Value of Weather and Climate to an Area

By far the most significant weather and climate variations occur from day to day, week to week, and month to month, and it is these relatively short-term variations that decision-makers of weather- and climate-sensitive enterprises have to contend with. The value that can be placed on these climate resources in terms of the variations that occur in the short-term weather requires, however, the identification of *activities in the areas that are affected* directly or indirectly by weather changes, and an analysis of the manner in which a given change results in gains or losses to such activities.

The problems associated with the use of economically important climate indices in econoclimatic models is discussed in detail in Chapters V and VI, but an essential part of these models is the calculation of commodity-weighted weather and climate indices based on the contribution of various areas to the economic activities (such as human population, energy sources, or corn production) in a region or a nation. But, how should this be done?

First, the value of weather and the climate to areas must be determined, and this requires the identification of those activities directly or indirectly affected by both the weather and the climate, *and* by weather and climate *information*; secondly, it is essential to assess meaningful commodity-weighted weather and climate indices for these activites. It must be emphasized, however, that for most purposes, traditional climatological information – produced several weeks *after the event* – is of limited operational use (whatever its educational or historical value may be, or its value in 'non-operational' decision-making). To contribute significantly to national economic planning, national meteorological services must therefore provide decision-makers with *real-time* weather and climate information, appropriately commodity-weighted by areas, and this must be available on a time-scale which provides sufficient lead-time for decision-making.

E. Responses to Weather and Climate Variations

'Is it time we learnt a bit more about the weather?' asked *The Economist* (9 April 1977) in an article on 'Weather: magnificent ignorance'. The journal rightly pointed out that 'the effort to get to grips on man's magnificent ignorance about the weather is now increasing' but noted that more money needed to be provided by national governments in order to answer the more pressing problems. However, perhaps *The Economist* could have been even more specific, and expressed concern not only about the 'magnificent ignorance' of weather, but also about the 'magnificent ignorance' of *weather information*. The reason for this – as has already been noted – is that while it is comparatively easy to make assessments of the *general* relationship between weather factors and production or consumption, or in some cases prices, the more *precise* relationships necessary for operational decision-making are much more difficult to formulate. Indeed, although it is obvious that various human activities respond to and are thus 'sensitive' to weather and climate variations on a wide range of scales, it is at times extremely difficult actually to assess these relationships or sensitivities.

These difficulties arise first because very little research has been done on assessing climate sensitivity, and second, what research has been done is often regarded by the many critics as being superficial, and by the uninformed as being obvious anyway. But while the concept of sensitivity may be obvious, a measure of this sensitivity, or more importantly a scale to measure these sensitivities, is far from obvious. For example, is the raisin industry of Australia more or less sensitive to the weather and the climate than, say, coffee production in Brazil, or the area of wheat sown in the Soviet Union, or the quality of wool production in Uruguay, or the tourist industry in Jamaica? Further, how would one relate the climate sensitivity of the raisin industry of Australia to the climatic sensitivities of *all* other climate-sensitive activities within Australia, or any other country?

The problem of assessing the sensitivity of human activities to weather and climate is heightened by a lack of proven methods for sensitivity analysis. The critical importance of such assessments lies in the need to know more about society's vulnerabilities in order to allocate funds and energies in problem-solving. The climate sensitivity of a sector of a nation, or of a nation as a whole, is also closely related to the *total* information available about such sectors and nations.

Although it is usually considered that the developing nations are *more* sensitive to weather and climate than the developed nations, the true situation in several cases is probably the reverse. That is, while developing nations may indeed be more *vulnerable* to weather and climate variations, the sensitivity of economic sectors to such variations in the developed nations is usually more important in an absolute sense. That is, wheat production in, say, Canada or the USSR may well be not only more sensitive (in the economic, social, and political sense) to a weather and climate variation than, say, the rice production industry in Sri Lanka or Bangladesh, but may also be more important in absolute terms. It must be emphasized, however, that few if any studies have shown the true 'value' of weather and climate sensitivity, and with the present knowledge only an educated guess or a 'feel for the situation' is possible.

It is evident that an awareness of weather and climate sensitivity is needed by resource managers, government officials, planners, industry analysts, and others involved in social and economic decision-making. Clearly, several specialized government agencies (agricultural, economic, and intelligence) are already involved in decision-making in politically sensitive areas which involves the use of weather and climate information. But, while the future of all national economies is dependent to some extent on weather and climate, the question remains whether researchers and decision-makers *really* know how to assess the climate sensitivities of various social and economic activities. Further, do they know how to take these sensitivities into account in their decision process? It is suggested that in only a few cases (e.g. agriculture and energy use) do we possess a quantitative description of weather and climate sensitivity, and the potential costs of possible future climate variations. Further, even in these sectors we do not know how important these costs are with regard to the total economy, especially if the latter is stressed by adverse social behavioural or political conditions.

The question remains, therefore, where does the meteorological profession go from here? The issues are clear. First, there is a continuing need to assess and present the impacts of weather and climate in terms of production figures, costs, or other similar measures which can be used directly by economists, agriculturalists, planners, and politicians. Second, it is important for national meteorological services to appoint and actively encourage personnel who have a background that will allow them to become 'development meteorologists'.

F. CLIMATE SENSITIVITY

1. Climatic sensitivity: An overview

During the 1970s, individual nations and the international community became aware of and accepted the view that there are some limits to the availability of natural resources and food. Partly as a consequence of this, world conferences on population, energy, food, water, human settlement, desertification, and climate have taken place. Among other things this led Barbara Ward to remark in *The Economist* (13 August 1977) that 'the 1970's have, in fact, been a decade of remarkable innovation within the UN system'. She continued:

> Virtually every subject of real substance in planetary existence has been brought into the open, discussed and published in a way unique in human history. The transnational impact of certain problems (for instance, the high interdependence of winds, seas and climates in the natural environment or the similar interdependence of the world grain and fuel market . . .) has been recognised in the only way in which recognition could have been secured so speedily – by the participation of virtually all governments in open conference.

Climate variations – including 'changes' in the climate – are paramount in any consideration of weather and climate sensitivity, but several aspects of these terms need to be considered before any worthwhile analysis can be attempted. They include answers to the questions: what is climate, what is a climate change, what is a climate variation; is there evidence for climate change or climate variation in terms of specific changes in temperature, rainfall, soil moisture, etc.; if so, how is the climate changing; what are the possible causal mechanisms; can these mechanisms be modified by people; are the effects of a climate change or variation predictable, and at what degree of accuracy? Finally, and of prime importance, what are the economic, social, strategic, and political implications of climate variations and climate changes?

Weather and climate variations obviously have economic consequences (which may be extremely significant), but they also affect human activities in many other pervasive ways. It is therefore to be expected that people should try to find ways of adjusting to and even altering the weather and climate. But it is equally important to realize that modification of tomorrow's (or next month's) weather and climate may be very dependent on the weather and climate that is forecast for tomorrow (or next month), particularly if decisions as to whether or not to modify are going to be based on what is predicted. Moreover, it is important to recognize that *to the user* accurate weather and climate forecasting and successful weather and climate modification may well have similar impacts, since what matters is what actually happens. For example, the retail store manager or the cattle farmer is not concerned with *why* it rains, but only that it does or does not rain.

It is equally important to realize that longer-range (and more accurate) weather and climate forecasts, as well as the ability to modify tomorrow's weather, can be coupled with many disadvantages. This is because if *everyone* including decision-makers know what next week's or next season's weather and climate is going to be, or if everyone is going to have a hand in modifying what the weather forecaster and the climatologist promises, then there will clearly either be conflict resulting from the different needs of people, or no economic advantage to those who 'pay' for information about what is going to happen in the future.

A key factor in these matters is the question of the sensitivity of activities, economic sectors, areas, and populations to weather and climate variations and to changes in the climate. These include an examination of weather and climate sensitivity analyses, including those of the important sensitive commodities, and an analysis of both weather and climate information and how such information may be packaged for different user groups. Some specific examples of weather and climate sensitivity from both a commodity and spatial viewpoint are also considered.

How then does one begin in a study of climate sensitivity? Often, such a study is already dictated by circumstances or events; that is, there is a previous commitment or preference to examine the relationship between climate and a particular crop, region, or population. However, the questions of what activities to study, in which places, and affecting whom, are frequently not predetermined. The sections which follow attempt to assist the impact assessor by surveying which activities, places, and populations analysts have found important to study in the past, and the methods they have used to identify climate sensitivity.

2. Characterizing climatic sensitivity

There is not likely to be any single best method for characterizing a society's climate sensitivities. However, several early, related steps might be considered, including an analysis of the uses and the users of weather and climate information. This would provide an idea of which groups or activities value climate knowledge and thus may be exhibiting sensitivity. Second, a review of the appropriate literature related to climate impacts could be made. This would lead to an implicit ranking of activities as to their sensitivity to climate influences.

Illustrations of several additional analytical methods are now examined. These include analyses of communications media and information content, reviews of national income and product accounts, examinations of seasonality, and correlation analyses. These methods all enable one to take readily available data and perform tentative analyses aimed at identifying the climate signal in social and economic activities.

The climate impact assessor – using his experience and knowledge – first completes an initial or occasionally an in-depth assessment of relative or absolute sensitivities; the task then becomes one of expressing these values in a usable form. The standard approach to this is the calculation of quantitative indices that combine the climate state and the economic activity into a 'single number' that can be treated as an independent variable. This 'single number' characterizes the climate sensitivity for the activity concerned and can be thought of as an 'econoclimatic indicator' or 'index'. It should be noted, however, that the typical response of many climatologists to interpreting regional and national economic indicators in terms of weather and climate is usually extremely conservative – or more kindly, it is just not understood.

But key econoclimatic indices can now be computed and made available not only regionally and nationally, but also, if required, on an international basis. Such indices, if used with appropriate economic and political understanding, can give very useful decision-type information. They are also clearly far more valuable (for real-time decision-making) than the usual non-weighted station-by-station tabulation of climate information issued by most national meteorological services. Specific examples of the value and use of econoclimatic indices are discussed in Chapter V.

3. Identifying and assessing the sensitivity of various activities to weather and climate variations

The whole range of weather and climate services all depend upon a knowledge of the weather and climate sensitivity of various activities. But what are the weather- and climate-sensitive activities, and what part do weather and climate variations play in these activities? Consider, for example, the following: (a) What is a significant weather and climate variation? (b) How much variation is needed before it becomes important in producing effects or in affecting decisions? (c) Are such significant weather and climate variations the same from one area to another, one day to the next, one season to the next, and from one type of activity to another? (d) What specific effect does a significant weather and climate variation have on weather- and climate-sensitive activities such as the restaurant business, tourism, retail sales, unemployment, ice cream sales, airlines, or petrol (gasoline) sales?

If we had answers to all the foregoing questions, then (a) we would be in a better position to evaluate weather forecasting and climate programmes; (b) those involved in weather- and climate-sensitive activities could operate more efficiently; (c) the economics of weather and climate modification programmes (either deliberate or inadvertent) could be more precisely determined; and (d) adjustments to tomorrow's weather and the climate of the next decade would be less of a chance, and much more a matter of choice.

Although some research has been undertaken to answer such questions, much has still to be accomplished. Thompson (1966) was one of the first to consider some of these questions, and at about the same time a staff writer of the *Winnipeg Tribune* made a study of the cost to the average Manitobian of a winter (Tremayne 1967). In this article Tremayne said that Manitobians indulge in a perverse sort of luxury during the winter, boasting that they can take the worst the elements throw at them. But as Tremayne remarks, 'It costs us plenty for the privilege of enduring the ravages of winter.' Using information gained from numerous sources, Tremayne estimated that an average winter costs Manitoba's 959,000 residents $246,846,800 or $257 per person (in 1967 dollars). What, one may well ask, is the cost in other parts of Canada, or the United States, or the Soviet Union, and would better use of weather forecasts, or more accurate weather forecasts, or more useful weather and climate information, lessen these costs?

Identifying climate sensitivities often varies with the scale of activities involved, and in many industries, normal analysis by management may reveal the climate-sensitive points. However, the assessment is more difficult at the national and international levels where a variety of compounding factors determine resource activities and productivity. For example, extracting the impact of climate variations from the *signals* of national economic growth and decline is a formidable challenge. Further, assessing national vulnerability to climate variability may be especially difficult in newly developing countries where baseline data are poor and rapid social and economic changes are occurring.

There are few well-developed methods for making an initial assessment of an economy's climate sensitivities, but basic steps can be listed. These include (a) a review of the literature related to climate impacts which may lead to an implicit ranking of activities and their sensitivity to climate influences; (b) an analysis of uses and users of weather and climate information, which could provide an idea of which groups value climate knowledge and thus may be exhibiting sensitivity; and (c) a quantitative analysis aimed at identifying the weather and climate signal in social and economic activities.

Each of these steps is now discussed in detail in sections 4, 5 and 6 below. However, in many nations the best guide to weather and climate sensitivity is not obtained from any of these steps, but *may* be obtained by a critical analysis of the relevant agricultural, economic, and business journals, and by a careful appraisal of the nation's newspapers. Indeed in many nations – both developed and developing – the media (newspapers, journals, radio and television) often provide the *only* indication of the importance *in a real-time sense* of the weather and the climate. In particular, a media analysis is particularly important in two kinds of situations – first, in data-poor nations or areas, where what little data does exist are available for analysis only several months (and in some cases several years) after the event; and second, in those data-rich nations or regions where decisions are made on a day-to-day basis (and sometimes even on an hour-to-hour basis) and where the prices of commodities are paramount. In both cases the financial, economic, and agricultural sections of the daily press (or if available the more specialized newspapers such as the *Wall Street Journal* or the *Financial Times*) publish valuable information provided the reports on the 'economic climate' of sectors, regions, and nations are read with a critical eye.

The identification of the weather- and climate-sensitive aspects of any activity or area, the extent to which appropriate meteorological and climatological advice can be given, and the form in which it should be conveyed, clearly require close collaboration between meteorologists, climatologists, and the users of weather and climate information. This is to ensure complete understanding of both the operational and the weather- and climate-related problems. In complex organizations such as large industries or construction works, an appropriate analysis by the management may reveal the weather- and climate-sensitive points; however, joint discussion between management and the atmospheric scientist will always be found to be beneficial.

One specific pioneering study in this area of assessing sensitivity has been made by the Center for Environmental Assessment Services (1980) of the US Department of Commerce. In this study the sensitivity of elements of the gross national product (GNP) to widespread anomalous weather events was examined. A summary of their findings is given in Table IV.1, this table being compiled from an analysis of the weather and climate sensitivity of various economic and social sectors in the United States as reported in the *New York Times* during a ten-year period. The survey showed that a major *increase* in the GNP can result from an unusually hot summer (specifically in the GNP elements of personal consumption expenditures on electricity and

Table IV.1 Sensitivity* of Gross National Product elements to widespread anomalous weather – United States

GNP elements	*Weather: unusually* hot summer	cold winter	dry summer	storm/rain	snow	mild
1. Personal consumption expenditures						
(a) Gasoline and oil	–	–	–	–	–	++
(b) Electricity	++	+	?	++	?	–
Natural gas, fuel oil, coal	?	++	?	++	+	–
(c) Furniture and appliances	–	–	–	?	–	++
(d) Food at home	++	+	++	+	++	– –
Food away	– –	– –	?	?	– –	++
(e) Apparel	–	+	?	?	–	+
(f) New and used cars	–	–	–	–	–	++
(g) Housing	–	– –	?	?	–	++
(h) Transportation	–	–	?	–	–	++
(i) Other	?	?	?	?	?	?
2. Non-residential fixed investment	?	?	?	?	?	?
3. Residential	–	–	–	?	–	++
4. Change in business inventories	+	+	+	+	+	– –
5. Net imports	+	++	+	+	+	– –
6. Government purchases						
(a) Federal	+	+	+	+	+	–
(b) State and local	+	+	+	+	+	–

Source: After Maunder and Ausubel (1985). Adapted from Center for Environmental Assessment Services (1980).
Note: * Weather-related changes in consumption: + = increase; + + = major increase; – = decrease; – – = major decrease.

food at home), and an unusually cold winter (specifically in the GNP elements of personal consumption expenditures on natural gas, fuel oil, and coal; also in net imports), whereas a major *decrease* in the GNP can be expected as a result of unusually mild conditions (specifically in the GNP elements of personal consumption expenditures on food at home, changes in business inventories and net imports). While the sensitivities shown in Table IV.1 must be regarded as tentative, they do give an indication of how the gross national product of a nation may be affected by widespread anomalous weather and climate conditions.

Specific weather and climate events have also been studied by the Center for Environmental Assessment Services. For example, as detailed in Chapter V, there was a series of reports (see Center for Environmental Assessment Services 1981) during the 1980 summer heat wave and drought in the United States updating the mounting economic losses that were finally estimated at more than $US20,000 million. Six months after the last special report on these events was issued, official statistics were released confirming the earlier weather-based estimates. To compare the impact of weather and climate of this event with previous similar events, a report was also prepared (Center for Environmental Assessment Services 1982) on the 1976/77 winter in the United States which indicated that the economic losses during that winter were almost twice those of the 1980 summer heat wave and drought. These media-based studies of a national economy are a useful method of placing specific weather and climate events in perspective.

4. Review of weather- and climate-sensitive studies

There are two overlapping sets of studies that constitute implicit, if not explicit, surveys of activities or places that are highly sensitive to weather and climate. The first set consists of studies of the users and the value of the information that 'national' meteorological services generate. The second includes studies connected with integrated assessments, scientific symposia, or governmental and intergovernmental conferences.

Typical of the impact studies related to 'national' meteorological services are the early assessments of Mason (1966) for the United Kingdom, and Tolstikov (1968) for the Soviet Union. Such studies emphasize the importance of meteorological conditions and information for agriculture, and point out their role in a variety of other activities, such as aviation. In addition, the value of national meteorological services was discussed at a New Zealand symposium (New Zealand Meteorological Service 1979). At this symposium topics examined included the relation of weather and climate to forest fires and forest management, agriculture (wool, wheat, potatoes), ship operations, electric supply, design and siting of power stations, boating and sport.

Although the national meteorological service studies generally focus on 'weather' rather than 'climate', they still may be viewed as a first-order approximation of the climate sensitivity of human activities, and clearly, the range of activities affected by short-term climate variations will be close to those affected by the weather. As an example, Table IV.2 presents a summary of annual dollar and percentage losses due to adverse weather in the United States as estimated by Thompson (1977). This showed that agriculture is an order of magnitude more sensitive to weather than any other activity in both relative and absolute terms.

Very few *routine* climate impact assessments are made, but the

Table IV.2 Losses due to adverse weather in the United States

Activity	*Overall losses* US$ × 10^6	% *of annual gross revenue*
Agriculture	8,240	15.5
Construction	998	1.0
Manufacturing	597	0.2
Transportation (rail, highway and water)	96	0.3
Aviation (commercial)	92	1.1
Communications	77	0.3
Electric power	45	0.2
Energy (e.g. fossil fuels)	5	0.1
Other (general public, government, etc.)	2,531	2.0
Totals	12,684	

Source: After Maunder and Ausubel (1985), from Thompson (1977).

Assessment and Information Services Center (AISC) (formerly the Center for Environmental Assessment Services, CEAS) of the US Department of Commerce is one organization that *does* make climate impact assessments on a weekly and monthly basis. This they do for the United States, plus several other countries and regions of the world. The monthly impact assessments for the United States cover eight broad categories of societal activity as follows (see Center for Environmental Assessment Services, 1980):

(1) *construction* – including housing starts, commercial and industrial construction, property damage, soil erosion, and land use;
(2) *economics and commerce* – including employment, banking, business, trade, manufacturing, and industry;
(3) *energy* – including utilities, supply and consumption of the different energy types, and alternative energy sources;
(4) *food and agriculture* – including food, fibre and orchard crops, forestry, and fisheries;
(5) *government and taxes* – including executive, legislative, and judicial branches; federal, regional, state, and local;
(6) *recreation and services* – including travel, vacation activities, and sport;
(7) *society* – including fatalities, injuries, health, air pollution, education, crime, and population movements; and

(8) *transportation and communications* – including highway, railroad, air, and mail delivery.

In addition to the central government groups concerned with the operation and provision of weather- and climate-related services, the second major source of information on weather and climate impacts is a rather miscellaneous research literature, much of it stemming from conferences. The agendas of such conferences and the contents of the reports provide a *de facto* judgement about what is sensitive to climate. Table IV.3 compares these individual agendas with the pioneer study by the author (Maunder 1970a) to suggest both the convergence of opinion on climate sensitivity and the idiosyncratic judgements as well. The impact studies cited include the monograph on *Economic and Social Measures of Biological and Climatic Change* of the Climate Impact Assessment Program (CIAP 1975), the Aspen Institute Conference on *Living with Climatic Change* (Aspen Institute 1977), the 1979 Conference on *The Impact of Climate on Australian Society and the Economy* (Commonwealth Scientific and Industrial Research Organization (CSIRO) 1979), the *Proceedings of the World Climate Conference* (World Meteorological Organization (WMO) 1979), the American Association for the Advancement of Science research agenda on the *Environment and Societal Consequences of a Passive CO_2-Induced Climate Change* prepared for the US Department of Energy (1980), the Assessment and Information Services Center ongoing climate impact assessments (CEAS 1980), and the Scientific Committee on Problems in the Environment (SCOPE) study on *Climate Impact Assessment: Studies of the Interaction of Climate and Society* (Kates *et al*. 1985).

This overview of climate impact studies suggests a broad consensus on the sensitivity to climate variation of agriculture, water resources, building construction, transportation, and energy activities. In addition, there is repeated concern about impacts on insurance, governmental expenditures, and recreation. However, Table IV.3 may suggest more orderliness and conscious selection than really exists. Any agenda represents to a certain extent networks of available researchers and not necessarily systematically identified topics. Moreover, impact studies and conferences have often focused on geographic regions, especially drought prone ones or high latitudes (Parry 1983), where life is obviously more sensitive to climate.

An alternative to the identification of large sectoral or regional sensitivity is to focus sharply on specific topics where climate impact analysis appears most useful for applications. Topics from a Massachusetts Institute of Technology (1980) conference on *Climate and Risk* illustrate some of these sensitivities. They include, for example: extreme wind speeds and structural failure risks; weather hazard probabilities and the design of nuclear facilities; application of climatology to air force and army operational planning; impacts and use of climate information in the hail insurance industry;

importance of climate and climate forecasting to offshore drilling and production operations in the petroleum industry; seasonal climate forecasts and energy management; evaluating farming system feasibility and impact using crop-growth models and climate data; the utilization and impacts of climate information on the development and operations of the Colorado River system; and snow management and its economic potential in the Great Plains.

While it is clear that there have been some attempts to be reasonably comprehensive from a sector point of view, few of the efforts have been internally consistent in assessing sector impacts. Indeed, differing methods and definitions of climate variation have often been employed within a conference or study. Most studies examine impacts at the national or regional level, but other levels may also be important. For example, municipalities may have their expenditures for education, welfare, police, fire, sanitation, and sewage services affected by climate variation (Sassone 1975). Households may also be sensitive in terms of income and expenditures for food, housing, transportation, clothing, and medical care (Crocker *et al.* 1975), and at the individual level, nutrition (Escudero 1985) and migration (Warrick 1980) may be major indicators of climate sensitivity. But few researchers have been systematic in their selection of indicators of sensitivity and units of exposure for comparative examination; it could therefore be fruitful to undertake studies of relative sensitivity to a given climate variation stratified in a variety of ways, such as by urban, suburban, and rural families, or by different firms in a particular region, or similar firms in different regions. Table IV.4 suggests a framework for defining sensitivity to climate, using several available studies as a guide.

Making sensitivity judgements based on studies of the value of national meteorological services or conference agendas has an important deficiency in that their dominant relationship is to industrialized societies. For example, while a few papers from the World Climate Conference relate to developing country situations (for example, Burgos (1979), Fukui (1979), and Oguntoyinbo and Odingo (1979)), the balance of analysis and methods development has been restricted to industrialized nations. Thus, although there is considerable evidence that primary activities (agriculture, pastoralism, water resources) may be even more sensitive to climate in developing countries than in industrialized countries, careful surveys of what the *real* sensitivity is remain to be done.

5. Analysis of information uses and value

Identifying weather and climate sensitivities by a 'willingness to pay' criterion is potentially a very useful approach. However, few studies have shown what the true value of weather or climate information is. Indeed, as Mason (1966) correctly pointed out many years ago, 'the economic value of a special

Table IV.3 Topics covered in eight major climate impact studies[a]

Sensitive area[b]	*Maunder 1970a*	*CIAP 1975*	*Aspen Institute 1977*	*CSIRO 1979*	*WMO 1979*	*DOE/ AAAS 1980*	*CEAS 1980*	*SCOPE*[c] *1985*
1 Agriculture	X	X		X	X	X	X	X
2 Forests and forestry	X	X		X	X	X	X	
3 Pastoral activities				X		X	X	X
4 Fish and fisheries	X	X			X	X	X	X
5 Ecosystems						X		
6 Environmental conservation								
7 Water supply, demand		X	X	X	X			X
8 Energy supply, demand	X	X	X	X	X		X	X
9 Manufacturing operations, location of plants	X	X		X			X	
10 Offshore operations					X			
11 Mining (extractive industries)				X				
12 Transportation – water, air, rail, highway	X		X	X			X	
13 Construction	X		X	X			X	

14 Materials weathering		X						
15 Aesthetic costs		X						
16 Trade	X						X	
17 Public expenditures		X					X	
18 Communications							X	
19 Insurance	X			X				
20 Financial planning and institutions			X				X	
21 Recreation and tourism	X			X			X	
22 Sea level rise, coastal zones						X		
23 Health – mortality, morbidity	X	X			X	X	X	X
24 Migration							X	
25 Social concerns, crime	X	X				X	X	X
26 Military planning and operations								
27 Political systems and institutions	X	X				X		
28 Legal systems and institutions	X					X		

Source: After Maunder and Ausubel (1985).

Notes: [a] To be checked here, topic must be treated explicitly or extensively. (See Bibliography for studies cited).

[b] List includes two topics (6 and 26) not covered in studies listed, but covered in other studies.

[c] See Kates, Ausubel, and Berberian (1985).

Table IV.4 Sample of impact studies in framework for defining sensitivity to climate

	CLIMATE VARIATION				*IMPACT LEVELS*						*INDICATORS USED*		
						social system unit of exposure					*quantitative*		
Study	*change*	*variability*	*season*	*extreme*	*biophysical*	*globe*	*nation/ region*	*locality*	*firm*	*household*	*monetary*	*non-monetary*	*qualitative*
CIAP (1975)	X				X	X	X	X	X	X	X		
WMO (1979)		X			X		X	X				X	
DOE/ AAAS (1980)	X				X		X						X
Seifert and Kamrany (1974)				X	X			X				X	X
National Defense University (1980)	X				X	X	X				X	X	
Garcia (1981)				X	X	X	X	X			X	X	X
Chambers *et al.* (1981)			X		X			X		X	X	X	X

Source: After Maunder and Ausubel (1985).
Note: Studies are placed in framework according to their emphases.

service for a particular customer can usually best be judged by the customer himself, but he may be reluctant to divulge this for fears that the charges will be increased'. Further, one cannot usually apply normal commercial criteria based on the price the customer is willing to pay because many of the basic meteorological and climatological services are free. However, in those countries where only a *basic* service is provided free of charge, normal commercial criteria may well be appropriate in assessing the 'true market price' for weather and climate services.

Several attempts have been made to place a dollar value on the various types of weather and climate information. In essence, such analyses require looking at a national economy through the 'joint eyes' of a meteorologist and an economist. Essentially, this means that a critical analysis of national economic data must be done in order to rank those parts of a nation in terms of their sensitivity to weather and climate *and* their contribution to national productivity.

In looking at sensitivities from the viewpoint of the value which either the provider of information or the user of information places on a particular product, four factors need to be considered: (1) the weather and climate sensitivity of various sectors of an economy; (2) the relative economic importance of these sectors; (3) the separation of weather and climate information into weather and climate forecasts, real-time weather and climate information, and past weather and climate information; and (4) the difference between the actual value of presently available weather and climate information, and the potential value of weather and climate information which could be provided.

As an example of the thinking required – and it must be emphasized that by the very nature of the problem this must be relatively incomplete – consider the argument put forward by Mason in the mid-1960s in determining the value of weather information to wheat and barley in the United Kingdom:

> Wheat and barley, our most important cereal crops, with an annual value of £200m (1966 values) are very weather-sensitive. A judicious choice of sowing time, avoiding short cold spells, and the intelligent use of 1–3 day forecasts and finespell warnings during harvesting operations, may improve yields by several per cent. In a wet summer, complete harvesting of the crop might well be impossible without continual meteorological advice to help take advantage of every available dry spell and to make full use of drying machinery. If the weather service contributes only 1 per cent to the total yields, this amounts to £2m in a year.

Similarly, the views of Russo (1966) who made a pioneer study of the United States building and construction industry are pertinent. He estimated the total weather losses as ranging from 1.1 to 11.3 per cent of the value of construction. More specific information from the study by Russo showed that in 1964 the total value of the US building and construction industry was $88 billion (this represented 10 per cent of the *then* US gross national product), of which $40 billion (45 per cent) was spent in what Russo described as 'weather-sensitive' areas. For example, with respect to highway construction, the total annual volume had a (1964) value of $6.6 billion of which Russo estimated that 73 per cent was potentially weather sensitive ($1.7 billion in perishable material, $1.6 billion in on-site wages, $0.8 billion in equipment, and $0.7 billion in overhead and profits). The total dollar loss (for all building and construction) due to the weather was quantitatively estimated by Russo (1966) at a minimum of $3 billion annually (representing 1.1 per cent of the then total value of building and construction, or 2.5 per cent of the total value of the weather-sensitive sectors of this industry), and a maximum of $10 billion (11.3 and 25.2 per cent respectively of the total values noted above). Russo noted that 'this wide range . . . results from a highly speculative estimate of the decreased construction volume due to seasonal weather effects'. This comment highlights the real difficulties in giving anything but approximate values of the weather-related losses of any weather-sensitive sector, even in highly developed societies, these difficulties relating directly to the problems associated with identifying and ranking weather and climate sensitivity.

6. Quantitative analyses

An initial quantitative approach to identifying climate sensitivities might simply be to look for seasonal variations in production and consumption data. For example, it is well recognized that some economic activities follow a recurring 'seasonal pattern' during the year; that is, the retailer is aware that there will be increased business in early spring and early winter, for which he must plan his purchasing and personnel changes. Similarly, the contractor buys materials and hires additional workers for the increased construction activity that inevitably comes in many countries during the summer months. Farmers' expenditure also rises in the spring and autumn because of planting and harvesting costs. In addition bankers recognize these and other patterns of 'seasonal' activity and they plan for an uneven deposit inflow and demand for loans during the year. In the same manner wage earners in industries with high degrees of dependence on 'seasonal' activity are well aware that their income will not flow evenly during the year.

A correlation analysis may also suggest sensitivities in the comparison of climate and economic data. For example, in one of the few studies of weather and specific retail trade in a large store, Linden (1962) related – in a very simple but telling way – sales of women's winter coats in New York departmental stores to the average monthly temperature in September and October. The analysis showed that about 20 per cent of the September –December sales of coats occur when the average September temperature is about 18°C, and that 30 per cent of the same sales occur when the average

Figure IV.1 Weighted* number of days of soil water deficit: The New Zealand sheep farm, 1949/50–1984/85

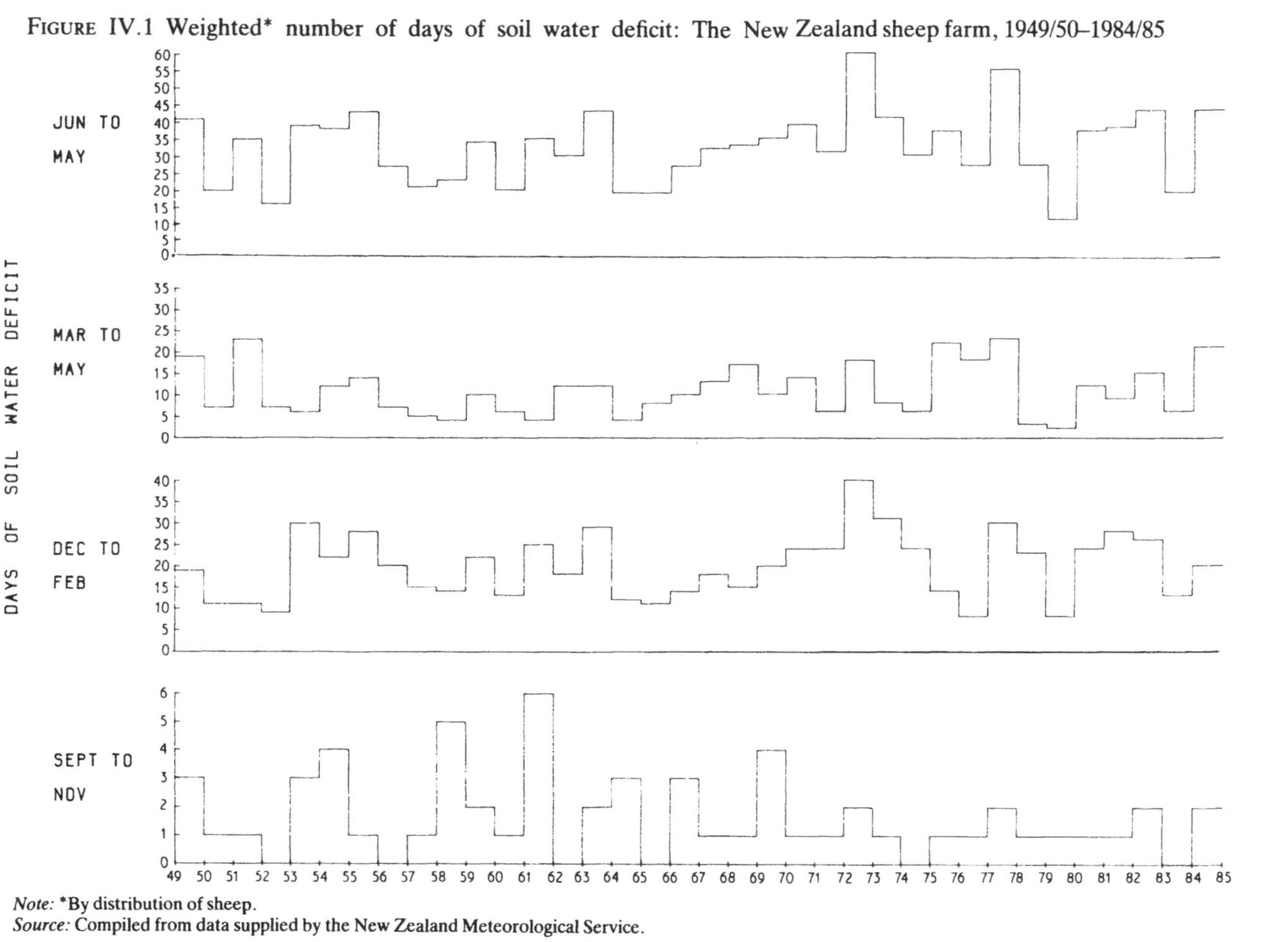

Note: *By distribution of sheep.
Source: Compiled from data supplied by the New Zealand Meteorological Service.

FIGURE IV.2 New Zealand wool production, sheep numbers, and wool clip per sheep: 1960/61–1984/85

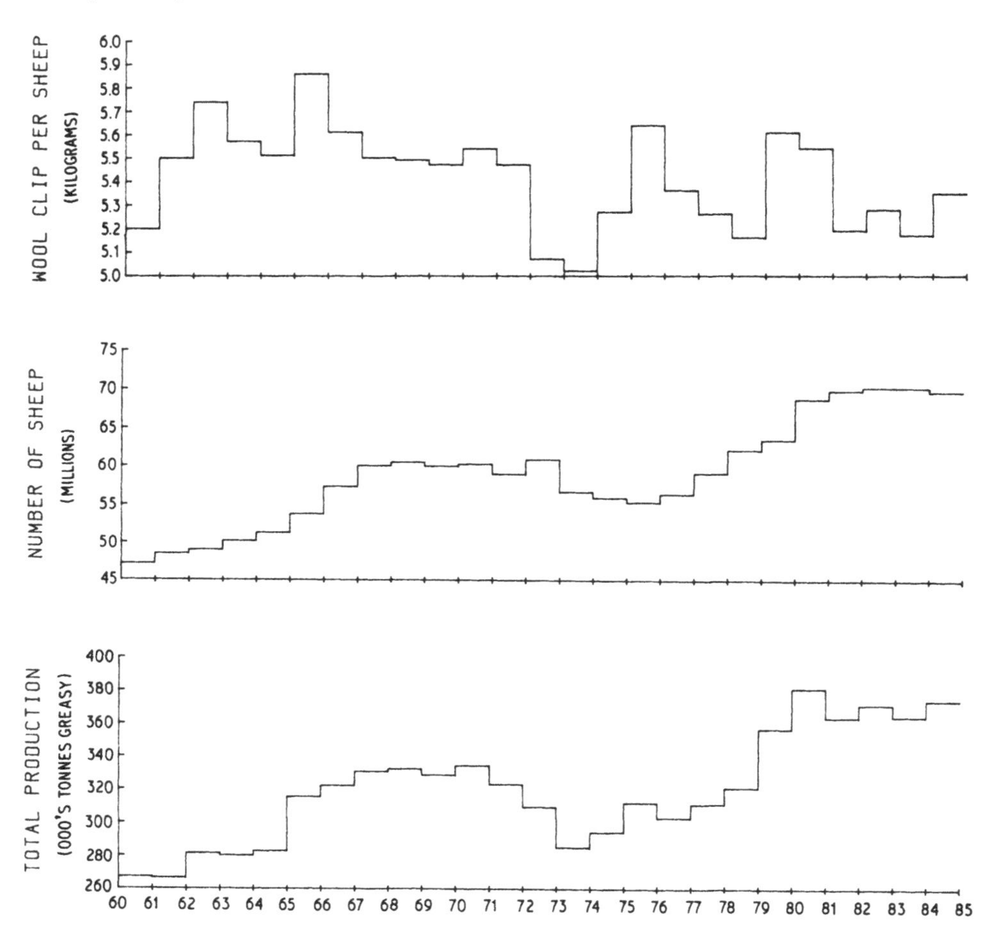

Source: Compiled from data supplied by the New Zealand Meat and Wool Boards' Economic Service.

October temperature is about 13°C. Of course, more detailed correlation analyses can also be made, and a good idea of climate sensitivity was obtained from an analysis in New Zealand of the climate data (specifically days of soil-water deficit) and the economic data (specifically sheep numbers, wool production and wool production per sheep shorn) shown in Fig. IV.1 and Fig. IV.2. Such an analysis, which is discussed in detail in Chapter VI (see Maunder 1980a), used a step-wise regression model to identify statistically the most important weather and climate factors in the New Zealand sheep industry, and the significant months or combination of months. Similarly, Palutikof (1983a) used a multiple regression analysis to evaluate the impact of a severe winter and a hot, dry summer on British industry (Table IV.5). This analysis showed that severe winters increased activities in the utility industry by 17.5 per cent over the same month for the preceding season, but reduced clothing and footwear production by 9.6 per cent, while a summer drought reduced utility activities by 6.4 per cent, but produced increases in clothing and footwear production of 3.0 per cent.

In some correlation-type analyses, however, the impact signals are weak, or hidden by larger, non-climate fluctuations. In such cases it is useful to focus on particularly anomalous climate episodes (such as major droughts, floods, cold spells) and search for concomitant variations in social and economic activities. However, if no apparent variation is found, it is reasonable to assume that the subject under study exhibits little sensitivity to that particular type of climate variation. Such a 'reasoning from extremes' is particularly useful in data-poor areas, and in some cases may be the only reasonable way of quantitatively analysing climate sensitivity.

Of course, while measures of sensitivity to climate are often economic,

Table IV.5 Performance of United Kingdom industries

Industry	*Average deviation per month from mean of preceding season*	
	1962–3 winter	*1975–6 drought*
Bricks, cement, etc.	−14.4	−0.9
Timber, furniture	−14.3	−1.8
Clothing and footwear	−9.6	+3.0
Paper, printing, and publishing	−5.6	−2.2
Mining and quarrying	−4.7	−2.3
Shipbuilding	−4.3	−0.9
Engineering and electrical goods	−3.7	−2.1
Non-ferrous metals	−2.7	−4.3
Drinks and tobacco	−2.4	+4.4
Ferrous metals	−2.3	−10.0
Food	−2.0	+0.7
Metal goods (not elsewhere specified)	−2.0	−4.6
Chemicals	−1.6	+2.1
Leather goods	−0.6	+0.6
Pottery and glass	−0.3	−3.9
Textiles	0.0	−0.6
Vehicles	0.0	−1.3
Coke ovens, oil refining, etc.	+3.3	−0.3
Utilities	+17.5	−6.4

Source: After Maunder and Ausubel (1985), from Palutikof (1983a).

other indices are also available. For example, measures can be offered in terms of the numbers of people affected by a climate variation (see Burton *et al.* 1978; Warrick 1980) in terms of dietary levels, or in terms of the patterns of land ownership (see Jodha and Mascarenhas 1985). However, the primary appeal of economic indicators, especially the monetary ones, is their ease of intercomparison, and the most extensive effort so far to assess relative economic sensitivities of different sectors and impacts to climate variation is that of CIAP (1975). The Climate Impact Assessment Program (CIAP) sought to formulate mathematical relationships between long-term climate change and many of the economic activities mentioned above. Results of one calculation are presented in Table IV.6. While the accuracy of these estimates has been questioned (see Ausubel 1980), the results are 'surprising' and, as such, worth noting. In particular, the *overall* effects on wages and health far surpass others, including agriculture and water resources, in importance. Other studies of note include those of Eddy *et al.*

Table IV.6 Estimates of economic impacts of a hypothetical global climatic change (−1°C change in mean annual temperature, no change in precipitation)

Impact studied	*Annualized cost – 1974 (millions of US dollars)*
Corn production (60% of world)	+21
Cotton production (65% of world)	−11
Wheat production (55% of world)	−92
Rice production (85% of world)	−956
Forest production	
(a) US	−661
(b) Canada	−268
(c) USSR (softwood only)	−1383
Douglas fir production (US Pacific Northwest)	−475
Marine resources (world)	−1431
Water resources (2 US river basins)	+2
Health impacts (excluding skin cancer) (world)	−2386
Urban resources (US)	
wages	−3667
residential, commercial and industrial fossil fuel demand	−176 lower bound. −232 upper bound.
residential and commercial electricity demand	+748
housing, clothing expenditures	−507
public expenditures	−24
aesthetic costs	+219

Source: After Maunder and Ausubel (1985), from Climate Impact Assessment Program (1975).

(1980) who have employed input–output and other economic models to assess the effects of contemporary climate variability on a range of sectors and spatial scales.

G. THE IMPACT OF CLIMATE

1. An overview

There are many interconnected components that are involved in climate impact studies (Fig. IV.3). Specifically, society and nature combine to influence societal structures which in turn are related to (1) impacts on specific human activities, (2) socio-economic impacts, (3) perception of impacts, and (4) the public demand for action and/or adaption to climatic variations.

FIGURE IV.3 The interconnected components that are involved in climate impact studies

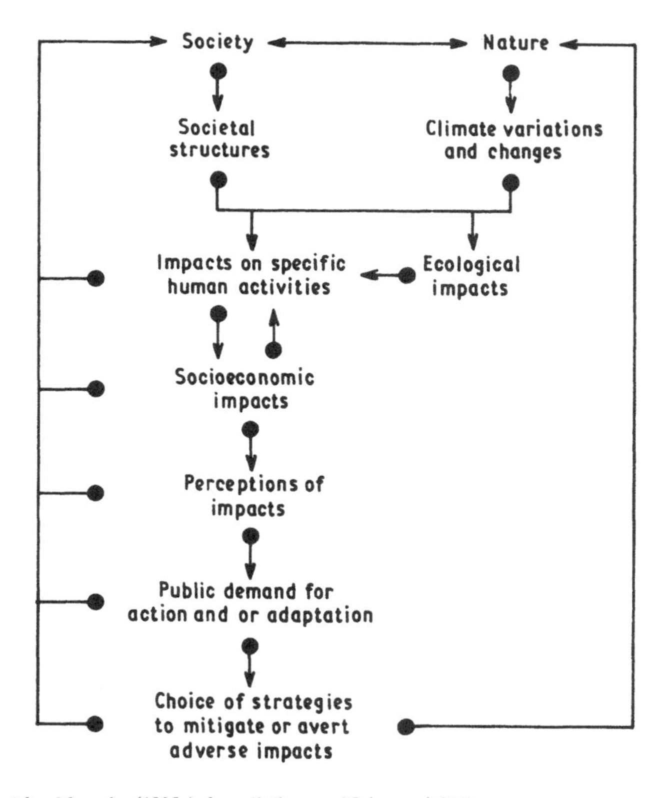

Source: After Maunder (1985a), from Kellogg and Schware (1981).

A political framework must of course be placed over the various interconnections, for it is clear that the strategies adopted in, say, a centrally planned economy will be different from the strategies adopted in a developing country or in a developed market economy. For example, if a climate variation or change is forecast, it is possible for a host of decision-makers in a developed market economy, such as Australia, to make independent decisions on the likely impact on themselves, and hence the strategies that they would adopt to minimize such impacts. In this case some guidance would come from government agencies, but the Australian wheat grower would finally make a decision as to his best course of action. In contrast, in a centrally planned economy, governmental agencies would have a much stronger influence on the strategies that would be adopted.

Studies of climate/society interaction, such as those discussed elsewhere in this chapter, provide the first clues as to climate sensitivities among human activities in various settings. As already noted, previous research has identified three important concepts: (1) that the atmosphere is an important natural resource which may be perceived, tapped, modified, despoiled, or ignored, and whose availability may be forecast; (2) that information concerning the atmosphere such as the past weather and climate, the present weather, and the future weather and climate is also an important resource; (3) that given an understanding of physical-biological-sociological-political-strategic interactions with the atmosphere, and given sufficient information about atmospheric events, people can at times use their management ability to improve the economic and social outcome of many weather-sensitive activities.

These concepts were first identified and developed during the 1964–70 period through the pioneering work of Sewell (1966), Sewell *et al.* (1968), McQuigg (1964, 1970), Maunder (1965, 1968a, 1970a), and Maunder and Sewell (1968). Other more specific studies were also made about this time, such as: (a) the study by Musgrave (1968) on measuring the influence of weather on housing starts in the United States; (2) the study by Johnson and Haig (1970) on agricultural land price differences and their relationship to climatic modification; (3) the studies by Maunder, Johnson, and McQuigg (1971a, b) on the effect of weather on road construction; (4) the edited review of climatic resources and economic activities by Taylor (1974).

These early studies on climate sensitivity encouraged many other studies, but it would be true to say that as we enter the late 1980s there are still very few case studies which give the dollar impact, or a measure of the sensitivity in dollar terms, of a *specific* weather or climate event, particularly in so far as the impact at a regional or national level is concerned. Further, since it is necessary first to have some measure of these dollar impacts before even a preliminary study of climate sensitivity can be attempted, it is obvious that much of the thinking on climate sensitivity must remain tentative. However, a review of the recent research in this area provides a valuable background.

These studies include (a) the research by Eddy *et al.* (1980) on the economic impact of climate; (b) the work by the Center for Environmental Assessment Services (CEAS) (1980) on the impact of climatic anomalies on the gross national product of the United States; (c) the various studies associated with the World Climate Programme (1980–85); (d) the agricultural climatic impact assessments for the year 2000 commissioned by the United States National Defense University (1980); (e) the work on managing climatic resources and risks published by the US National Research Council (1981); (f) the area and commodity specific studies by Maunder (1977a, b, c, 1980b); and (g) the 'ideas' studies of Maunder (1979, 1982a, 1983) and Mather *et al.* (1981). Two of these studies are now considered in more detail.

2. Climate impacts research: Case studies

The work of the Oklahoma Climatological Survey (see Eddy *et al.* 1980, 1981, 1982, 1983, 1984, 1985) on *The Economic Impact of Climate* is of considerable importance for its strong economic base. This series of publications summarizes the work of several workshops and includes various aspects of climate impact. In the foreword to the series, editor Eddy says that use is first made of current economic modelling practice in order to infer the optimal methodology to be employed in analysing the effect of climate on the United States economy. He comments that this analysis permits significant questions to be answered concerning the sensitivity and responsiveness of the economy to climate fluctuations. He notes that water and food, and energy supply and distribution, are all functions of climate and play an important role in any econometric model.

The Introduction to Volume I of the series, written by Eddy, emphasizes that in examining that particular subset of human activities which comprises the production and consumption of goods and services, the construction of any model to describe such a system must first address the problem of scale. Specifically, this is the problem of aggregation in space and time, as well as with respect to economic sectors, in that high resolution (low aggregation) enables a study of cause and effect but may lead to an unstable and very complex model (even if all the data required could be obtained). In contrast, high aggregation (low resolution) gives a more stable model with sufficient data for validation, but much of the detailed cause-and-effect information is masked.

The questions to be asked and the type and scale of the model required to produce the right answers are ultimately linked; however, of prime importance is finding out what kind of questions concerning economic–climate relationships should and can be asked. Of particular interest,is the ranking of such questions with respect to the benefits and costs associated with obtaining answers to them. Such a ranking is directly related to climatic sensitivity.

There are several ways to undertake such an investigation. One can begin by considering a regional input–output economic model having enough structural detail to permit a fairly explicit relationship between weather and climate and the economically related activities of people. Further, if the region to be considered is at least the size of a small nation or a state or province of a larger nation, a certain amount of model stability can be expected, as well as – and this is *most* important – some hope of obtaining the required economic data. Alternatively, or in addition, aggregating to the national level through the use of a multi-regional input-output modelling technique may be considered, as well as the technique of sectoral aggregation employed in constructing national input-output models.

The real economic 'health' of a region depends primarily on its 'basic' industries; that is those industries which sell their product principally *outside* of the region. But not only is this segment of final demand influenced by weather *remote* from the region, so also is the price which is derived from supply/demand relationships. Thus a vital factor in the input-output 'recipe' is influenced by weather and climate from both *within* and *outside* of the region. The input-output technique also permits a study of the differences in impact on a regional economy. For example, weather and climate impose a demand for utilities such as heating or cooling on a regional scale, and on time intervals of a few hours to a few days. This demand cannot, however, be aggregated safely *beyond a season* without losing the space-time location of the economic impact. On the other hand, climate affects the supply of food being produced within a region the size of a state, and often on a subregional scale. In this case, the critical time scales are the few weeks associated with planting, flowering and harvesting; there are also 'events' such as blizzards, droughts and floods to be considered, and the interactions between weather, pests and diseases.

A further important consideration noted by Eddy *et al.* in their studies is that the economic impact of the climate includes not simply the passive response of production and consumption to climate variations, but also how climate information may be used both to stabilize the economic system and to optimize the benefit/cost ratio with respect to planning strategies and actions taken in individual sectors. A specific application of using climate information is in the food production sector. Significant fluctuations in food production occur from year to year because of weather and climate factors, such fluctuations being in respect of both the total amount of production and the sub-totals of production in the geographical (and economic) regions. More importantly, runs of several years of weather which are 'bad' on the global scale for producing food must be expected from a climatological point of view, and runs of weather which are 'bad' for food production in one region but 'good' in another region are also to be expected climatologically. However, since both overproduction and underproduction can cause instability within an economy, the ability to be able to anticipate such occurrences in order to provide for corrective measures is highly desirable.

As an example of this, it is evident that a central marketing agency (either government controlled or a private sector company) can use stockpiles of grain or meat or wool to add to the current production in times of shortfall (thereby keeping prices below a critical level), and to subtract from current production (that is increasing reserves) in times of high production, thus keeping prices from dropping below a critical level. Clearly, this economic stabilizing activity is in part the result of weather and climate influences, and is a good example of how climate sensitivity not only crosses both regional and national boundaries, but can also be a critical factor in marketing, political, and even strategic decision-making.

The second econoclimatic study to be considered in this review is the work of the Center for Environmental Assessment Services (CEAS) (1980). The Center (which is part of NOAA, which in turn is part of the US Department of Commerce) makes routine weekly, two-weekly and monthly climatic impact assessments of the United States and many foreign countries. The objective in preparing these reports is to qualitatively assess the impact of major climate and other natural events on eight broad categories of United States societal activities. The impact assessments consider those unusual or abnormal meteorological or geophysical events (that is, unusual in time, location, intensity, frequency, or persistence) that are likely to impact on societal or economic activities in a special and significant manner. The reports are based on the sensitivity of the gross national product to widespread anomalous weather as given earlier in Table IV.1. The information given in this table is believed to be the first of its kind in the world and was compiled by CEAS from a critical analysis of the weather and climate sensitivity of the various economic and social sectors in the United States as reported in the *New York Times* during a ten-year period.

A further noteworthy contribution of CEAS and other groups who are working in this area is the concept that 'although last month's weather or climate is history, the measurement of its economic impact is not'. For example, as previously noted, economic 'indicators' take time to compile and are usually not available until one to several weeks after the fact, while economic 'statistics' take months and in some cases a year or more to compile. But weather and other environmental satellites, combined with traditional data-collection systems, provide worldwide measurements of the weather within minutes or hours of the events. Furthermore, provided appropriate analyses have been made, these weather data can be converted in real-time into economic impacts and can therefore become estimates of that part of the economy that is affected by the weather and the climate. Accordingly, when weather and climate is a *limiting factor* its impact is known immediately. Thus a forecast of economic data – *which at the time the forecast is made do not actually exist or more correctly have not been compiled, collected, or analysed* – can be made using real-time weather and climate data. Moreover, because climate is *outside* the economic system, its direct effects are easier to measure than the effects of other economic variables.

H. CLIMATIC SOLUTIONS TO PROBLEMS

A key aspect of climate and socio-economics is 'what is your problem?' followed by the even more important question, 'is there a *climate* solution to this problem?' That is, would an appropriate analysis of weather and climate information be of assistance in providing a better understanding of this problem? Consider for example the problem of agriculture (Fig. IV.4). In general the environment can be said to influence production. More specifically one can define the biological response to a variation in the environment. For example, very dry conditions may decrease wool production but increase wool quality, while colder than average temperatures may decrease corn production but increase potato production. The significant question then follows: 'what does the farmer, farm adviser, agricultural producer board official, or government Secretary of Agriculture, do?' As shown in Fig. IV.4 one can either do nothing and therefore accept the economic, social, and political consequences, or one can act.

In the case of agriculture at least four options of 'doing something' are available. These include relocation of the agriculture to a more suitable area (which is a real choice if a longer term climate change is envisaged), or changing the agricultural operations by using, for example, a different species of plant or animal. One can also stop production (such as deciding not to plant wheat or rice), or one can modify the environment by growing a shelter belt of trees, by providing irrigation water, or by some form of deliberate weather modification. In all cases, the choice of 'do something' would hopefully improve production and/or quality, and as a result allow more useful management decisions to be made. In all cases, the role of managers, with their differences in willingness and ability to assess risk, becomes crucial.

A typical example of these risks are those associated with a drought situation, as *Time* (28 March 1983) so vividly reported from Australia.

> Even the region's hardiest inhabitants find the eerie silence unnerving. Under the blazing sun, the quiet that grips vast stretches of eastern Australia is accentuated by the whine of the hot, blast-furnace-like winds that bear down on farmlands long since stripped of vegetation, whipping up the bone-dry topsoil into whirling clouds of reddish-brown dust. Here and there, the superheated stillness is broken by the squawking of white cockatoos and black crows, wheeling and circling overhead, searching the parched land for dead and dying animals below.

FIGURE IV.4 Patterns of responses leading from environmentally induced production events to a management decision

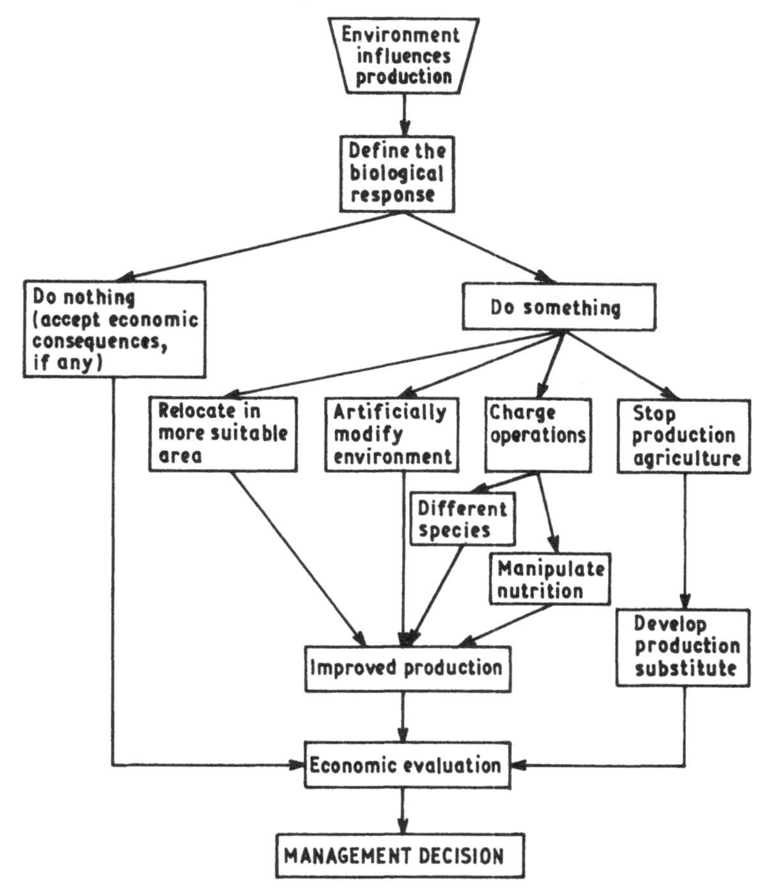

Source: After Maunder (1985a), from National Academy of Sciences (1976).

These comments describe the extremely severe drought conditions prevailing in Australia in March 1983. Droughts of course occur in many areas, and what follows is an examination of a typical drought situation in New Zealand – written in 1982 *as it was happening* – and the kind of decision-making that faced farmers at that time. The strategies a farmer could take in any drought situation depend of course on a number of factors and clearly these will vary from area to area, and farming type to farming type, but the specific case* of a Canterbury farmer in New Zealand provides a typical real-life situation.

* As reported by John Goulter in 'Disaster country', *Christchurch Star*, 1 September 1982.

One property with an area of 210 hectares is a sheep farm in which all the income comes from converting grass into wool and lambs. This farm had increased its stock numbers up to 2500 by January 1982, but because of the drought $NZ30,000 had to be spent in providing supplementary feed. This included $NZ17,000 on barley which was brought in as a grain feed to carry the ewes over the winter (June–August). The farm paid another $NZ6000 to graze sheep on a nearby farm, and the balance of $NZ7000 went in 'invisible' costs like selling lambs at low prices. The owners of the farm in this case are established farmers and they were able to carry these costs. As a result of their decision to purchase feed, they were able to close off most of the farm over the winter and, as of September 1982, with the drought reaching a peak, the farm's pastures were not eaten out and the ewes (which were fed on barley) would have gone into lambing in good condition. A key question relating to this farm is 'what would have happened if $NZ23,000 had *not* been spent on supplementary feed?' and further, 'now that $NZ23,000 has been spent, what is the correct strategy if the drought continues?'

In another farm in the area the farmer spent $NZ25,000 on supplementary feed, mainly to purchase 9000 bales of hay which were fed to the 2025 sheep from February 1982. On this farm feeding out would not normally have started until late autumn (May), and as with the first farm the strategies available to the farmer were relatively restricted. In this case a possible long-term strategy would be the use of irrigation. However, the cost of a full-scale irrigation scheme of about $NZ150,000 is a cost the farmer considers he cannot justify.

Many questions arise from the brief factual comments given on these two typical farms. Most important, they emphasize that decisions have to be made – *even the decision to do nothing*. Further, in the absence of a reliable long-range weather or climate forecast, a careful analysis of the climatological record and probabilities – together with appropriate advice from agricultural and financial advisers – may well provide the only real guide to the correct strategy.

I. MEASURING THE COSTS OF THE WEATHER AND THE CLIMATE

Any attempt to estimate the economic effect of a significant weather or climate variation in any country is beset with difficulties. For example, in assessing the cost of the 1969/70 drought to New Zealand (Maunder 1971a), it was necessary to piece together material from a variety of official and semi-official documents. Such a task has real difficulties; nevertheless, in this case it was possible to give two estimates of the costs of the drought, a 'high' cost of $NZ47 million (in 1969 dollars) which is the probable maximum cost of the drought, and a 'low' cost of $NZ24 million, which is the probable minimum cost of the drought. Many assumptions had to be made in this study but it is believed that the costs were of the correct order of magnitude, and that such estimates were (and are) potentially useful to political decision-makers.

There is, in fact, increasing interest in assessing both the costs of weather and climate events, and the benefits obtainable through a better use of the weather-economics relationship. As is noted several times in this book several studies have shown that proper meteorological and climatological advice can play an important role in the decision-making involved in weather-sensitive enterprises. Included in this advice is the availability of commodity-weighted weather and climate indices. The problems of developing such indices through econoclimatic models are discussed in detail in Chapter V but an essential part of such models is the calculation of commodity-weighted weather indices which are based on the contribution of various areas (regions) to the total 'economic activity' (such as energy consumption or rice production) of a nation.

The primary purpose in developing commodity-weighted weather and climate indices is to provide a means by which a vast amount of weather and climate data can be conveniently expressed as an index, which is both representative of the data and also meaningful to economic activities. The application of such indices to economic activities, however, is not easy, and anyone designing a model to be used in a study of the impact of weather events on human activities finds out very quickly that very little is known in quantified terms that can be used in this way. Indeed the basic handicap in applying weather and climate information to nation-wide economic activities in almost all countries is the almost complete lack of economic indicators for 'short' time periods. Two known notable exceptions are those associated with energy consumption and certain aspects of agricultural production, and it is perhaps not by accident that much of the work in the application of meteorology and climatology to economic development is associated with these activities, as is discussed in detail in Chapter VI and VII.

J. NATIONAL SENSITIVITY TO WEATHER AND CLIMATE EVENTS

The economic and social impact of a specific national weather or climate event such as the 1980 heat wave and drought in the United States has been assessed in a study by the Center for Environmental Assessment Services (1981), an agency of the US Department of Commerce. The direct effects of the 1980 heat wave and drought are estimated in this report to have been $US11,000 million for major crop losses, $1000 million for livestock and poultry losses, $1000 million for increased federal, state, and local government expenditures, $800 million for increased power consumption (principally for additional air conditioning), $600 million for increased requirement for health services, plus a further $3000 million for other sectors. While these dollar losses are very high, they are also significant on a *per capita* basis, representing in total nearly $US100 for every American.

The overall sensitivity of nations to weather and climate is of course much more difficult to assess, especially if *all* the nations of the world are considered. One attempt to do this has been made by the author (see Maunder 1986a) based on the use of each nation's exports as an indicator of weather (and possibly short-term climate) sensitivity. It is emphasized that this is an approach to only *one* aspect of national weather sensitivity, but it does give a means of ranking the nations of the world in terms of their respective sensitivities to the variable atmospheric resource all nations have to live with. This analysis considered the proportion of exports of each of 133 countries in terms of whether they were food/beverages, non-food agriculture, fuels/minerals/metals, machinery equipment, or other manufactures. These components were then weighted by factors of 10, 7, 2, 1, and 3 respectively (the factors being based on a subjective assessment of each component's sensitivity to weather and climate) to give what could be called an 'export climate sensitivity index'. In effect, a total weight of 23 was used, of which 10/23 (about 44 per cent) was allocated to the 'high' climate-sensitive food/beverages export sector, but only 1/23 (about 4 per cent) was allocated to the 'low' climate-sensitive machinery equipment export sector. For example, Fiji, with mid-1970 export values of 75 per cent for food/beverages exports, 1 per cent for non-food agriculture, 7 per cent for other manufactured goods, 14 per cent for fuels/minerals/metals, and 3 per cent for machinery equipment, had an 'export climate sensitivity index' of 809 (that is 10×75, plus 7×1, plus 3×7, plus 2×14, plus 1×3), whereas the United States had a comparable index of 367 (made up of 10×19, plus 7×5, plus 3×29, plus 2×6, plus 1×43).

The analysis considered mid-1970 data from 133 countries (see Table IV.7) and showed that the 'export climate sensitivity ranking' at that time varied from 1000 in Gambia (rank 1) to 198 in Japan (rank 133). Some of the major agricultural exporting countries had relatively high rankings (for example, Argentina (31), New Zealand (43), Australia (61), United States (83)), whereas several of the oil exporting countries had relatively low rankings (for example, Venezuela (126), and Saudi Arabia (130)). Obviously the ranking and the 'export climate sensitivity index' will vary according to the specific year being studied since the composition of exports from most countries varies considerably from year to year, these variations being in many cases closely associated with the 'seasonal' climate in the specific country. The index will also clearly vary if a different set of weightings is used, and this may well be appropriate in a more rigorous analysis. The weightings will of course vary considerably from region to region, but if a nation-by-nation comparison is to be made (as has been done in this case), there are advantages in using the same weighting system for each country. By so doing, a useful climate sensitivity ranking is obtained which may be

Table IV.7 Sensitivity ranking of export climates

Country	*Exports (%)*					
	Food Beverages	*Non-food agricult.*	*Fuels Minerals Metals*	*Machinery Equipment*	*Other manufact.*	*Export climate sensitivity index*
Weight	10	7	2	1	3	
Gambia	100	0	0	0	0	1000
Mauritius	96	1	0	1	2	974
Burundi	93	5	0	0	1	968
Uganda	90	7	2	0	0	953
Niger	91	5	0	1	2	952
Western Samoa	92	1	0	0	7	948
Malawi	91	2	0	2	5	941
Somalia	86	8	0	4	2	926
Ethiopia	83	12	3	1	1	924
Belize	82	9	0	0	8	907
Kampuchea (Cambodia)	72	24	2	1	1	896
Burkina Faso (Upper Volta)	68	28	0	1	3	886
Cameroon	73	19	5	1	3	883
Cuba	83	1	16	0	0	869
Benin	71	18	0	3	8	863
Madagascar	79	5	9	2	5	860
Ivory Coast	71	17	4	2	6	857
Burma	65	26	7	1	2	853
Honduras	73	11	6	0	10	849
Dominican Republic	79	0	3	1	17	848
Mali	48	52	0	0	0	844
Paraguay	61	25	1	0	13	826
Iceland	77	1	16	1	5	825
Tanzania, U/Rep.	61	25	5	0	9	822
Sri Lanka	65	17	6	0	13	818
Costa Rica	74	1	0	3	21	813
Fiji	75	1	14	3	7	809
Colombia	70	6	4	3	17	804
Ghana	69	9	19	0	3	800
Nicaragua	56	26	1	1	16	793
Argentina	69	6	1	9	15	788
Mozambique	57	23	10	5'	5	771
Guatemala	59	13	0	2	26	761

Table IV.7– *Cont.*

Country	*Exports (%)*					
	Food Beverages	*Non-food agricult.*	*Fuels Minerals Metals*	*Machinery Equipment*	*Other manufact.*	*Export climate sensitivity index*
Weight	10	7	2	1	3	
Sudan	24	74	1	0	0	760
Yemen, Arab Republic	34	56	0	1	9	760
Kenya	65	5	18	2	10	753
Senegal	65	4	12	4	15	751
Chad	24	71	0	1	4	750
Togo	67	2	25	2	4	748
Thailand	59	12	9	3	17	746
Papua New Guinea	62	9	1	13	16	746
El Salvador	59	10	2	3	26	745
New Zealand	46	29	6	5	14	722
Panama	63	0	28	2	8	712
Brazil	62	2	10	12	14	708
Turkey	51	16	8	1	23	708
Central African Repub.	43	29	0	0	24	705
Rwanda	81	3	35	0	1	704
Uruguay	45	21	0	2	31	692
Afghanistan	36	36	17	0	11	679
Philippines	51	7	17	2	23	664
Guyana	50	2	34	0	13	621
Angola	45	8	25	1	21	620
Haiti	42	10	13	3	32	615
Cyprus	47	1	7	8	37	610
Peru	44	6	49	0	1	583
Ireland	40	2	3	15	40	555
Malaysia	19	39	27	7	9	551
Barbados	39	0	13	12	35	533
Egypt	19	30	26	0	25	527
Australia	31	13	33	5	18	526
Greece	33	3	14	5	45	519
Pakistan	29	6	5	2	58	518
India	32	3	9	6	50	515
Ecuador	37	2	59	0	2	508

Denmark	34	5	5	27	29	499
Jordan	35	1	24	13	27	499
Lebanon	30	6	4	19	41	492
Morocco	31	2	46	1	20	477
Bangladesh	8	30	0	0	62	476
Mexico	33	9	29	7	22	461
South Africa	22	10	27	7	34	453
French Polynesia	22	0	0	5	73	444
Portugal	16	10	4	15	56	421
Netherlands	22	4	22	19	34	413
Sierra Leone	18	1	20	0	61	410
Israel	15	4	2	10	70	402
Hungary	23	3	7	33	34	400
Spain	21	2	6	26	45	397
Indonesia	12	15	71	1	1	371
Congo	10	14	60	2	14	362
Jamaica	20	0	71	2	6	362
United States	19	5	6	43	27	361
Yemen, People's Dem. Rep.	15	7	70	3	6	360
Tunisia	15	2	49	1	32	359
Laos People's Dem. Repub.	5	28	36	30	1	351
Korea	11	2	2	17	68	349
Malta	11	0	6	12	71	347
Singapore	10	14	32	24	20	346
Poland	11	2	13	14	60	344
Syria, Arab Repub.	6	18	66	2	8	344
Taiwan	12	1	2	21	64	344
Canada	12	12	26	33	18	343
Yugoslavia	12	8	10	32	38	342
France	15	3	6	37	39	337
Finland	4	16	6	26	48	334
Liberia	4	19	72	1	3	327
Belgium	10	2	11	24	53	319
Zaire	12	3	78	0	7	318
Norway	11	4	32	29	24	303
Hong Kong	4	2	1	16	77	303
Austria	4	7	5	28	56	295
Italy	8	2	7	34	50	292
USSR	8	10	25	22	35	288
Mauritania	8	3	88	0	1	280
United Kingdom	7	2	10	37	44	273
Bahrain	7	1	28	28	37	272
Sweden	3	11	6	44	36	271

Table IV.7– *Cont.*

Country	*Exports (%)*					
	Food Beverages	*Non-food agricult.*	*Fuels Minerals Metals*	*Machinery Equipment*	*Other manufact.*	*Export climate sensitivity index*
Weight	10	7	2	1	3	
Bermuda	1	0	42	1	56	263
Switzerland	4	1	3	33	58	260
Suriname	5	2	92	0	2	254
Chile	4	3	88	1	4	250
Nigeria	6	0	93	0	1	249
Gabon	0	8	91	0	1	241
New Caledonia	0	0	51	3	45	240
Bolivia	3	2	93	0	3	239
Germany, Fed. Republic	5	1	5	48	41	238
Trinidad & Tobago	4	0	92	0	4	236
Czechoslovakia	3	2	7	51	37	220
United Arab Emirates	1	0	96	1	2	219
Algeria	2	0	97	0	1	217
Iran	1	1	97	0	1	214
Kuwait	1	0	88	3	8	213
Qatar	0	0	99	0	1	211
Netherlands Antilles	1	0	97	0	2	210
Venezuela	1	0	98	0	1	209
Iraq	1	0	99	0	0	208
Zambia	1	0	98	1	0	207
Bahamas	0	0	96	0	3	201
Saudi Arabia	0	0	100	0	0	200
Oman	0	0	100	0	0	200
Libyan Arab Repub.	0	0	100	0	0	200
Japan	1	1	1	56	41	198

Source: 'Export climate sensitivity index' compiled from data on mid-1970s exports from the countries listed. Original export data compiled by the World Bank and published in *World Tables*, second edition, 1980 (published for the World Bank by the Johns Hopkins University Press, Baltimore and London).

updated each year. If this is done, trends in the world 'rankings' of 'export climates' may be assessed.

Naturally, there are many overtones to such national climate sensitivity indices, notably the need for nations to import food, as well as their ability to export food, but it is considered that the ranking and sensitivity index provides a first estimate of one aspect of the weather and climate sensitivity of nations. Nevertheless, a relevant question related to these assessments is their value and use. Clearly, they are of little specific value to a decision-maker, but it is considered that international agencies may find the yearly rankings of some interest, particularly if such agencies are involved in climatically related activities in the form of financial or food assistance programmes.

K. NATIONAL INCOME AND PRODUCT ACCOUNTS: A NEW ZEALAND EXAMPLE

A more thorough and orderly process of assessing climate sensitivity can be obtained through a critical reading of national income and product accounts. But, although the trained eye looking at an economy may well be able to differentiate between those sectors which are more weather and climate sensitive than others, it is not always easy. As an example, consider the climate sensitivity of New Zealand by examining the various components of the economy as they appear in the *New Zealand Official Year Book*.

An analysis has been prepared (see Annexe to this chapter, p. 124) which gives for New Zealand an indication of the scale of weather and climate sensitivity. Specifically it reflects the results of an 'econoclimatic view' of the 1981 *New Zealand Official Year Book* and should provide an example as to how a similar 'econoclimatic view' could be applied to the equivalent statistical surveys of other countries. The approach used in the New Zealand example involves an overview of the economy and then a more detailed look at key elements. A few examples will indicate the typical step-by-step process that is required to make an initial assessment of national climate sensitivity. In the case cited the items noted as being climate sensitive are listed in the order they appear in the *New Zealand Official Year Book*.

First consider transport, an important component of any economy. However, it is not transport as such that should be considered but its various sectors, and specifically the climate-sensitive aspects of each sector. In New Zealand, the gross expenditure on railways (using data from the 1981 *Year Book*) was $404 million, this including $26 million (or 6 per cent) spent on fuel, most of which is imported petroleum products. It is considered that it is this, the fuel element of railway transport in New Zealand, which is sensitive to weather and climate information, since any reduction in fuel oil used – as a result of better weather and climate information for rail transportation – will be a benefit to the nation, through a reduction in the need to import fuel oil.

While it may be equally argued that other aspects of railway expenditure are also weather and climate sensitive, and this may well be the case in other countries, an analysis shows that in New Zealand few other sectors of railway operations are as weather and climate sensitive. Similarly, in the case of shipping, the key weather and climate factor that can be identified (as noted in the Annexe) is the loading and unloading of containers. This is because most of the 200,000 such containers moved in a year contain weather- and climate-sensitive commodities such as dairy produce, meat, wool or fruit products. The optimal movement of these mainly perishable commodities is crucial to New Zealand's export competitiveness and thus it is logical to include container movements in the list of weather and climate sensitivities.

Second, in terms of agriculture one of the key factors in assessing 'absolute' rather than 'relative' climate sensitivity is to consider agriculture production in terms of its dollar value. In the case of New Zealand, wool accounts for about 19 per cent of the gross value of agricultural production, compared with 3 per cent for vegetables. Thus although some aspects of the vegetable sector (such as the transport of vegetables to market) are much more weather sensitive than many aspects of the wool industry (such as the effects of severe frost on the quality of wool), in monetary terms the value of the weather and the climate on the total national agricultural scene will in normal circumstances be associated with the wool sector rather than the vegetable sector. Similarly, since 62 per cent of New Zealand's agricultural income is from pastoral products (i.e. wool, dairy products, and meat) it is evident that the key climate factors in New Zealand will be very much related to the state of the nation's pastures. Naturally in other countries this would be quite different, especially where field crops or horticultural type crops are the principal agricultural earners.

A third aspect is government expenditure, especially expenditure or 'subsidies' from public funds. The *New Zealand Official Year Book* lists two climate-related subsidies – the first concerning adverse events (such as drought conditions); the second, a fertilizer transport subsidy. This latter subsidy can at times be highly weather sensitive, since at that time it reflected the 'official' concern of the New Zealand government about farm income variations, in that a subsidy on fertilizers is often given *after* a relatively poor climatic (and hence poor income) season, in order to encourage farmers to fertilize their pastures during the following season. It should be noted that the natural response following a 'poor' season is for fewer dollars to be spent on fertilizer; the government fertilizer subsidy is therefore, in effect, a means of smoothing not only the irregularities in the application of fertilizer to New Zealand pastures, but also the climate-induced variations in farmers' incomes.

Fourth, in the energy sector, the key factor is the heavy reliance of New

Zealand on imported oil, which supplies 50 per cent of New Zealand's total energy requirements. In contrast, of the 22 per cent of the total energy from primary electricity, a very high proportion (86 per cent) comes from hydro-electric generation. The total energy situation in New Zealand is therefore related to two main factors – the need to use imported oil, and the natural availability of relatively cheap hydro-electricity. The first factor depends to a considerable extent on the amount of water available for hydro-electric power production (as electricity is generated from imported fuels when the local water supply is low), and the general overall state of the New Zealand economy. The second factor essentially involves (a) the adequate supply of water during the critical summer and autumn periods in the hydro-electric dams; and (b) the severity of the winter which dictates the proportion of electricity needed to be generated by the more expensive oil and gas. Both factors are therefore sensitive to an adequate supply of water during the critical summer and autumn periods at hydro-electric stations, and the severity of the winter, the latter being a critical determinant of the proportion of electricity to be generated by the more expensive oil and gas.

Finally, one may consider the components of the gross domestic product (GDP). As shown in the Annexe, under item (14), 11 per cent of New Zealand's gross domestic product comes from the agricultural sector and 6 per cent from food manufacturing. Other important weather- and climate-related market production groups include transport and storage (6 per cent), construction (5 per cent), energy (3 per cent), and wood products and forestry (3 per cent). These components comprise one-third of New Zealand's gross domestic product, and it is evident that they are much more weather and climate sensitive than other sectors. Hence, they must be considered the key weather and climate components of national income accounts.

As previously noted, the weather- and climate-sensitive sectors of other nations will be in many cases quite different from those of New Zealand. However, it is considered that an analysis similar to that described for New Zealand using appropriate national 'Year Books' would very clearly point to the significant weather- and climate-sensitive sectors. It should however be noted that most of what has been described relates to a 'feel for the situation' rather than the results of any in-depth analysis. The data illustrate in fact only the 'economy' side of the 'weather–economic mix', and it is clear that the 'weather' side of the mix is equally important. Regrettably, however, few in-depth studies have been made of the total mix – particularly on a national scale – and the analysis described of the New Zealand economy, while not being rigorous, may well be in the absence of more relevant studies the only realistic approach to the problem.

L. CLIMATE SENSITIVITY – THE LINK WITH POLITICAL REALITY

> Dear reader, I fear that a new multidimensional, multidisciplinary, cross-cutting subject of the utmost importance and concern, namely socioeconomic climatology, is about to be launched on an undeserving world and, if this book is a foretaste of things to come, let it be a warning.

These comments by Sir John Mason writing in the *Quarterly Journal of the Royal Meteorological Society* in July 1981 highlight the very great difficulties facing the economic climatologist. Mason was reviewing the book *Climatic Constraints and Human Activities* (Ausubel and Biswas 1980). The book is a report of a task force convened by the International Institute of Applied Systems Analysis (IIASA) in Vienna and the following are three typical extracts from the book.

> Climate is a classic multidisciplinary problem. It is at one extreme a subject for basic geophysical field research and computer modelling and at the other an applied question of immediate importance to policy makers concerned with disaster relief, agricultural policy, and so forth.
>
> It is not difficult to see that in some senses the atmosphere may be regarded as a commodity or a bundle of economic goods, but it remains to be seen whether some sort of 'atmospheric sector' may be useful in examining either the relationships between an atmospheric sector and the rest of the economy, or the relationship between subsectors within an atmospheric sector.
>
> Meteorologists usually talk as if climate is simply a set of statistics, rather than anything tangible which can be regarded as a resource. Yet, if an economist were to describe weather modification to a meteorologist as an attempt to bring about an enhancement or redistribution of climate resources, it would be an acceptable statement. From an economic point of view, climate is matter and energy organized in a certain way.

Despite these comments, or possibly because of them, the book at least as far as Mason is concerned 'does not inspire confidence'.

In contrast, the present author reviewing the same book in *Weather and Climate* in February 1981 took quite a different viewpoint.

> This is an important book because it is concerned with what many would regard as the non-scientific aspects of weather and climate. It is in fact a refreshing change from the dynamics and physics of the atmosphere which although important scientifically have, at least in terms of 'better' weather forecasts, a relatively poor 'track record' over the last decade. But let us not deceive ourselves in thinking that all is easy when the non-scientific aspects of the subject are considered. It is not, and terms like education, marketing, awareness, relevance, and value are often much more difficult for both meteorologists and users of meteorological information to come to terms with, than the more familiar terms – at least to most meteorologists – of vorticity, divergence, convection, radiative processes etc.

Mason, however, clearly has some difficulty with the concept of the atmosphere as a resource, a fact which he delightfully demonstrates in the following additional comments from his review:

> It seems that there is more between earth and sky than is dreamt of in our philosophy; perhaps we should study the 'situational dynamics' and extend our models to include the 'socioeconomic dimension' and think of the atmosphere simultaneously as 'a global common', 'a modulator of the supply of climate', 'a living resource' (including the birds!), and climate as 'a depletable resource of matter and energy captured from the atmosphere'!

Perhaps the real truth lies somewhere between the views expressed by Sir John Mason in his review and those expressed by the present author in his review of the book. It may therefore be appropriate to quote Sir Brian Pippard, Professor of Physics at Cambridge University, who in 1981 gave an inaugural lecture at the 150th anniversary meeting of the British Association. *The Times*, reporting on this meeting (see also Pippard 1982), said that Professor Pippard questioned whether methods applied in the laboratory by the physicist, biologist, chemist and mathematician have any potential value for the politician and sociologist in resolving everyday problems. Sir Brian went on to say:

> It would be no bad thing if at prizegiving ceremonies it was whispered in the ears of mathematicians and scientists in their hour of triumph that they had succeeded because they had chosen to tackle relatively straightforward problems: and that if politicians and social reformers are not so obviously successful, it is because they have challenged problems of enormously greater complexity.

Professor Pippard then illustrated the connection between the problems associated with weather forecasting and the difficulties in making political predictions with a series of models showing the instabilities that cause earthquakes, electronic interference, and the disorderly flow of liquids and gases. These he said were examples of events leading to chaotic behaviour. And since the world around us also shows all the signs of being chaotic, he wondered if lessons could be drawn.

This section has focused on some of the difficulties associated with the mix of the economic and meteorological systems, but has also highlighted some of the challenges. But, in this very difficult area of defining and measuring weather and climate sensitivity, it is important that the *applied nature* of the problem again be emphasized. For example, the sensitivity of the commodity markets to weather and climate information is a clear indication that in the economic world, weather and climate sensitivity is a reality. There is also increasing climatic realism in the very difficult areas of disaster relief and agricultural and energy policies.

Finally, in considering weather and climate sensitivity and the linkages with political reality, it is essential to re-emphasize that appropriate planning must be evolved. But, if this is really to be accomplished, the politician and the planner must become more weather and climate orientated, for only then will optimum use be made of the climatic resources of the 1990s. Central to this planning, as has already been recognized by the World Meteorological Organization and several national meteorological services, is the need for a much more comprehensive monitoring and analysis of the world's climate, both to detect and to predict changes. The sensitivity – in economic, social, and political terms – of nations, sectors of nations, and commodities to weather and climate variations and changes must also be better understood. Indeed, the need for such understanding offers the most important challenge facing the meteorological and climatological community as we move towards the twenty-first century.

ANNEXE
Weather and climate sensitivity: A New Zealand example

Note: All data are from the *New Zealand Official Year Book* (1981), and refer to the most recent period *then* available. The percentages shown are the percentage of the relevant totals.

1. RAILWAYS

		Expenditure ($ million)	*%*
Gross expenditure		404	
of which *FUEL*	=	26	6

2. SHIPPING

CONTAINER TRAFFIC		*Number*
Unloaded	=	107,000
Loaded	=	113,000

3. ROADS

	Expenditure ($ million)	*%*
Highway maintenance	51	30
Highway construction	36	21
Grants to local authorities	70	41

4. AIR

AIR NEW ZEALAND

	Million
International: Passenger/km flown	4,430
International: Cargo tonne/km flown	152
International: Total revenue (tonne/km)	573

5. AGRICULTURE

GROSS AGRICULTURAL PRODUCTION

	$ million	*%*
Wool	851	19
Dairy produce	688	15
Cattle	668	15
Sheep and lambs	565	13
Agricultural services	199	4
Crops and seeds	184	4
Fruit	130	3
Vegetables	121	3
Poultry and eggs	112	2

SUBSIDIES FROM PUBLIC FUNDS

Season	*Adverse events relief ($ million)*	*Fertilizer transport subsidy ($ million)*
1970/71	3.5	7.1
1971/72	0.3	9.0
1972/73	0.2	12.5
1973/74	1.0	11.6
1974/75	0.2	8.2
1975/76	0.3	9.2
1976/77	0.4	12.7
1977/78	0.2	16.1
1978/79	6.1	23.4
1979/80	0.3	28.7

6. MANUFACTURING

	Turnover ($ million)*	*%*
Meat export works	792	7
Co-operative dairy companies	713	7
Brewers	142	1
Wool fibres/spinning/weaving	141	1
Canning/preserving fruit and vegetables	117	1

* Total of all sales and other income.

7. BUILDING AND CONSTRUCTION

	Turnover ($ million)	%
Building – non-residential	900	34
Construction	766	29
Building – residential	446	17

VALUE OF BUILDING PERMITS

	$ million	%
Houses	379	33
Alterations/additions	154	14

8. ENERGY

TOTAL ENERGY

	Petajoules	%
Imported oil	178	50
Primary electricity	72	22
Coal	55	15
Natural gas	37	10
Indigenous oil	14	4

ENERGY USERS

Industry	39%
Transport	34%
Households	16%
Commerce	11%

ELECTRIC POWER GENERATION

Hydro	86%
Geothermal	5%
Other*	9%

* *Fuel used* Light oil 3,994 tonnes
Heavy oil 17,408 tonnes

ELECTRICITY GENERATION

	% by Oil and Gas
1976	0.7%
1977	6.8%
1978	5.5%
1979	3.4%
1980	0.6%

9. TRADE/DISTRIBUTION

	Sales ($ million)	%
Wholesale	9,776	55
Retail	6,548	37
Restaurants and hotels	904	5
Personal and households	441	3

WHOLESALE TRADE

	Sales ($ million)	%
Dairy products	576	6
Meat and meat products	556	6
Wool, hides, skins	508	5
Fresh fruit and vegetables	202	2

RETAIL TRADE

	Sales ($ million)	%
Petrol stations	546	8
Supermarkets	501	8
Dairies	308	5
Chemists	213	3
Butchers	210	3
Greengrocers and fruiterers	175	1

RESTAURANTS AND HOTELS

	Sales ($ million)	%
Licensed motels/hotels	355	39
Licensed taverns and clubs	182	20
Cafeterias, coffee bars, etc.	109	12
Take-away foods	107	12

PERSONAL AND HOUSEHOLD SERVICES

	Sales ($ million)	%
Motor repairs	142	32
Panel beaters/spray painters	85	19

10. EXPORTS

INDIVIDUAL COMMODITIES

	Value ($ million)	%
Wool	931	18
Beef	504	10
Lamb	486	9
Butter	361	7
Milk powder etc.	220	4
Casein	113	2
Cheese	106	2
Mutton	98	2
Leather	72	1
Lamb pelts	64	1
Fish	62	1
Tallow	43	1
Apples	36	1
Kiwifruit	35	1
Sheep pelts	34	1
Sausage casings	34	1

MAJOR GROUPS

	Value ($ million)	%
Meat and meat preparations	1,192	23
Wool	931	18
Dairy products	799	16
Pulp, paper, paperboard	248	5
Hides, skins and pelts	180	3
Fruit and vegetables	129	3
Tallow	43	1
Sausage casings	34	1

11. IMPORTS

	Value ($ million)	%
Petroleum products	901	17
Vegetables and fruits	52	1
Sugar	43	1

12. FOODSTUFFS EXPORTED

	% of total production
Tallow and animal fat	86
Fish	83
Processed milk	83
Butter	75
Beef	67
Veal	66
Cheese	63
Pulses	63
Mutton and lamb	60
Onions	56
Shellfish	46
Apples	37
Berry fruits	21

13. LOCAL GOVERNMENT

SALES OF COMMODITIES AND SERVICES

	$ million	*%*
Electric power boards	370	41
City and borough councils	262	29
Harbour boards	100	11

14. GROSS DOMESTIC PRODUCT

MARKET PRODUCTION GROUPS*

	Value ($ million)	*%*
Agriculture	2,378	11
Food manufacture	1,266	6
Transport/storage	1,199	6
Construction	934	5
Electricity/gas/water	700	3
Wood products' manufacture	342	2
Forestry	176	1

* Considered to be most weather and climate related.

V Assessing the impact of weather and climate

A. ECONOMIC AND WEATHER/CLIMATE DATA SOURCES

1. The data problem

Attempts to find associations between nationwide economic and weather/ climate data present many problems. In particular there is the incompatibility of the two raw data sets since economic information is usually related to areas and weather and climate data to places. It is therefore usually necessary to transform one of the data sets and, from the nature of the information, the weather and climate data are generally the easier to adjust. Nevertheless, the economic data banks that do exist are still of only limited value in assessing weather and climate impacts, since most economic data is either too embracing in its composition, or it is for large areas or relatively long time periods.

For example, it is difficult if not impossible to find official economic data on the sales of, say, ice cream on a daily basis for a small area, nor is it usually possible to find *any* official economic data available sufficiently soon after an event to be useful in any real kind of decision-making. Economists lead you to believe otherwise, but except for a few notable economic indicators – such as prices for commodities or shares, the availability of economic information in any decision-making time framework is minimal, and cannot in any sense be compared with the many thousands of weather observations made every few hours at places all over the world and available to meteorological decision-makers (primarily in the form of weather forecasters) within minutes of the events taking place.

2. Sampling the national scene

Although the basic unit for statistical purposes in many countries is the county or its equivalent, in the United States (as well as some other countries) many indices of national economic activity published in business journals or in government reports are based on *sampling* and are *not* related to counties or states. For example, *Business Week* reports national weekly data on such items as electric power consumption (supplied from the Edison Electric Institute), and intercity truck tonnage (supplied from the American Truck Association). In each case it is considered that such data when used in any weather impact study can only be realistically associated with appropriate commodity-weighted weather information for the nation, that is the United States, as a whole. Similarly, many aspects of New Zealand's economic activity are not at all associated with counties or with county data, and some economic indicators such as meat, wool, and dairy production, retail trade, manufacturing, and prices are based on sampling which is designed to give a reasonably true indication of *national* activity in these economic sectors.

It therefore follows that for the many national economic indices produced from various sample surveys, weather and climate information from, say, 200 first-order weather stations in the United States, or from 50 synoptic stations in New Zealand can provide, if appropriately weighted, a useful indication of the 'national atmospheric climate'. But it must be emphasized that the economic data used in weather-impact studies cannot usually be reduced to a spatial scale that is less than the nation as a whole. For example, as previously noted, data on US weekly electric power consumption published in *Business Week* are *designed* to give a nationwide index of national activity, and should not therefore be considered as estimates of *regional* activity, nor are the national data a summation of regional data but rather a carefully designed sample to give a national picture.

3. The use of commodity-weighted data

As discussed in the next section, commodity-weighted weather and climate indices are now available in New Zealand for 120 different weightings (from human population to skiing), and for a wide variety of climatic parameters (rainfall, rainfall as a percentage of normal, growing degree days, days of soil moisture deficit, etc.). It should be noted, however, that to date few countries produce and market commodity-weighted weather and climate indices. From both a political and an economic point of view this is both surprising and disturbing, and if meteorologists and climatologists really believe that they have a product worth marketing, the assessment and promotion of commodity-weighted weather and climate indices must be actively encouraged.

A prime reason for this kind of research into weather economics is to enable *nationwide* economic activities to be related to *nationwide* weather. For example, an early study made by the author on weather and retail trade in the United States (Maunder 1973a) noted that specific and useful aspects of nationwide weather are published weekly by the US Department of Commerce (Weather Bureau) and the US Department of Agriculture in their *Weekly Weather and Crop Bulletin*, together with the five-day nationwide weather forecasts issued by the National Weather Service. Various kinds of weekly economic indices are also available for the United States; among these are the *Weekly Retail Sales* published by the US Department of

Commerce.* However, these and many other similar sets of weekly data are available only on a national basis, and although it may well be very desirable for weekly *regional* data to be available, the fact that they are not necessitates the calculation of national weather indices for the same periods of time as are available for the economic data.

A further complicating factor is that most if not all indicators of national economic activity are *not* designed to be related to weather conditions; furthermore, the publication *Weekly Retail Sales*, which gives estimated weekly retail sales for the United States, is restricted to thirteen major *store types*, and no information is published relating to weather-sensitive commodities such as ice cream, soft drinks, women's winter and summer clothes, refrigerators, air-conditioners, automobile tyres, beer, iced tea, paint, umbrellas, etc. It is clear that the unavailability of the right kind of data makes any analysis difficult, and it is fully appreciated that it would make things much easier for both the climatologist and the economist if such information were readily available. However, several nationwide companies in the United States and Canada advised the author at that time (that is in the early 1970s when the original report was prepared) that they either did not have information on the daily or weekly sales of weather-sensitive commodities, or that they were not permitted to release such information.

It is appropriate at this stage also to refute the false assumption often made that government economic statisticians can provide data suitable for use in econoclimatic studies. Regrettably, this impression is far from the truth, and although the availability of data varies from country to country, the plain fact is that government economic statisticians can only on rare occasions provide the economic climatologist with the information he or she desires. There are exceptions, but unfortunately the basic position is as stated by McQuigg (1972a) more than a decade ago:

> There are not as many well-documented, quantified relationships between weather events and economically important activities as one would think existed. One of the great difficulties that has led to this situation is the lack of homogeneous samples of operational data comparable to the reasonably complete body of meteorological observations that exist in many countries.

B. Commodity-Weighted Weather Indices: The Why, What and How

1. An overview

A key factor in providing appropriate guidance to decision-makers is the commodity weighting of weather and climate information, for, as is very

* Publication of this series ceased in 1979. The original study by the author was carried out in the early 1970s using retail trade data for the period 1966–69.

obvious in applying such information, all things, and all areas, are *not* equal. For example, a forest industry is basically and in most cases *only* interested in the weather and climate conditions pertaining to the forested areas, and a wheat industry in the weather and climate conditions pertaining to wheat areas. Similarly a meat and wool industry is interested in the weather and climate in the beef cattle and sheep areas, and an electricity industry in the weather pertaining to the populated and industrial areas with respect to consumption, but in other respects to energy source areas and transmission routes.

Unlike many national organizations which observe, collect and process data in *non-real-time*, national meteorological and climatological services observe, collect, and process data in *real-time*. Accordingly national meteorological and climatological services can compute key climate and (if appropriate analyses are made) *econoclimatic indices* on both a regional and a national basis, and also if required on an international basis. A key problem associated with the development of such indices, however, which must be re-emphasized is that most commodity type information refers to *large* areas of a country whereas weather and climate data are usually related to *points*. Accordingly, any study which endeavours to associate weather and climate conditions with economic activities such as housing starts in the United States, rice production in Thailand, corn production in Iowa, or energy consumption in Brazil, is faced with the incompatibility of the commodity/economic and weather/climate data. An essential task, therefore is to adjust the weather and climate data to fit the available economic data. One way to do this is to use the economic data collated on a county basis, and this method has been used in an operational manner in New Zealand for ten years; details are given below.

Similar information for some other countries is also being assessed, but except for some population-weighted temperatures and heating-degree data for the United States and Canada, it is believed that the comprehensive commodity-weighted aspects of the national indices produced for New Zealand are unique. An example of the potential of this kind of real-time weather and climate information is given in the May 1979 issue (p. 485) of the *Bulletin of the American Meteorological Society*. This report stated that the Earth Satellite Corporation had introduced CROPCAST–1979, the third generation of a computerized agricultural information system designed to give a near real-time view of world agricultural information and expected future change. The corporation reported that CROPCAST–1978 demonstrated 'excellent accuracies of forecast for all crops, worldwide'. In particular, CROPCAST–1978 placed weekly computer-generated maps for US corn, soybeans, wheat (spring and winter), Canadian spring wheat and USSR wheat (spring and winter) on agri-business desks in various world locations via computer display terminals. CROPCAST–1978 also provided twice-monthly written crop reports covering selected production, demand, supply, price and policy questions, daily telephone consultations, and daily computer-delivered crop weather factor analyses.

2. The New Zealand experience

As previously discussed the key factor in optimizing the marketing of weather information is the availability of appropriate information in real-time. The weather-sensitive sectors of New Zealand's economy are now generally well known, as well as the availability of real-time weather information specifically tailored to national commodity sectors, such as the wool industry, horticulture, or energy production. In particular, the availability of quite complex programming techniques has enabled the provision in real-time of highly relevant weather information to decision-makers. Moreover, the potential availability of such information to decision-makers and the public through videotex type systems now means that applied meteorological and climatological information of prime importance is available on demand. Such information can provide *indices of weather and climate activity* which are not only updated daily (and in some cases hourly), but are also available on a regional and a national basis.

Specifically, weather and climate indices weighted according to the distribution of 120 economic and agricultural parameters are currently analysed by the New Zealand Meteorological Service for several areas of New Zealand and for New Zealand as a whole. The actual weightings used in each analysis are based on the contribution of the 'geographical county' to the New Zealand total 'population' or 'area' (see Maunder 1972a and 1972d).

The 120 economic/commodity weightings – which are discussed in detail in the next section – include: human population, sheep population, dairy cow population, and beef cattle population, as well as the area in crops, land, plantations of timber trees, orchards, market gardens, potatoes, wheat, vegetable crops, hay and silage, apples, pears, cabbages, carrots, lettuces, vegetable onions, vegetable peas, sweetcorn, outdoor tomatoes, oats, barley, maize, threshed peas, field onions, grassland irrigated, irrigated fruit, and irrigated vegetables. In addition, weightings for the following commodities are computed and used operationally: hydro-electric capacity, shorn sheep and hoggets, shorn lambs, the area of grassland established, the area of grazing land, the total area of holdings, breeding ewes, wool production, expenditure on farms, pig population, cargo operations at seaports, and passenger operations at airports. Dairy cow population indices related to five specific areas used by the New Zealand Dairy Board are also computed, plus sheep population indices related to the eight classes of farms used by the NZ Meat and Wool Boards' Economic Service. Additional weightings have also been prepared which update some of these weights which were originally based on data for the late 1960s, and new weightings have also been provided for deer farming, tourism, skiing, and biomass production.

(a) ECONOMIC UNITS

The basic economic unit for statistical purposes in New Zealand is the county, and although some counties extend over more than one homogeneous or 'geographical' region, the counties in New Zealand can be used as the basis for a regional breakdown of economic activities. Indeed, in many instances the county is the only data source which can be used for a regional division of New Zealand.

In 1970 when the original analysis was made, there were 110 counties in New Zealand, and it was deemed appropriate to use economic data from these, together with the climate data applicable to one or more climatological stations in each county, as the basis for the formulation of weighted weather indices. It should be noted that the counties used are the so-called 'geographical counties'; that is, they are the administrative counties plus the population and or production within any cities, boroughs, or independent town districts which are included within the boundaries of the 'geographical county' as defined by the Department of Statistics. For example, the Hutt geographical county at the 1966 census had a human population of 296,156, which included 41,284 in the Hutt administrative county, as well as the 131,555 in Wellington City, 22,190 in Porirua City, 19,084 in Upper Hutt City, etc. The advantage of using the geographical county is that the total New Zealand population (of animals, or land, or people in urban areas, etc.) can be subdivided into 110 areas.

(b) REPRESENTATIVE WEATHER/CLIMATE STATIONS

As of 1970, there were 260 climatological stations in New Zealand, and the original study (Maunder 1972a and 1972d) selected from these stations one representative climatological station for each geographical county. There were also over 1500 rainfall stations which could have been used, but it was decided to confine the weather data (with a few exceptions) to that obtainable from the climatological station network. An important factor in this decision is that the procedure now described can also be applied with minor adjustments to other climate elements such as maximum temperature, minimum temperature and sunshine. In many counties there was only one climatological station available, and in these cases the one station was used in the analysis. In a few cases there were no climatological stations within the county boundaries, and in these cases the economic data in the county concerned was combined with an adjacent climatologically similar county in which a climatological station was located.

There were also cases where more than one climatological station was available. In these counties, consideration was given to the significance of the county from a population or area viewpoint and in fifteen cases the weightings of two or more climatological stations were used. A complete list of the counties and the climatological stations used in the analysis is given in the original paper (Maunder 1972a).

(c) ASSESSMENT OF COUNTY/ECONOMIC INDICES

The basic consideration in the calculation of the weightings of the climatological data for each geographical county is the contribution of the geographical county to the New Zealand total population or area. For example, recent data show that the Waimate geographical county has 0.3 per cent of the New Zealand human population, 1.4 per cent of the land area, 1.9 per cent of the sheep population, 0.1 per cent of the dairy cow population, 0.7 per cent of the beef cattle population, and 5.0 per cent of the crop area. Data showing the relative significance of six geographical counties for twenty-three economic parameters is shown in Table V.1, which indicates that there are relatively large differences in the economic base of various areas of New Zealand. For example, the effect of similar weather variations in say, the Waikato and Vincent counties will be significantly different in so far as the

Table V.1 Relative significance* of geographical counties in New Zealand

	Geographical county					
Economic parameter	*Waikato*	*Cook*	*Horowhenua*	*Mackenzie*	*Waimate*	*Vincent*
Human population	1.0	1.3	1.0	0.1	0.3	0.3
Sheep	0.8	1.9	0.4	1.3	1.9	1.2
Dairy cows	5.5	0.3	1.8	0.0	0.1	0.1
Beef cattle	1.3	4.1	0.6	0.5	0.7	0.4
Crop area	0.4	0.9	0.5	1.6	5.0	0.8
Orchards	0.6	3.5	0.1	0.0	0.3	9.1
Market gardens	1.3	3.0	4.9	0.0	0.6	0.2
Potatoes	1.0	0.4	7.1	0.1	2.3	0.1
Wheat	0.0	0.1	0.0	1.5	7.3	0.7
Grazing land	0.1	0.2	0.1	9.6	3.0	8.2
Breeding ewes	0.7	1.8	0.4	1.2	2.0	1.2
Sheep/hoggets shorn	0.9	1.9	0.4	1.3	1.9	1.2
Hydro capacity	0.3	0.3	0.0	7.5	1.3	0.5
Apples	0.1	0.5	0.1	0.0	0.5	1.0
Pears	0.1	0.4	0.1	0.0	1.2	1.7
Cabbages	1.3	1.6	7.0	0.0	0.9	0.1
Lettuce	1.5	2.3	6.1	0.0	3.5	0.0
Sweet corn	2.1	49.6	0.3	0.0	0.1	0.0
Outdoor tomatoes	0.8	11.4	2.8	0.0	0.0	0.0
Sheep/beef units	0.9	2.6	0.4	1.1	1.6	1.0
Wool production	0.5	2.0	0.3	1.4	2.1	1.3
Biomass (maize)	3.0	1.2	1.4	1.2	2.1	0.0
Biomass (beefs)	2.4	1.2	1.0	0.7	1.8	0.6

Source: After Maunder (1985a).
Note: * Percentage of New Zealand total.

weather is associated with the activities of people, or the production from dairy cows, beef cattle, sheep or crop land, or the capacity for hydro-electric power production. It is logical, therefore, that the assessment of the weather and climate effects on economic activities in New Zealand (or any other country) in any month (or week, or season) should take into account the relative importance of *activities* in various areas.

(d) Formulation of monthly climate indices

The assessment of the commodity-weighted weather and climate indices now described is based on the weighting of the weather and climate data applicable to the county climatological or weather station (or stations). The weighted values or deviations from the average are then summed for various regions, for the North and South Islands, and finally for New Zealand. A specific example is now given for rainfall in March 1969 – a particularly dry month over much of New Zealand – but the method described can also be applied to other climate factors such as temperature or sunshine, and for other months, seasons, years, or for that matter *any* period.

The rainfalls at the 109 climatological stations were first expressed as a percentage of the normal rainfall for the specific stations. For example, in March 1969, Kaitaia Aerodrome had 5 per cent of its normal rainfall, Rukuhia 47 per cent, Ruatoria 2 per cent, Milford Sound 172 per cent, Christchurch 14 per cent, and Waimate 40 per cent. These percentages were then weighted by multiplying them by the county contribution of the New Zealand total population or area. For example, in the case of Waimate County, the Waimate climatological station rainfall percentage of 40 per cent was multiplied by 0.3 for human population (see Table V.2) for the land area, and by 1.9 for sheep population. A similar procedure was followed for the other geographical counties, except in those cases where two or more climatological stations were used. In these latter cases, the county population or area was subdivided so that each climatological station was associated with an appropriate proportion of the total county weight.

(e) Formulation of regional and national indices

The weighted county indices are then combined into regional or national indices using the following formula:

$$\text{Regional or national index I} = \frac{\Sigma R_i E_i}{\Sigma E_i}$$

where: R_i is the rainfall (or other climate element) for county i expressed as percentage of normal; E_i is the percentage of the national or regional economic parameter in county i; and I is an index which ranges in terms of rainfall expressed as a percentage of the average from 0 to over 200. This particular index may therefore be considered an index of wetness or dryness. The summation is over all counties in the region, or the nation.

Table V.2 Weighted rainfall computations for the South Canterbury/North Otago region of New Zealand for May 1983

		Weightings					
*Climatological station**	*Rainfall % of normal*	*Land area A*	*Land area B*	*Human population A*	*Human population B*	*Sheep population A*	*Sheep population B*
Lake Coleridge	129	1.1	142	0.1	13	1.7	204
Ashburton	77	1.2	92	0.8	62	2.4	185
Lake Tekapo	108	2.7	292	0.1	11	1.1	119
Timaru Airport	46	0.8	37	0.4	18	1.1	51
Timaru	48	0.3	14	1.2	58	0.7	34
Waimate	60	1.4	84	0.3	18	1.9	114
Tara Hills	146	1.2	175	0.1	15	0.7	102
Oamaru	54	1.2	.65	0.9	49	1.0	54
Region		9.9	901	3.9	244	10.8	863
Weighted Index (B/A)		$\frac{901}{9.9} = 91$		$\frac{244}{3.9} = 63$		$\frac{863}{10.8} = 80$	

Source: After Maunder (1985a), updated from Maunder (1972a).
Notes:
* These climatological stations are located in and assumed to be associated with various geographical counties. For example, it is assumed that 1.7% of the total New Zealand sheep population is climatologically associated with the Lake Coleridge climatological station.
A = Percentage of the New Zealand total of the economic parameter located in the geographical county and associated climatological station(s).
B = (Column A) × (the rainfall at the climatological station expressed as a percentage of normal).

An example of the computation of the regional index is given in Table V.2 for the South Canterbury/North Otago area (for New Zealand regions, see Map, p. 140). Table V.2 shows that in May 1983 the weighted regional rainfall indices were 91 (per cent of average) for the land area, 63 (per cent) for the human population, and 80 (per cent) for the sheep population. As of 1985, similar weightings for over 120 commodities are available for a wide variety of climatic elements.

Similar computations to those shown in Table V.2 for the South Canterbury/North Otago region were made for twenty-one regions in New Zealand. In addition, the weighted indices for the sixty-four climatological stations in the North Island and the forty-five climatological stations in the South Island were combined to obtain rainfall indices for the North and South Islands respectively. Finally, the weighted indices for all 109 climatological stations were combined to obtain the weighted rainfall index for New Zealand.

The resulting weighted rainfall indices for March 1969, for twenty-one regions in New Zealand, are shown in Table V.3. The corresponding rainfall

Table V.3 Rainfall indices weighted according to the significance of various economic parameters. Data for March 1969

	Economic parameters					
Area	*Land area*	*Human population*	*Sheep population*	*Dairy cows in milk*	*Beef cattle population*	*Crop area*
1. Northland	15	11	17	16	16	14
2. Auckland	15	17	13	18	16	19
3. Coromandel	11	12	10	11	10	16
4. Waikato	27	35	26	27	26	26
5. Bay of Plenty/Taupo	34	27	32	23	32	31
6. Gisborne	10	18	12	14	12	17
7. Hawkes Bay	31	31	30	32	31	29
8. Waitomo	42	40	41	38	40	45
9. Taranaki	25	25	24	24	25	26
10. Wanganui	40	29	39	33	39	38
11. Manawatu	42	46	36	48	38	41
12. Wairarapa	32	35	34	29	34	31
13. Wellington	27	27	27	27	27	28
14. Nelson	12	10	11	8	12	10
15. West Coast	104	95	95	82	108	85
16. Marlborough	15	14	14	18	15	13
17. North Canterbury	27	20	27	27	26	29
18. South Canterbury/ North Otago	70	57	59	56	62	53
19. Coastal Otago	101	96	105	103	103	106
20. Central Otago	117	117	110	113	111	107
21. Southland	124	124	127	127	126	128
North Island	28	25	29	25	27	31
South Island	79	57	81	60	75	69
New Zealand	58	35	52	27	36	62

Source: After Maunder (1972a).

indices for the North and South Islands, and for New Zealand, are also shown. Similar commodity-weighted indices are available for rainfall from January 1950, for mean temperatures from January 1954, for sunshine from January 1967, for days of soil water deficit from January 1949; and for heating degree days with base 18°, 15°, and 12°C, and growing degree days with base 5°, 10°, and 15°C, from January 1950.

The commodity-weighted data provide useful specific information on the climate of a specific period. For example, in February 1976 the weighted rainfall index for New Zealand as a whole varied from 66 (i.e. 66 per cent of normal) for grazing land, to 259 (i.e. 259 per cent of normal) for vegetable crops. Thus it was well below normal (rainfall-wise) for the sheep and beef

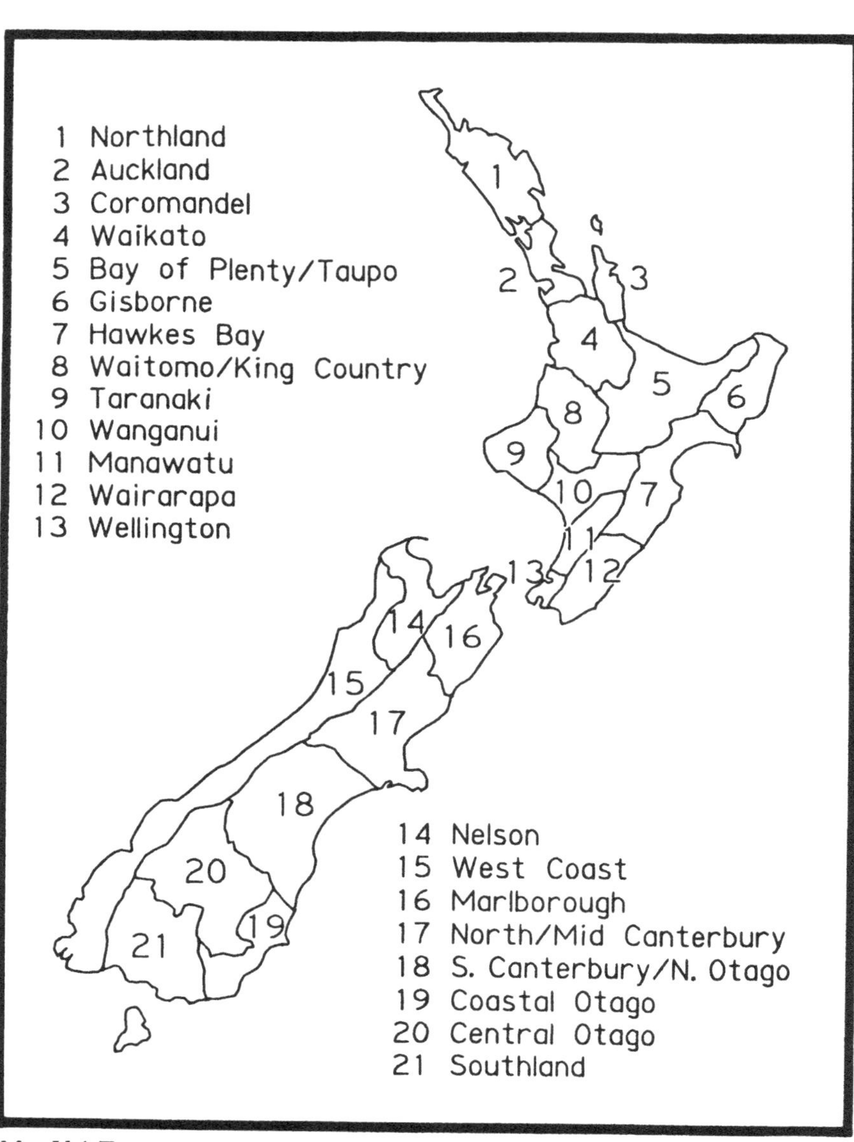

MAP V.1 Twenty-one generalized regions of New Zealand

farmer using grazing land associated with the meat and wool industry, but very wet for those involved with the vegetable industry. The weighted indices may also be used for regional comparisons: for example, in February 1976 for the commodity sheep, the rainfall indices varied from 295 (i.e. 295 per cent of normal) in the east of the North Island, to 24 (i.e. 24 per cent of normal) in the north of the South Island.

Apart from their use in operational models, such as those discussed later for various economic and agricultural industries, the weighted values provide a real alternative to the traditional method of portraying weather and climate information which treats all areas as being equal. Accordingly, it is no longer necessary or indeed acceptable to say that last month (or last week) temperatures (or rainfalls) were above normal. What is required is a statement that last month (or week), temperatures, weighted according to the distribution of a specific population (whether they be people, animals, crops or their productivity) were above (or below) normal. In this way, the vast quantity of weather and climate information can be used operationally to assist decision-making processes.

(f) WEIGHTED SOIL WATER DEFICIT INDICES: DAILY TO SEASONAL VALUES

A simple water balance analysis of *daily* rainfall data at representative climatological stations can also be used to calculate indices (both *weighted* and *actual*) of 'soil water deficits' (Coulter 1973). In such an analysis the number of *days of soil water deficit* can be calculated from daily rainfall data by assuming that evapotranspiration continues at the average monthly

Table V.4 Soil water deficit index computations – New Zealand: January 1971

Stations (selected)	*Days of deficit (a)*	*Weight* (b)*	*(a) × (b)*
Kaitaia	0	1.9	0
Dargaville	19	4.1	78
Rukuhia	11	5.1	56
Tauranga	9	4.1	37
New Plymouth	0	2.4	0
Gore	2	0.4	1
New Zealand	–	100.0	445

$$\text{Weighted average} = \frac{445}{100} = 4.5 \text{ days}$$

Source: After Maunder (1977b).
Note: * County percentage of New Zealand total of dairy cows.

potential evapotranspiration rate (calculated by an appropriate formula) until 75 mm of soil moisture has been withdrawn. Thereafter, *days of soil water deficit* are counted until there is a day with rainfall in excess of the daily potential evapotranspiration.

By way of example consider the computations involved in assessing the number of *days of soil water deficit* for the 'New Zealand dairy farm'. The relative significance of selected counties in New Zealand in terms of dairy production (among other things) is shown in Table V.1, and the specific application of the dairying weightings in computing 'New Zealand dairy farm' commodity-weighted *days of soil water deficit* for January 1971 is given in Table V.4. In this the number of days of soil water deficit (a) is weighted by the percentage of the total dairy cow population in the county concerned (b), and the weighted days of soil moisture deficit indices for each month (i.e. (a) × (b)) are then combined into a national index using the following formula:

$$\text{Soil water deficit index I} = \frac{\Sigma W_i E_i}{\Sigma E_i}$$

where: W_i is the actual number of days (in a given period) with soil water deficit for station i, E_i is the percentage of the New Zealand dairy cow population in county i, and I is an index which ranges from 0 to 31 days (if months are considered), or 0 to 90 days (if a season is considered), or other ranges for other periods.

The calculated weighted soil water deficit indices may be considered to give an indication of 'excessive' dryness. The abridged computations of the

Table V.5 Weighted* soil water indices (days of soil water deficit) – New Zealand: February

Region	*1967*	*1972*	*1973*	*1976*	*1982*	*1985*
Northland	0	14	21	11	9	6
Waikato	0	17	19	1	12	2
Bay of Plenty	0	11	24	0	8	0
Taranaki	0	1	16	0	8	0
Manawatu	0	12	22	0	10	0
Marlborough	15	21	24	8	22	6
North Island	0	12	20	2	11	2
South Island	13	12	19	6	13	9
New Zealand	1	12	20	2	11	3

Source: Updated from Maunder (1977b).
Note: * Weighted by distribution of dairy cows.

New Zealand deficit index for January 1971 are given in Table V.4. The relevant *dairy cow* weighted soil water deficit indices for selected regions and New Zealand for February 1967, 1972, 1973, 1976, 1982, and 1985 are given in Table V.5, which indicates very clearly a marked variation from one region to another and from one February to another. Using similar methods, weekly, monthly, and seasonal indices of the number of *days of soil water deficit* have been computed for 120 different weightings and for twenty-four regions of New Zealand. An example for the weighting 'dairy cows' for the Waikato region for the significant January to April months of the dairy season is given in Fig. V.1.

(g) Human population weighted temperature indices

In addition to computing various climate indices for a wide range of commodities for different periods, temperature indices weighted by the distribution of the human population are computed three times a day by the New Zealand Meteorological Service. Such indices are useful in a variety of ways. For example, during June 1976, relative extremes of temperature occurred in New Zealand, and this is shown clearly by the population-weighted temperatures based on the average of the 8 a.m., noon, and 5 p.m. observations for 2 June and 21 June. On 2 June 1976, the New Zealand population-weighted average temperature was 15.3°C, with 17.2°C being recorded at noon. By contrast, on 21 June 1976 – possibly the coldest day in New Zealand for at least 100 years – the national weighted average temperature was only 4.9°C, with 5.9°C being recorded at noon. Thus there was a difference of over 10°C in the New Zealand weighted average temperatures on these two June days, which, although not large by 'continental' standards, is extremely large for New Zealand since this difference is equivalent to the *normal* January–July difference in the mean monthly temperatures. Knowledge of such population-weighted temperatures on a national basis has obvious implications and value to energy producing and distributing authorities, since a variation of 1°C in the temperature is equivalent to a 2 per cent variation in electric power consumption.

Similar population-weighted temperature indices or heating degree day indices are computed in the United States and Canada. In the United States they are combined in a very effective way with the monthly and seasonal temperature forecasts to forecast the expected demand for energy, details of which are given in Chapter VI.

(h) The importance and application of weighted weather and climate indices

Weighted weather and climate information can be assessed – like all weather and climate information – in either real-time or non-real-time. The difference is *significant*, and relates to the time between when the event happened and when information about that event is known. The technique of

FIGURE V.1 Weighted* soil water deficit indices (days of soil water deficit) – Waikato

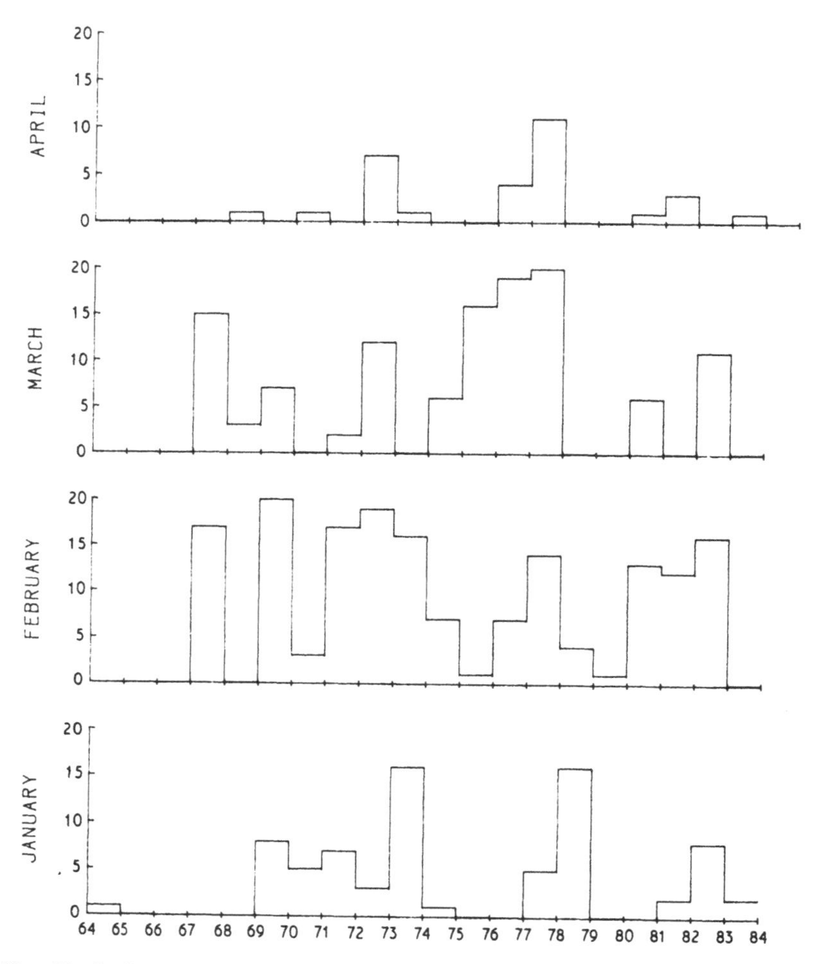

Note: *By distribution of dairy cows.
Source: Compiled from data supplied by the New Zealand Meteorological Service.

monitoring and applying real-time weather information has – as previously noted – been developed to a considerable extent in New Zealand. Specifically, such real-time information is now available in New Zealand for combinations of 8 regions, 120 commodities, and 31 climate parameters, the data being available for *any sequence of days* as well as for the traditional calendar month.

A typical example of the kind of weighted climate information that is available in New Zealand is shown in Table V.6. Further examples of

Table V.6 Typical (unweighted) climatic data for four stations in New Zealand, and weighted climatic data for two regions, and six national commodities (Data covers the 100-day period ending on 2 July 1984)

'Weighting'	*Rainfall percentage**	*Temperatures*** *maximum*	*minimum*
Station			
Auckland	68	0.0	+0.1
Napier	31	+0.8	−0.2
Wellington	69	+1.0	+0.9
Alexandra	90	+1.5	+0.4
Area			
Waikato/Bay of Plenty	68	+0.3	+0.4
Canterbury	47	+1.0	0.0
Commodity			
Beef cattle	64	+0.5	+0.3
Sheep	65	+0.6	+0.3
Cows	67	+0.3	+0.3
Wheat	57	+0.8	0.0
Vegetables	43	+0.7	−0.2

Source: After Maunder (1984a).
Notes:
* Percentage of the average for the 100-day period.
** Deviations from the average for the 100-day period.

real-time weighted weather information for New Zealand are given in Tables V.7 and V.8. Specifically Table V.7 gives *national* weather indices for four key aspects of the New Zealand economy: (1) rainfall in the hydro-electric generation areas; (2) minimum temperatures in the urban areas; (3) rainfall in the agricultural areas; and (4) maximum temperatures in the agricultural areas. This information relates to the previous 7 days, 2 weeks, 4 weeks, 3 months, 6 months, and 12 months to 10 October 1979. Similarly, Table V.8 gives rainfall and temperature information for the *national wool farm*. Information such as given in these tables can be provided by the New Zealand Meteorological Service for any sequence of days (1 to 365).

In the case of the agricultural industry, a convenient and useful way of condensing a vast amount of weather information into a compact index is to use the number of 'days of soil water deficit', or 'days below wilting point'. As previously explained these may be calculated from the balance between the daily rainfall and estimates of the daily potential evapotranspiration. The weighted number of days of soil water deficit for New Zealand for the critical March–May period for the thirty-six seasons 1949/50 to 1984/85, using six different weightings, are shown in Fig. V.2.

Table V.7 National weighted weather indices – New Zealand

Period to 9 a.m. 10 Oct. 1979	*Hydro-elect. rainfall (1)*	*Urban temperatures (2)*	*Agriculture rainfall (3)*	*Agriculture temperatures (4)*
Last 7 days	90	−0.5	138	−1.2
Last 2 weeks	153	0.0	142	−0.9
Last 4 weeks	122	−0.1	116	−0.6
Last 3 months	113	+0.3	115	−0.1
Last 6 months	102	+0.3	105	+0.1
Last 12 months	111	+0.3	112	+0.1

Source: After Maunder (1979).
Notes:
(1) Rainfall (weighted by the significance of hydro-electricity generation stations and their catchment areas) expressed as a percentage of the average for the period stated.
(2) Minimum temperatures (weighted by the significance of urban areas) as a difference from the average for the period stated.
(3) Rainfall (weighted by the significance of agricultural areas) expressed as a percentage of the average for the period stated.
(4) Maximum temperatures (weighted by the significance of agricultural areas) expressed as a difference from the average for the period stated.

Table V.8 National weather indices: wool production – New Zealand

Period to 9 a.m. 10 Oct. 1979	*Rainfall (1)**	*Maximum temperature (2)**	*Minimum temperature (3)**
Last 7 days	125	−1.5	−0.5
Last 2 weeks	141	−0.9	+0.1
Last 4 weeks	115	−0.6	0.0
Last 3 months	116	−0.1	+0.3
Last 6 months	102	+0.2	+0.3
Last 12 months	111	+0.1	+0.3

Source: After Maunder (1979).
Notes:
* Weightings according to the distribution of wool production as follows:
(1) Rainfall expressed as a percentage of the average for the period stated.
(2) Maximum temperatures expressed as a difference from the average for the period stated.
(3) Minimum temperatures expressed as a difference from the average for the period stated.

FIGURE V.2 Weighted* number of days of soil water deficit in the autumn (March–May): New Zealand, 1949/50–1983/84

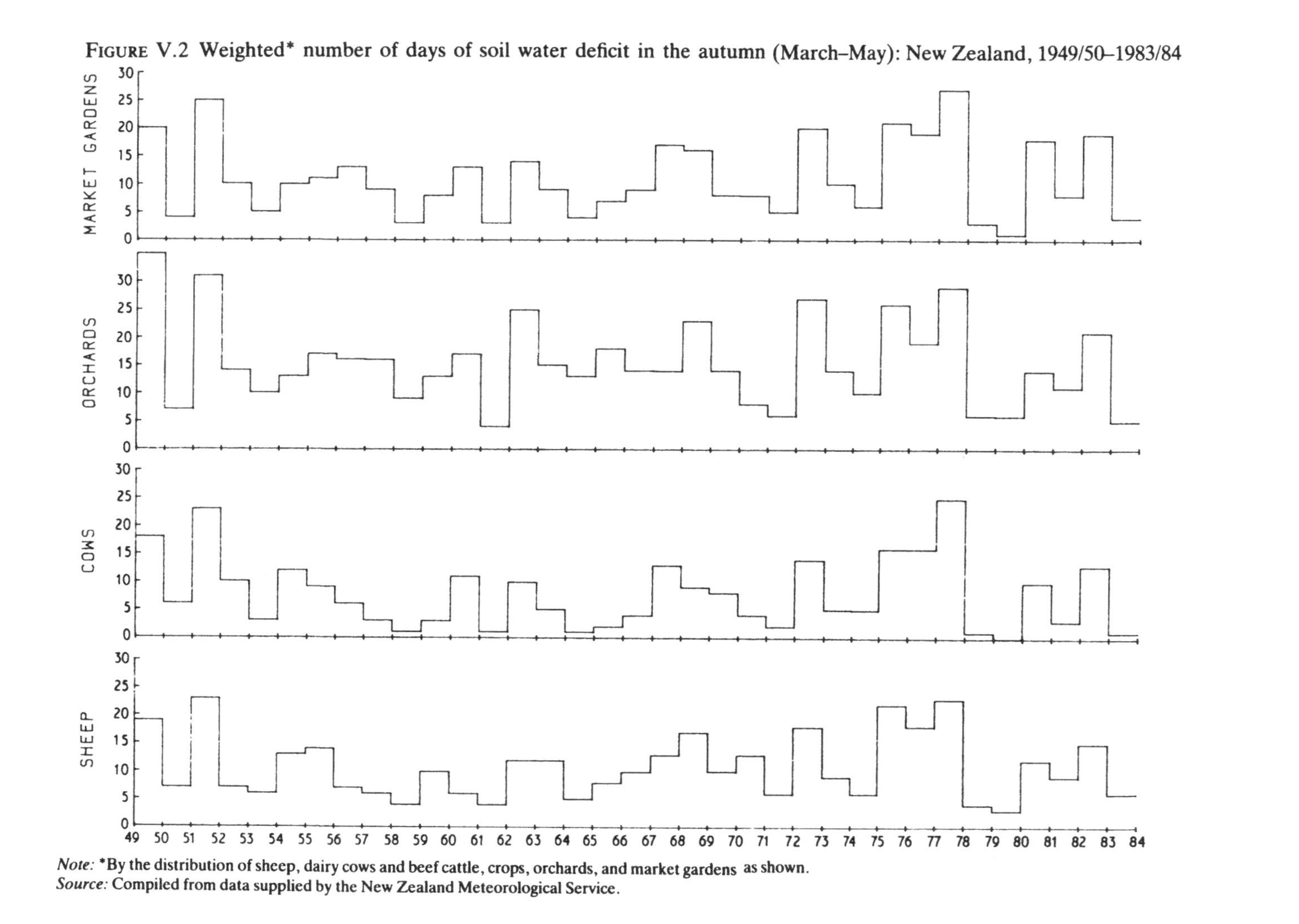

Note: *By the distribution of sheep, dairy cows and beef cattle, crops, orchards, and market gardens as shown.
Source: Compiled from data supplied by the New Zealand Meteorological Service.

These data are for New Zealand as a whole, and can therefore be said to represent the 'national sheep farm', the 'national dairy farm', the 'national cattle farm', the 'national crop farm', the 'national orchard', and the 'national market garden' respectively. The indices show just how variable the New Zealand agricultural climate is in the autumn period; noteworthy are the very high values for several of the commodities in the seasons 1951/52, 1968/69, 1972/73, 1975/76, 1975/77, and 1977/78.

As previously pointed out, such weighted weather information can be assessed in either real-time or non-real-time. This difference – which is important – relates to the time between the last event happening and the time that information about that and previous events is known. For example, if at, say, noon today it is possible to obtain the total rainfall from fifty places in a country for the last forty-five days *to today*, then such information may be considered to be real-time information. In contrast, 'traditional' climate information for, say, the month of May is in many countries only available after 'mailed-in-records' have been processed in a central office. In reality of course there is no reason (other than economic) why all weather and climate information should not be available in real-time. Indeed this may well become the norm in a large number of countries within a few years as the cost of collecting and processing 'mailed-in-records' becomes too expensive.

Although the computations involved in the commodity-weighted weather indices given in this chapter are simple and straightforward, and the methodology and arguments for developing such indices have been publicized for over fifteen years (see, for example: Maunder 1970a, 1972a, 1979, 1981a), it is still surprisingly and regrettably traditional for most national meteorological services to publish weather and climate information only for places. In some cases the information is mapped, but a considerable deficiency still exists in that the basic weather and climate point data cannot, even when mapped, be directly used in association with commodity/economic data. Further, even when spatial values are published, the values (with only a few exceptions – such as the New Zealand data already discussed and the human population weighted heating degree day data published in the United States and Canada (see Warren and Le Duc 1981) – are generally simply 'station averages' and consequently bear little if any relevance to a *real* commodity.

3. The United States example

The methodology used in the examples discussed for New Zealand can of course be translated to any other area, and one example of this translation – to the United States – is now examined. Translating the New Zealand station/county model to the United States would mean using weather data from about 3000 non-real-time stations (in the United States), and/or data from about 1500 real-time stations. Neither alternative is possible. Nevertheless meaningful commodity-weighted weather indices can be produced in the United States using real-time or near real-time basis data from the few hundred weather stations which *are* available.

Table V.9 United States weighted temperature departures (°F)

Commodity	*Weeks ending* *Oct. 22 1978*	*Sept. 2 1979*	*Jan. 25 1981*
Employees in transport/public utilities	−1.2	+3.1	+5.0
Electric energy: production	−1.2	+2.7	+4.6
Electric energy: installed capacity	−1.2	+2.8	+4.6
Electric energy: residential sales	−1.3	+2.8	+4.5
Gas: residential sales	−0.6	+3.4	+6.5
Gas: commercial sales	−0.3	+3.1	+6.7
Gas: industrial sales	+0.4	+2.1	+4.7
Interstate highway mileage	−0.3	+2.6	+6.2
Farm products sold	+0.2	+2.6	+8.9
Corn: value	−0.5	+3.5	+11.4
Wheat: value	+1.6	+2.7	+12.5
Cotton: value	+1.4	+0.8	+1.6
Soybeans: value	−0.7	+2.9	+9.7
Cattle production	+0.5	+2.4	+8.8
Hogs and pigs: population	−0.3	+3.3	+12.7
Broilers production	−2.2	+2.0	+1.1
National forests – acreage	+3.1	+1.5	+9.9
New housing units: value	−0.1	+2.4	+5.3
Retail trade: sales	−1.2	+3.2	+5.0
Federal aid (Medicaid)	−1.6	+3.5	+4.9
Coal production	−2.7	+3.3	+4.6
Human population (black)	−2.2	+3.0	+2.6
Human population (white)	−1.2	+3.2	+5.1
Households	−1.2	+3.1	+4.9

Source: After Maunder (1985a).

As an example, commodity-weighted temperature departures from the average – for the United States as a whole – for the weeks ending 22 October 1978, 2 September 1979, and 25 January 1981 are given in Table V.9. This shows very clearly that within each of these weeks there were wide differences in the 'response' of different commodity groups to temperature. For example, in the week ending 25 January 1981 the United States weighted temperature departures varied from a very high +12.7°F (7.1°C)* for hogs and pigs, to a very small positive departure of +1.6°F (0.9°C) for cotton. These weighted values provide information that is in contrast to the traditional synoptic viewpoint in which all points (and areas) are generally considered to be 'meteorologically equal'. Clearly, however, all points (and areas) are not equal from an economic, social, or political viewpoint, and it is considered essential that such differences must be taken into account if meaningful econoclimatic explanations or analyses are to be undertaken.

C. National Commodity-Weighted Weather and Climate Indices

1. The setting

Irrespective of any major long-term climate trends which may occur, it is clear that the most significant variations for human activities are the week-to-week, month-to-month, and season-to-season fluctuations. Decision-makers must contend with these short-term variations in regard to a wide range of activities such as electric power production, natural gas consumption, farm production, and export trade. There is clear evidence that better use of weather and climate information can produce benefits in the programming and location of many aspects of an economy; to this end there have been a number of attempts at a national level to match weather and climate data sets with those for various economic activities. As previously discussed, these involve transformation of one or both of the data sets, usually the former.

It is, of course, perhaps not surprising to find that the concept of 'weather adjusting' national economic indicators is usually considered to be beyond the scope of *both* the meteorologist and the economist. Indeed many people consider that such adjusting is meaningless. But despite this, economists still 'statistically smooth' weekly, monthly and quarterly data, label the product 'seasonally adjusted', and claim that this accounts for variations in seasonal activities *including* the weather. Perhaps one should not need to emphasize that such seasonally adjusted data have *no* relationship to weather as distinct from the *average* climate, but as has been explained in the previous sections weather information *suitably weighted* and *specifically adjusted* to take into account various economic distributions such as population, energy consumption, corn production, and energy production is now available, and *can be used* for 'weather adjusting' various economic activities.

At this stage it is important to re-emphasize, particularly for conservative climatologists, that (for example) a horticultural industry is basically interested in weather conditions pertaining to horticultural areas, a cotton

* The original study using data for the United States was in degrees Fahrenheit.

industry in the weather conditions pertaining to cotton areas, a wool industry in the weather conditions pertaining to sheep areas, and a natural gas industry in the weather pertaining to the (human) population areas with respect to consumption, but in other respects to energy source areas and transportation routes. Using such information in appropriate analyses it is possible – as shown in the previous sections – to 'weather adjust' or at least 'environmentally monitor' indices of national economic activities on a nationwide scale. It is also potentially possible to use commodity-weighted weather information to provide a *forecast* of the tendency of national economic indicators one, two, or three weeks (or months, depending on the economic parameter) before the actual national economic indices are available, since weather information is available in real-time, whereas national economic indicators usually have a publication delay of at least two or three weeks – even in the most developed societies.

2. National weather and national consumption and production

One cannot of course presume that the variations in any nation's weather and climate in the recent past can explain more than a small part of the totality of the reasons for consumption and production variations. Nevertheless, like the Dow-Jones index of the New York Stock Exchange (which despite its limited base is considered by most investors to be a good indicator of stock market performance), it is believed that commodity-weighted weather and climate indices, particularly on a national scale, can be and indeed must be considered in relation to the vagaries of any nation's 'economic climate'.

As discussed in section B of this chapter, computer tabulations and graphical analyses of this type of information are produced for many time scales and for over 120 commodities by the New Zealand Meteorological Service, and they have many applications. For example, the tabulations provide information that is not available through the traditional maps or tables, and they can be utilized in any investigation which takes into account the widely varying land usage and production capability of people. In particular, they can be used in operational models – such as those discussed in Chapters VI and VII for the dairy industry, the meat and wool industry, the electricity industry, the retail industry and the transport industry. The weighted data also provide useful specific commodity–weather information on a national scale as shown in Table V.10 for twenty commodities and for four climate parameters. For example, in March 1982 the weighted rainfall index for New Zealand as a whole varied from 53 (53 per cent of normal) for wheat, to 112 (112 per cent of normal) for dairying. Thus the month had well below normal rainfall for the national wheat farmer, but was a little wetter than average for those involved in the dairy industry. In contrast, March 1983 in New Zealand was very dry (and much warmer than March 1982) for

Table V.10 Commodity-weighted climate indices for New Zealand: March 1983 and March 1982

Commodity	*Rainfall % of normal* March*		*Max. temp. departures** March*		*Growing degree days (base 5°C) March*		*Days of soil water deficit*** March*	
	1982	*1983*	*1982*	*1983*	*1982*	*1983*	*1982*	*1983*
Sheep	91	81	0.3	+0.5	317	318	7.0	12.3
Dairy cows	112	63	−0.1	+0.3	351	375	1.4	12.5
Beef cattle	100	59	−0.1	+0.9	335	358	2.8	14.4
Potatoes	77	75	+0.6	+0.6	331	328	11.3	13.8
Wheat	53	98	+1.4	+0.6	306	275	18.9	10.1
Vegetable crops	93	32	−0.3	−0.2	349	388	5.8	23.9
Breeding ewes	86	87	+0.4	+0.4	314	310	8.3	11.7
Apples	68	55	+0.3	+1.4	352	359	7.0	17.3
Lettuce	82	67	+0.3	+0.5	353	363	7.2	14.8
Oats	67	123	+1.2	−0.1	293	253	15.8	7.5
Barley	65	72	+1.0	+1.1	323	313	15.5	13.1
Maize	102	23	−0.8	+2.5	357	420	0.3	23.6
Wool	91	84	+0.3	+0.3	317	314	6.5	11.8
Sheep/Beef	92	78	+0.2	+0.5	323	327	5.8	12.5
Exotic trees	87	54	−0.5	+0.5	335	356	3.4	13.2
Lucerne	68	95	+0.9	+0.6	319	302	15.9	11.2
Biomass (maize)	92	66	+0.4	+0.8	334	342	8.4	12.9
Grazing land	71	116	+0.9	0.0	297	266	14.3	9.7
Crop area	66	97	+1.0	+0.5	309	287	15.0	10.8
Orchards	84	75	+0.2	+1.0	354	362	6.5	16.3

Source: After Maunder (1985a).
Notes:
* Percentage of the 1941–70 normal.
** Departure from the 1941–70 normal.
*** Based on water balance computations.

the 'national maize farm', and March 1983 was also very much drier than March 1982 for the 'national dairy farm' and the 'national beef cattle farm', if the number of 'days of soil water deficit' or 'days below wilting point' are considered.

It is considered that the weighted weather and climate indices described here for sequences of days provide a real alternative to the traditional method of portraying weather and climate information which treats all areas as being equal. Rather, what is required is information relating to the fact that during the last 'n' days temperatures (or rainfall, or soil temperature, or sunshine) – weighted according to the distribution of a specific population (whether people, animals, crops or their productivity) – were above (or below) normal.

Figure V.3 New Zealand rainfall indices:* 1982–1984

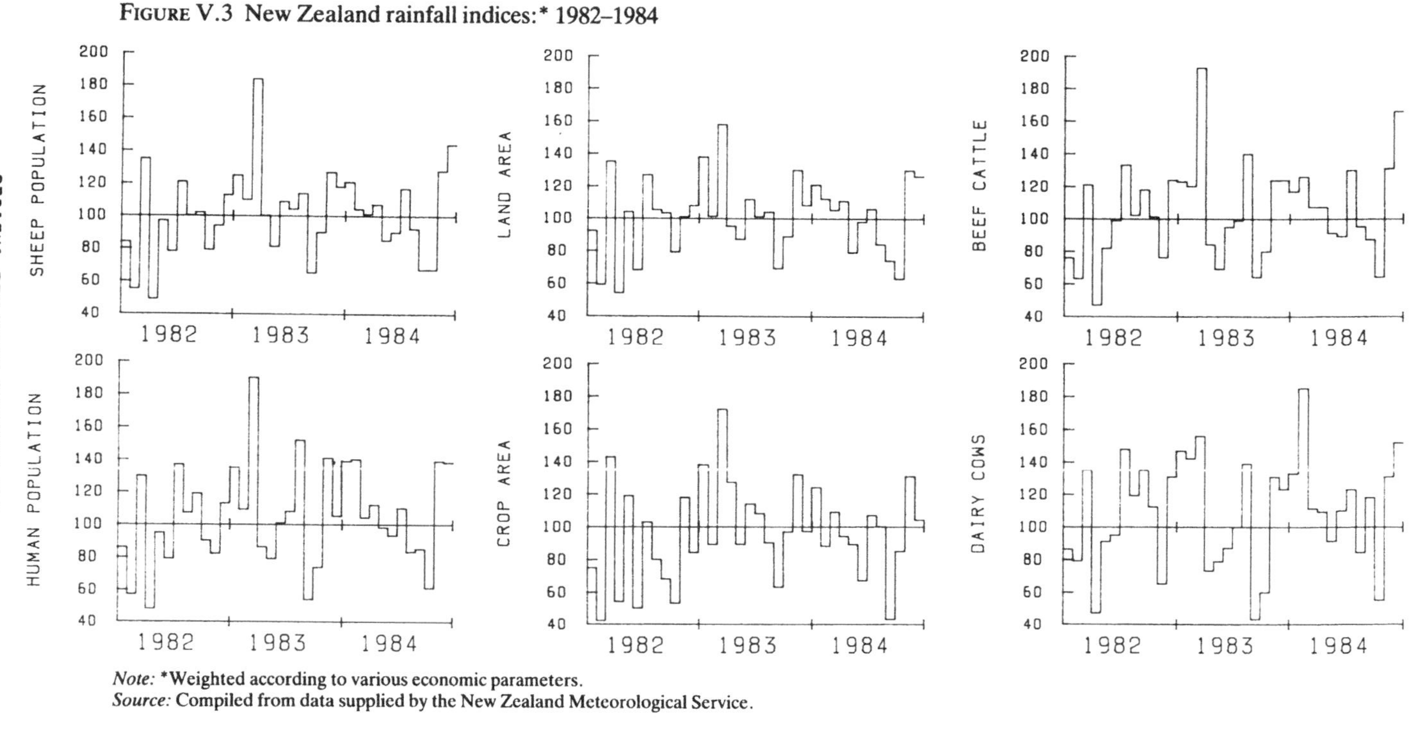

Note: *Weighted according to various economic parameters.
Source: Compiled from data supplied by the New Zealand Meteorological Service.

3. New Zealand national rainfall indices: 1950–84

The weighted rainfall indices for March 1969 for various New Zealand regions (given previously in Table V.3) show very clearly two important factors: first, the variation from one region to another, and second, and most important, the variations which occur if different economic parameters are considered. The New Zealand indices for this month indicated the relatively dry period which was prevalent at that time. However, it is pertinent to note that the dry period was most severe in the dairying areas (index of 27), and least severe in the crop areas (index of 62). Another interesting feature is that the dry period appeared to be more severe in the urban areas than the rural areas, for considering the total land area as a weight, the rainfall index was 58, whereas the index weighted according to the human population was only 35.

Commodity-weighted rainfall indices for New Zealand for 'the national dairy farm', the 'national sheep farm' and 118 other 'commodities' including people, energy, and transport have been computed for each month from January 1950. A sample of the weighted data is given in Fig. V.3 for the period January 1982 to December 1984, and the extreme values for 15 of 120 economic parameters available are also shown in Table V.11. It is clear that there is considerable variation from month to month, and it is believed that the type of information shown in Fig. V.3 is of considerable value in

Table V.11 Extremes in New Zealand rainfall indices:* 1960–84

Economic parameter	*Index*	*Highest month/year*	*Index*	*Lowest month/year*
Human population	235	Nov. 1952	19	Feb. 1973
Sheep	218	Nov. 1952	26	Feb. 1973
Dairy cows	239	Nov. 1952	12	Feb. 1973
Beef cattle	229	Nov. 1952	26	Feb. 1973
Crop area	237	Apr. 1968	30	Nov. 1974
Land area	199	Sep. 1970	27	Feb. 1973
Plantations	232	Nov. 1959	14	Feb. 1973
Orchards	258	Mar. 1979	21	Feb. 1973
Market gardens	240	Nov. 1952	19	Feb. 1973
Potatoes	252	Nov. 1952	24	Feb. 1973
Wheat	262	Apr. 1968	26	Nov. 1974
Vegetable crops	340	Mar. 1979	15	Feb. 1973
Seed, hay, silage	229	Nov. 1952	21	Feb. 1973
Area cultivated	225	Nov. 1952	25	Feb. 1973
Lambs shorn	203	Nov. 1952	23	Feb. 1973

Source: Updated from Maunder (1979).
Note: * Percentage of the 1941–70 normal.

assessing the overall 'wetness' or 'dryness' of New Zealand. For example, if the human population parameter is considered to reflect urbanized New Zealand, then an analysis of the rainfall indices weighted according to the sheep population should indicate the degree of 'wetness' or 'dryness' of the 'national sheep farm'. Such an analysis shows that during the 420 months from January 1950 to December 1984, there were 8 months in which the rainfall index exceeded 170 (or 170 per cent of the average). Thus about 2 per cent of the months on the 'national sheep farm' in the period 1950–84 could be regarded as being extremely wet. By contrast, in fifteen or about 4 per cent of the 420 months the rainfall index was less than 50 (or 50 per cent of the average) or very dry.

4. Weighted national agricultural soil water deficit indices

The use of traditional monthly data, particularly those related to rainfall, assumes an even distribution of the parameter throughout that period. *Daily* climate information can be used to overcome this deficiency, for example by computing the number of days in any period in which there is a 'soil moisture' or 'soil water' deficit. As explained on page 141, a 'day of soil water deficit' can be defined as occurring when the combined precipitation and soil moisture, assuming a maximum soil moisture capacity of 75 mm, is less than an assessment of the water need.

The weighted number of 'days of soil water deficit' for New Zealand for the thirty-five seasons 1949/50 to 1984/85 using three different pastoral weightings is shown in Table V.12. These data are – for the June–May season and the critical March–May period – for New Zealand as a whole, and can therefore be said to represent the 'national sheep farm', the 'national dairy farm', and the 'national beef cattle farm' respectively. The indices show just how variable the New Zealand climate is; noteworthy are the very high values in the drought seasons of 1972/73, 1973/74, 1977/78, which are in marked contrast to the 'green era' seasons of the mid-1960s. For example, the New Zealand indices weighted by the distribution of dairy cows give a *total* index of 13 days for the three consecutive seasons 1964/65, 1965/66, 1966/67, this being in very marked contrast to the *average* index during the six seasons 1972/73 to 1977/78 of 30 days, and the *total* index during the two consecutive drought seasons of 1972/73 and 1973/74 of 80 days.

5. Application of national commodity-weighted weather and climate indices

The development of commodity-weighted weather and climate indices is primarily aimed at studies which relate weather and climate events to national economic activities. Such research requires in most cases economic data for short time periods, of which there is an almost complete lack in New

Table V.12 Weighted* number of days of soil water deficit in New Zealand

Season	*June–May*			*March–May*		
	Sheep	*Cows*	*Cattle*	*Sheep*	*Cows*	*Cattle*
1949/50	41	40	43	19	17	19
1950/51	19	13	15	7	5	5
1951/52	34	29	31	22	23	20
1952/53	16	12	14	7	10	9
1953/54	39	36	14	6	2	4
1954/55	38	35	35	12	12	11
1955/56	42	20	33	13	9	10
1956/57	26	23	28	6	5	8
1957/58	20	19	29	5	2	8
1958/59	22	4	13	4	0	2
1959/60	34	13	27	10	3	7
1960/61	19	19	17	5	9	6
1961/62	35	22	31	3	0	1
1962/63	29	23	28	11	10	12
1963/64	42	26	39	11	5	10
1964/65	18	4	12	3	1	2
1965/66	19	4	14	8	1	4
1966/67	26	5	14	9	3	8
1967/68	32	29	35	13	13	16
1968/69	32	12	24	16	9	14
1969/70	33	34	30	9	7	7
1970/71	38	16	26	13	3	9
1971/72	31	19	24	6	2	3
1972/73	59	45	52	17	13	15
1973/74	40	35	37	8	4	7
1974/75	30	17	25	6	4	6
1975/76	36	20	25	22	16	17
1976/77	27	24	22	17	15	14
1977/78	55	41	51	23	24	24
1978/79	27	20	28	4	1	2
1979/80	11	3	8	2	0	1
1980/81	38	21	28	12	10	11
1981/82	38	21	30	9	3	5
1982/83	44	40	50	15	13	18
1983/84	20	7	15	6	2	3
1984/85	43	11	31	21	6	13

Source: Updated from Maunder (1980b).
Note: * Weighted by the distribution of sheep, dairy cows, and beef cattle as shown.

Zealand as well as in most if not all other countries. Two important series are, however, available in New Zealand and provide useful examples.

The first concerns the consumption of electric power where studies have indicated that a 1°C deviation of the temperature from the average for a particular period is associated with a 2–3 per cent change in the demand for electricity. In a recent analysis (see Maunder 1974a) a population-weighted temperature index was computed for each month for a number of years. These indices were then correlated with the pattern of electricity consumption in New Zealand (with appropriate allowance for the 'normal' climate, seasonality, trend, and workdays). The results indicated that a considerable portion of the variance in the amount of electric power used could be accounted for by the deviation of the weighted temperature index from the average for each period.

Such an analysis provided a valuable insight into the weather and climate sensitivity of an important sector of New Zealand's economy. This is particularly important since thermal power stations are used in New Zealand in a back-up role during periods of peak demand, and it is evident that the proper use of temperature forecasts can prevent the unnecessary generation of this more expensive power. Although it is difficult to forecast very cold periods more than two days in advance, useful 12–36 hour temperature forecasts can be made with sufficient accuracy to allow a worthwhile contribution to the efficiency of the electric power generation industry. Such forecasts are especially valuable when the forecast differences from the average for the specific period are weighted by the human population distribution. Further details of the use of weighted temperature data in the electric power industry are given in Chapter VI.

The second series that is readily available in New Zealand concerns the important dairy industry. This sector of the economy is particularly weather-sensitive and month-to-month variations in its production are of concern to many decision-makers involved in the marketing and transportation of dairy products. The relevant weighted soil water deficit indices computed for this industry have been successfully used to account for much of the monthly variations in dairy production. Predictions are now made in an operational manner in New Zealand using appropriately derived regression equations, and the results suggest that a high degree of accuracy can be achieved in predicting dairy production from weather and climate data in the form of dairy cow weighted soil water deficit indices. Further details of the dairy production application models, as well as other pastoral production application models, are given in Chapter VII.

D. Climate productivity indices: National and international

1. National climate productivity indices: A New Zealand example

In many countries the weather and climate have long been taken for granted, and other factors are readily advanced to explain the vagaries of a nation's 'economic climate'. But while this term may well be used correctly if one meaning of climate is taken (that is the prevailing temper or environmental conditions characterizing a group or period), it is evident that the *atmospheric* component of the economic climate must also be taken into account. One of the reasons for 'explaining' the vagaries of a nation's 'economic climate' only in terms of economics, rather than in terms of *both* economics and climate, is the ready availability of nationwide economic indicators, the absence of suitable nationwide climate indicators, and the difficulty in understanding such indicators even if they are available. It is, for example, difficult to ascertain from the (US) *Weekly Weather and Crop Bulletin* the *overall* state of the weather as it has affected United States grain production or meat prices. This publication is generally excellent in its station-by-station, and even state-by-state analysis, but it is difficult to use the data provided to give a national indicator of climate conditions.

In view of these difficulties, particularly in terms of the impact of weather and climate on a nation, on international trade, on commodity prices, or on food shortages and surpluses, it is considered that a 'climate productivity' type index could provide decision-makers with a good 'single index' of the 'weather–business' scene. The availability of real-time weather and climate information from around the world, and the availability of a few reasonably credible 'weather–business' models, also means that for the first time a world and/or hemispheric 'weather–productivity' or 'climate–productivity' index is a reality.

The importance of agriculture and forestry to New Zealand's economy has long been stressed by numerous observers, scientists and politicians. Nevertheless, even in New Zealand the climate of the nation's grassland, cropping areas and forests has in general been taken for granted, and factors other than the weather and the climate have usually been advanced to explain the vagaries of New Zealand's 'economic climate'.

In the belief that real-time information on the agricultural climate of New Zealand is not only important but also essential, if the business community and the farmer are to maximize New Zealand's agricultural potential and at the same time minimize costs, an index called an 'Agroclimatic Confidence Index' or more properly an 'Agroclimatic Productivity Index' has been devised. The index compares the past with the present weather to give a single-figure indicator of what could be called the *true* economic climate of the 'national farm'. It is based on a combination of the number of growing degree days and the days of soil water deficit, appropriately weighted with regard to key agricultural commodities and the importance of different months in the production season.

There are two aspects of any national econoclimatic indicator which need to be considered. First, for the information to be useful to decision-makers it must be available in real-time, or near real-time; second, the information must be available on a national basis, and if used at a political level, it should preferably be available in the form of a single index. The Agroclimatic Productivity Index computed for New Zealand is essentially an economic weighting of the standard commodity-weighted weather and climate indices which have been previously discussed. That is, the commodity weightings of sheep, dairy cows, beef cattle, lambs shorn, etc. (the main components of New Zealand's agriculture) are economically weighted according to their contribution to the total New Zealand agricultural production. In reality this means giving a weight of 30 per cent to the weighted weather and climate indices for wool (sheep/hoggets shorn 25 per cent, lambs shorn 5 per cent), plus weights of 25 per cent for mutton and lamb production (sheep population 5 per cent, breeding ewe population 20 per cent), 25 per cent for dairying (dairy cow population 25 per cent), and 20 per cent for beef and veal production (beef cattle population 15 per cent, calves born population 5 per cent). The combination of the *economic* weightings with the 'normal' commodity weightings (i.e. the weighting of weather and climate data based on the distribution of a commodity such as crop production or animal numbers) produces in effect a double weighting – one for the significance of the individual commodity, and one for the significance of that commodity to the nation as a whole.

Two aspects of the weighted weather are considered in computing the Agroclimatic Productivity Index: first, 'days of soil water deficit' (or specifically 'days below wilting point'); second, growing degree days (base 5°C). Each factor is in turn weighted according to the significance of the months being considered. For example, the influence of days of deficit is much more significant in mid-summer than in mid-winter. These weighted values are then further adjusted by the number of months being considered in the composite index. The final index should therefore provide a good 'one value' indicator of the overall state of the New Zealand agroclimatological scene, and hopefully the 'New Zealand Farm'.

The resulting New Zealand Agroclimatic Productivity Indices for the consecutive months from June 1977 to May 1985 are given in Fig. V.4, each index being a cumulative weighted value of the previous six-monthly period. The index during the eight seasons shown varied from 815 for the six months ending March 1978 (the peak of the 1977/78 drought) to 1200 for the six months ending February 1980. Comparing the average value for each of the whole seasons given, the figure shows that 1979/80 and 1983/84 were the 'best' of the six seasons shown, and 1977/78 and 1982/83 were the 'worst' of

FIGURE V.4 New Zealand agroclimatic productivity indices

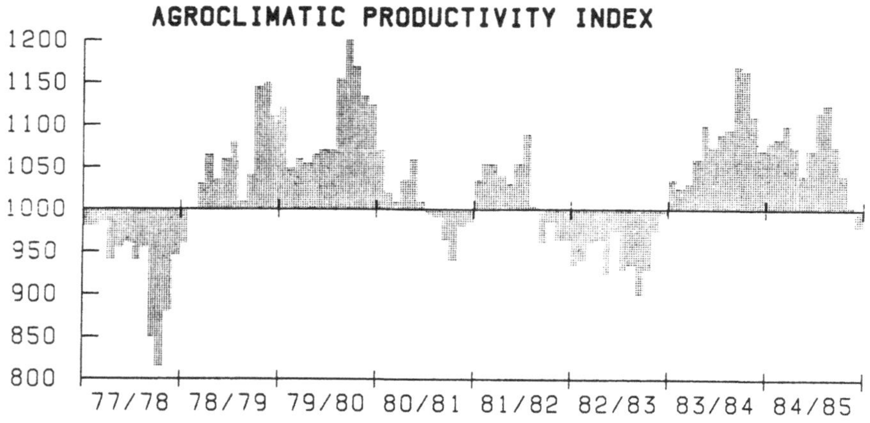

Source: Compiled from data supplied by the New Zealand Meteorological Service.

these six seasons. In addition, it shows that the change in the seasonal average of the Agroclimatic Productivity Index from 947 in 1982/83 to 1085 in 1983/84 was quite dramatic. The index is so constructed that a value of more than 1000 implies more 'confidence' in the pastoral farming sector of New Zealand (that is, overall it should be 'good' for farmers), while an index of less than 1000 implies that there should be less 'confidence' in the farming sector. It is not suggested of course that weather is all there is in deciding the general level of national farm productivity or confidence, but as a purely climate-based indicator it is a useful 'one glance' guide.

It is considered that agroclimatic productivity indices could also be developed for use in an operational manner for specific application to international commodities. That is, Agroclimatic Productivity Indices could be developed for nations as a whole, for commodities for selected nations, and for commodities for combinations of nations.

2. International climate/agricultural linkages

World production and the international trade of grains, fruit, and pastoral products are key elements in the economic structure of many countries. Indeed the balance or imbalance between exports and imports of some crops constitutes one of the most important causes of concern in the world today. Obviously the weather and climate conditions in the exporting countries are paramount in these considerations, as are the weather and climate conditions in the importing countries, in that 'good' agricultural weather in the importing country may reduce the need to import certain agricultural commodities. Similarly, variations in the weather and climate of the major exporting countries clearly have an effect on the price of those exports and the ability of the importing country to pay for such commodities.

The principal exporters and importers of wheat, corn, and rice are given in Table V.13, and Table V.14 shows the principal producers of pastoral products. An analysis of the information in these tables shows clearly several significant features: notably (1) the United States, Canada, and Australia as exporters of wheat, and China, USSR, Japan, and Brazil as importers of wheat; (2) the United States and Argentina as exporters of corn, and the USSR and Japan as importers of corn; (3) Thailand and the United States as exporters of rice, and Indonesia as an importer of rice; (4) the United States and USSR as producers of beef and veal; (5) the USSR and New Zealand as producers of sheep meat; (6) Australia and New Zealand as producers of wool; and (7) France and West Germany as producers of butter.

Naturally the ability of all nations to anticipate their agricultural production, and hence their capacity to export and their need to import, is of paramount importance. It is evident that the advantages (either economic, social, political, or strategic) that weather-based forecasts of such information could give some countries are wide ranging, and involve decision-makers at several levels and in several sectors such as production and processing, and marketing and transportation.

In the specific case of New Zealand, it is clear that New Zealand's success as an exporter of pastoral products has been highly dependent upon her ability to produce high-quality products at relatively low costs. However, the situation is increasingly becoming very different, in that the costs 'beyond the farm gate' have reached critical levels. It is considered that the meteorological and climatological input into minimizing these costs is a significant factor. But it is evident that if the close liaison between the meteorological and climatological system and the pastoral producing industry in New Zealand is to be strengthened, the significance of weather and climate variations to the pastoral industry *as a whole* must become fully realized. This involves the impact of weather on agricultural production both within and outside New Zealand, and as a contribution to the place of New Zealand in the international weather commodity scene, two hypothetical examples of the potential value of weather-based pastoral production forecasts are now given.

The specific example relates to the sensitivity of pastoral products in the international market from the viewpoint of one country (New Zealand) and gives an indication of the potential value of weather-based pastoral production forecasts. Hypothetical pastoral commodity information based on weather information to a specific date is given in Table V.15. By way of example the predicted indices for the commodities are given as 110 (i.e. 10 per cent above last season) for New Zealand, and 90 (i.e. 10 per cent below last season) for the rest of the world. Using the weightings shown (in this case

Table V.13 World trade of principal grains

000 metric tons

Country	*Wheat*		*Corn*		*Rice*	
	exports	*imports*	*exports*	*imports*	*exports*	*imports*
Algeria		1277		138		12
Argentina	3673		5091			2
Australia	9148			2		1
Austria	133	6	1	26		
Bangladesh		1490				370
Belgium		908		2272		202
Brazil		3510		697		198
Bulgaria	281	203	86	296	1	10
Burma			8		560	
Canada	12756			647		82
Chile		852		138	14	6
China		6845	70	2540	1390	50
Colombia		376		64	87	
Czechoslovakia		460		763		76
Denmark	170	25		262	1	12
Egypt		3409		568	168	2
Ethiopia		113				2
Finland	63	98		5		13
France	6724	380	1614	984	39	251
Germany, East	58	994		1591		41
Germany, West	743	1397	260	3099	51	175
Greece	116			933	17	
Hungary	656	21	392	138		27
India	450	1840		5	138	148
Indonesia		845	8	57		1767
Iran		1134		271		382
Iraq		834		52		267
Ireland	15	184		244		3
Italy	35	2985	7	3922	414	167
Japan		5748		9848	121	35
Kenya	2	56	61			4
Korea, South		1777		1707	18	187
Malaysia		450	1	402		285
Mexico	17	530		1470		19
Morocco		1346		59		2
Netherlands	701	1443	1487	4027	99	172
New Zealand	5	37	33			6
Nigeria		827		52		413
Norway	1	318		78		7
Pakistan		1246			973	
Peru		746	2	192		48

Philippines		695		96	57	33
Poland	15	2537		1850		85
Portugal		559		1577		72
Romania	1050	582	790	395		54
South Africa	118	1	2119	37	1	109
Spain	5	200	2	4068	51	
Sweden	615	18		32		23
Switzerland		357		251		25
Syria	2	99		37		86
Thailand		107	1918		2348	27
Turkey	800	51				27
United States	29460	16	48773	49	2238	1
USSR	1256	6616	152	11225	12	389
United Kingdom	165	3337	31	3588	54	205
Uruguay	32	37	10	1	116	
Venezuela		765		551	55	33
Yugoslavia	30	451	243	300		16

Source: Data reprinted by permission from the 1981 *Britannica Book of the Year*, copyright 1980 by Encyclopaedia Britannica, Inc., Chicago, Illinois.
Note: Data are averages for 1976–79 period.

national exports/production as a percentage of world exports/production) the world pastoral/weather index would, in this example, have varied from 90 for merino wool and 91 for cheese, to 100 for lamb and mutton and 107 for cross-bred wool. Such information may then be compared directly with the hypothetical New Zealand index of 110. The comparative advantages and/or disadvantages to a particular country – in this case New Zealand – of a specific season may then be assessed. Of course, in a real situation, weather-based forecasts of both national and international commodities including the important crops of corn, wheat, and rice would be progressively updated and monitored; such a procedure is currently followed by several meteorological consultant groups in the United States.

The implications of the type of information given in Table V.15 are further developed in Table V.16 which gives hypothetical production forecasts for a major pastoral-based commodity, wool. The weightings are the respective national production forecasts of wool expressed as a percentage of the world production (in some cases percentages of world exports rather than world production would be more appropriate), and the wool production forecasts are examples of the kind of weather-based wool production forecasts that could be produced. In this example, it is shown that *if* the 1984/85 wool production in New Zealand was expected to be 105 (i.e. 5 per cent above the 1983/84 season) and the 1984/85 or 1984 production in other wool producing countries was expected to be as indicated, then the world wool production index for 1984/85 would have been 96 (i.e. 4 per cent less than the previous

Table V.14 (a) World production of pastoral products* (000 tonnes 1978/9)

Country	*Wool***	*Butter*	*Beef/Veal*	*Sheep meat*	*Pig meat*
Argentina	166		3193	130	214
Australia	713	101	2134	492	199
Brazil			2200	44	850
Canada		102	1060	4	620
China	100				
Denmark		140			
France		571	1658	157	1609
Ireland		119			
Iran			88	377	
Italy			1027	50	931
Japan			403		1284
Netherlands		211			
New Zealand	357	268	549	498	43
Mexico			1054	55	440
Poland			709	20	1766
South Africa	103				
United Kingdom	48	162	1048	238	848
United States	47	451	11283	140	6075
Uruguay	73		354	36	16
USSR	472		6590	875	3727
West Germany		564	1435	26	2618

Source: Compiled from information supplied by the New Zealand Meat Board, the New Zealand Wool Board, and the New Zealand Dairy Board.
Notes:
* Only principal items and countries shown.
** Also see table below.

season), and the Southern Hemisphere (less New Zealand) wool production index would have been 92 per cent (i.e. 8 per cent less than the previous season). In reality, of course, specific weather- and climate-based forecasts of wool production could be made for all wool producing countries.

The specific implications of such information to the New Zealand Wool Board, Minister of Agriculture, Treasury, wool brokers, and the wool grower are many, and clearly the *volume* of the agricultural commodity (in this case wool) is only *one* factor in a whole range which, say, a buyer of wool has to be concerned about. In particular, the relative value of a nation's currencies is clearly reflected in the buying price of most agricultural products. Nevertheless, a key factor in almost all aspects of agricultural production is the actual physical volume available for use, and it is evident that the physical volume of production – which *is* weather related – must be taken into account. The atmospheric scientist must therefore alert the

(b) World production of raw wool (000 tonnes)

Country	*1982–83*	*1983–84*	*1984–85*
Argentina	162	162	153
Australia	702	729	773
Brazil	28	25	24
Canada	2	2	2
China	202	194	180
France	25	25	24
Iran	16	16	16
Italy	13	13	13
New Zealand	371	364	373
Poland	12	13	14
South Africa	113	108	105
United Kingdom	50	53	56
United States	49	47	43
Uruguay	82	82	75
USSR	474	484	485
West Germany	5	5	5
All other	606	613	616
Total	2912	2935	2957

Source: 'Wool Statistics, 1984–85', table 2, Commonwealth Secretariat International Wool Trade Organization and International Wool Study Group and New Zealand Wool Board.
Note: The 1984–5 New Zealand figure is from the New Zealand Wool Board and the total has been adjusted accordingly.

decision-maker in national economic planning of the value of such weather information, and to assist him to use such information in the best possible way.

3. International climatic confidence indices

It is evident that the national Agroclimatic Productivity Indices discussed earlier could also be developed for use in an operational manner for specific application to international commodities. Thus Agroclimatic Productivity Indices could be developed for nations as a whole, for commodities for selected nations, and for commodities for combinations of nations. Clearly, a global (or at least a hemispheric) Agroclimatic Productivity Index developed for corn, wheat, soybeans, wool, meat, coffee, etc., would have obvious political and trading overtones.

This further development of weighted climatic indices is discussed in a publication of McQuigg Consultants Inc. (1981). This development used the

Table V.15 Weighted international pastoral weather indices†

Commodity	*New Zealand*		*Rest of World*		*World*
	Weight	*Index*	*Weight*	*Index*	*Index*
Lamb and mutton	(50)*	110	(50)***	90	100
Beef and veal	(9)*	110	(91)***	90	92
Butter	(21)*	110	(79)***	90	94
Cheese	(7)*	110	(93)***	90	91
Dried milk	(11)*	110	(89)***	90	92
Wool					
Merino	(2)**	110	(98)****	90	90
Cross bred	(87)**	110	(13)****	90	107
Other	(11)**	110	(89)****	90	92

Source: After Maunder (1979).
Notes:
† Hypothetical (based on weather information up to . . .)
* NZ exports as percentage of world exports.
** NZ production as percentage of world production.
*** Rest of World exports as percentage of world exports.
**** Rest of World production as percentage of world production.

Table V.16 Weather-based world wool production forecasts*

Area	*Weighting***	*Marketing period forecasts* 1984/85	1984
New Zealand	16	105	
Rest of Southern Hemisphere			
Australia	35	93	
Argentina	9	89	
South Africa	5	85	
Uruguay	3	93	
Northern Hemisphere			
USSR	23		95
China	4		102
USA	3		108
UK	2		105

Source: After Maunder (1979).
Notes:
* Hypothetical.
** Percentage of world production.

Table V.17 International commodity-weighted agroclimatic indices: May 1981

		Rainfall	*Temperature*
		*percentage probability**	
USA	Corn	72	17
	Soybeans	84	2
	Winter wheat	96	49
	Spring wheat	23	56
CANADA	Wheat	17	82
USSR	Winter wheat	36	8
	Spring wheat	28	16
	Sunflower seed	16	4
	Corn	24	4

Source: After Maunder (1984b), from McQuigg Consultants Inc. (1981).
Notes: * 100 = very wet/very warm.
0 = very dry/very cold.

technique already described in detail for assessing the Agroclimatic Productivity Index of New Zealand. Table V.17 shows some climate values for various national commodities for May 1981 which could easily be developed into hemispheric commodity indices.

In terms of the impact of weather and climate on a nation, on international trade, on commodity prices, and on food shortages and surpluses, it is suggested that such 'agroclimatic productivity' type indices would provide many decision-makers with a good one-index indicator of the 'agro–weather –economy' scene. Further, it is clear that the availability of real-time weather information from around the world, and the availability of reasonably credible 'agro–weather–economy' models, means that for the first time a world and/or hemispheric 'weather confidence' or 'weather-sensitivity' or 'weather productivity' index is a reality.

Future implications of these econoclimatic developments are full of potential; for example, the system could fit neatly into the developing videotex systems which enable very detailed information to be made available to individual farmers with visual display units. This suggests that even further sophistication in the increasing professionalism of farm management will be needed. Alternatively a group of consultants could interpret the data with appropriate advice and recommendations and thus reduce decision-making on the part of individual farmers.

At the international level, such commodity-weighted weather information can be used – as previously noted – by a country to monitor and in some cases predict the agricultural production of another country. It is considered that the advantages (either economic, strategic, or political) that such information* could give a country are wide ranging, and involve decision-makers at the government, producer board, producer, and processing levels. Clearly the marketing advantages of knowing that domestic production of a commodity is doing comparatively better than that of one's competitor is a challenging prospect, but with the important corollary that astute competitors would also know when one's own country's production was likely to be less than expected.

* Mention should also be made of the semi-operational vegetation index which can be computed from an appropriate analysis of meteorological satellite data. See, for example, Yates *et al.* (1984), Tarpley *et al.* (1984), and Taylor, Dini, and Kidson (1985). The latter paper, in commenting on the use of such an index in New Zealand, states

> The aim of the investigation was to determine the relevance to New Zealand agriculture of a vegetation index which is readily available from locally received, meteorological satellite data. The results indicate that the potential exists to monitor farming areas . . . and to measure pasture growth for use by the agricultural community in daily management and forward planning. The areas involved are comparable in size (and location) to those recognized by . . . the Economic Service of the New Zealand Meat and Wool Boards . . . and also to those used by the New Zealand Meteorological Service in the compilation of agriculturally weighted real-time weather information and the forecasting of pastoral production.(Maunder 1979, 1980a)

VI The assessment process

A. Droughts

1. Drought impact pathways

The human consequences of a significant weather or climate variation such as a severe drought depend largely on the ways in which direct sector effects filter through the socio-economic-political fabric of a society. For example, reduced agricultural production or altered water supplies, or changes in hydro-electric output, are not the *ultimate* concerns, but rather the degree to which people *are affected* in terms of changes in income, effect on health, or community stability. That is, a reduced wheat yield does not – except in certain critical circumstances – directly affect a society, but an increased or decreased price for wheat, or an inferior substitute, does.

A framework for tracing the impacts of drought occurrence specifically in the Great Plains of the United States has been devised by Warrick and Bowden (1981) and is shown in Fig. VI.1. It depicts a variety of pathways

FIGURE VI.1 Hypothetical pathways of drought impacts on society

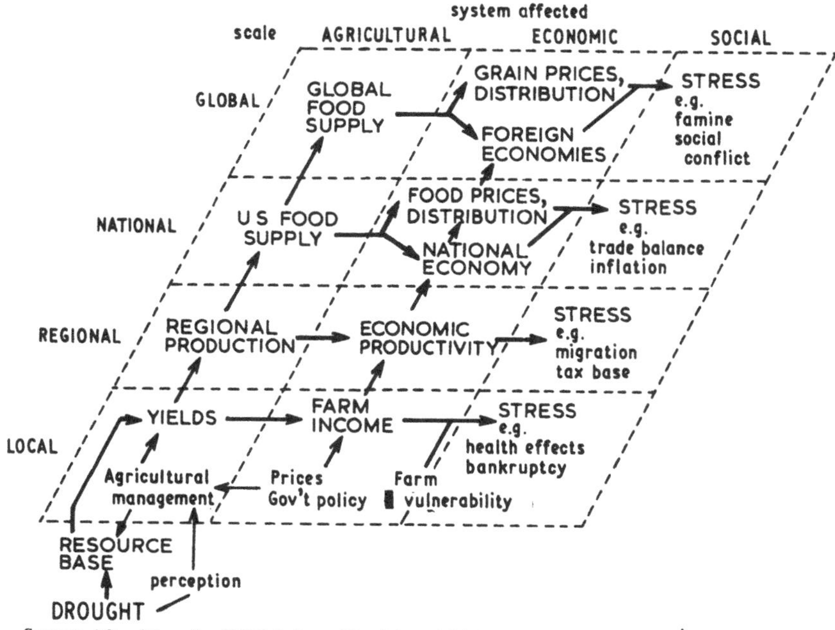

Source: After Maunder (1985a), from Warrick and Riebsame (1981), after Warrick and Bowden (1981).

that drought impacts could take, spanning spatial scales from local through regional and national to global, and the three major systems that could be affected (agricultural, economic, social). As shown, the initial perturbation originates from a meteorological event, and becomes an agricultural event or drought – as distinct from a meteorological drought – when agricultural production falls below some perceived threshold (which will vary widely). The agricultural drought will then translate into a drought impact when the stress is detected in the economic, social, and political sectors. Naturally, the degree to which the initial climate event is transformed into stress is influenced by a number of factors including market prices, government policies, farm stability, and the degree to which the drought is seen as a local, regional, or global problem. For example, while the drought impact in the Great Plains of the United States in the 1930s took an essentially horizontal pathway to local stress (as shown in Fig. VI.1), it is highly probable that a similar drought today (in the Great Plains and in many other areas) would take a much more vertical pathway.

2. Costs of drought: Problems of measurement

The costs of most meteorological events, and especially droughts, are difficult to estimate, and over two decades ago an issue of the Australian *Current Affairs Bulletin* on 'Drought' (Anonymous 1966) rightly pointed out that it is often very difficult to establish real cause-and-effect relationships between drought and total economic production. In addition, there are – as they say – other 'book-keeping problems'.

> The mind may well boggle at estimates of the cost of a drought calculated to include wool shorn from sheep never conceived and wheat not harvested from crops never sown. Even if it can be admitted that one can in some way lose something which never came into being, the calculation of values of the items becomes very involved.

A critical insight into many of the problems relating to the evaluation of drought conditions was made in an early study by Heathcote (1967). He noted in particular the 'negative' aspects of droughts, which include both spasmodic and incessant effects. Spasmodic effects are the relatively short-run and immediate results of drought. More specifically they include the immediate day-to-day reactions to the effects of drought, the reduction of surface and soil moisture volumes which can have a disastrous effect upon a range of economic activities from crop and fodder growth to water navigation and the rehabilitation aspect associated with the follow-up measures to restore losses and restart production. By contrast, the incessant effects result from the costs of preparing against the recurrence of drought. There are, according to Heathcote, four main effects: first, the abnormal service capacities to meet abnormal demands for water, such as a storage or flow capacity well in excess of normal use; second, the maintenance of a foreign credit reserve to cover the fluctuations of national production; third, the cost of drought research; and fourth, the reticent attitude to resource development in the drought-prone areas.

Several detailed studies of the costs of drought have been made, including that of Foley (1957) who discussed the major droughts in Australia from the earliest years of settlement until 1955. Foley's study is very comprehensive, and one point which he makes is as valid today as it was in 1957. He stated:

> Droughts such as experienced in the past are considered to be a characteristic of the climate, a feature which must be borne in mind in relation to its effects on the national economy.

How much of the reduction in production in a drought year can be directly attributed to a 'climate' variation from the so-called '*average*' is thus questionable. For example, it has been shown (Maunder 1970b) that in the 25-year period 1941 to 1965, the average percentage (by area) of the Australian continent with above-average rainfall was only 40 per cent and in only five of the 25 years (1947, 1950, 1955, 1956, 1960) did more than half of Australia receive more than the average rainfall.

The *economic* aspects of drought in Australia have also been discussed by a number of investigators including some pioneering but still very relevant studies by Cumming (1966) and Duloy and Woodland (1967). In this regard

Table VI.1 Financial impact of drought on properties in two areas of Australia

Area	*1960–61 to 1962–63*	*1964–65*	*1965–66*	*1966–67*
New South Wales Pastoral Zone		($A)		
Net farm income	15,629	8,834	−10,311	10,544
Cash balance	7,193	12,592	−1,792	536
Interest paid	803	729	932	1,090
Cash investment (inc. land purchases)	8,049	4,539	1,690	5,077
Queensland Pastoral Zone				
Net farm income	9,139	2,570	−7,902	9,134
Cash balance	7,933	4,292	1,735	4,275
Interest paid	1,050	1,401	1,863	1,522
Cash investment (inc. land purchases)	4,291	4,397	1,371	4,168

Source: After Maunder (1971c), from 'An economic survey of drought affected properties – New South Wales and Queensland: 1964–65 to 1965–66', *Wool Economics Research Report*, no. 15, Bureau of Agricultural Economics, Canberra.

it is pertinent to note that many assumptions have to be made in most economic studies of drought. For example, Duloy and Woodland pointed out in their 1967 study that a more satisfactory study of drought costs would enquire into the likely effects on receipts of output changes for different commodities, and would entail a detailed study of farm costs. Such a survey of the drought in Australia in the mid-1960s was completed by the Australian Bureau of Agricultural Economics (1969). The resulting financial implications of this drought are clearly indicated in Table VI.1, which shows that the largest absolute decline occurred in the pastoral zone of New South Wales, where net farm income (in 1960s dollars) dropped from a profit of $A8800 in 1964–65 to a loss of $A10,300 in 1965–66.

3. Beneficial effects of drought

As with most weather and climate variations – extreme or otherwise – the effects of drought conditions are not exclusively negative. Indeed in certain ecological contexts it has been shown that the effects of droughts are sometimes beneficial. For example, Perry (1962) found from a study of the drought of 1958–61 in Central Australia that it had caused little long-term damage to the vegetation and that the perennial grasses were probably in a healthier condition than after a more 'normal' succession of seasons, because the death of livestock had reduced the pressure on grazing.

In addition to these ecological benefits are the more direct economic benefits, for, to many people both inside and outside of drought areas, drought clearly means increased business, such as through the cartage of animals, water, and feed supplies. Moreover, it is clear that in many cases those receiving the benefits are in a region or even a country other than the one affected by the drought. For example, a drought in Europe's dairying area brings substantial benefits to the dairy farmer of New Zealand some 20,000 km away, in the form of additional outlets in the United Kingdom market for New Zealand's butter and cheese. Such transactions – in this case partly related to drought conditions – may of course have much wider international implications than the simple export or non-export of a commodity.

4. Drought and intense cold wave: An example

If we are to use the 'climatic income' which is available to us, we must begin to learn how to cope better with the wide swings of nature's pendulum. One specific and dramatic swing of this pendulum occurred in the United States in the winter of 1976/77 when a drought in the west was associated with an intense cold wave in the east. The direct losses of this climatic swing have been calculated by the Center for Environmental Assessment Services (1982) and show that the total 1977 dollar costs were about $27 billion, comprising $9 billion in the production sector, $7 billion in the foodstuff sector, $5 billion in the transportation sector, $4 billion in the retail trade sector, and $3 billion in the energy consumption sector.

As shown in Fig. VI.2 the impact was most severe during the six-week period from January through to mid-February. In assessing such losses, it is important to appreciate however that *nationally* it is the *net* loss (or *net* cost) to the key economic sectors that is important, irrespective of the fact that some sectors (both regional and national) may in fact 'gain' from the occurrence of severe weather.

5. Drought in New Zealand: A case study

The occurrence of a severe drought in New Zealand in the 1969/70 season resulted in a number of in-depth studies. Some details of this drought are now given, and emphasize the fact that even in a 'well watered' agricultural country like New Zealand, the impacts of a drought can be quite extensive, and can also extend well 'beyond the farm gate'. In particular, the economic consequences of this drought were of considerable political concern; indeed the Hon. R. D. Muldoon, the then Minister of Finance, in presenting the 1970 Budget to the New Zealand Parliament, stated:

FIGURE VI.2 Costs of the 1976/77 intense freeze and drought in the United States

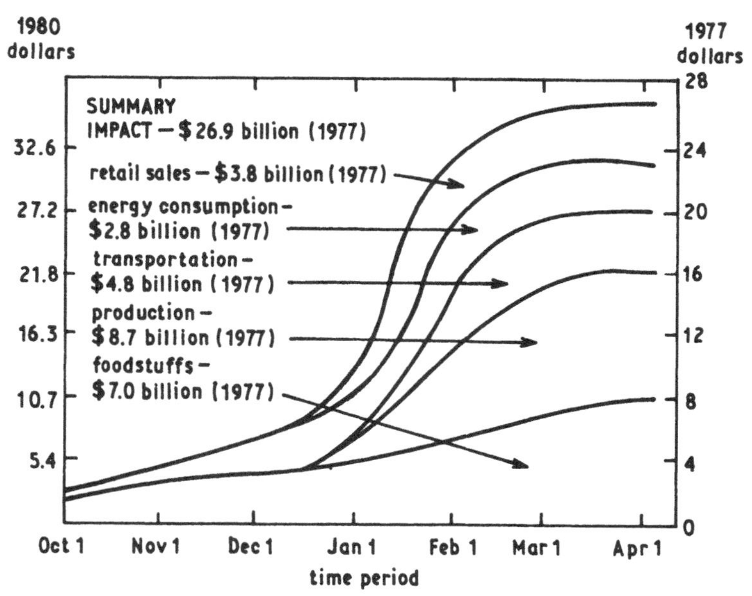

Source: After Maunder (1985a), from Joyce (1982), after Center for Environmental Assessment Services (1982).

Table VI.2 Summary of estimated 'costs' of 1969/70 drought in New Zealand

Item	*Reason*	*High cost ($)*	*Low cost ($)*
Livestock	Stock movements[1]	500,000	250,000
Wheat	Lower production[2]	5,000,000	3,000,000
Fertilizer	Subsidies[3]	6,750,000	2,700,000
Drought relief	Budget provision[4]	3,800,000	2,850,000
Rural lending	Budget provision[5]	6,000,000	2,000,000
Electric power	Thermal plants – fuel[6]	3,500,000	2,800,000
Forest fire losses	Additional losses[7]	150,000	112,000
Forest fire prevention	Additional costs[8]	140,000	75,000
Dairy production	Lower production[9]	15,000,000	6,800,000
Lamb production	Lower production[10]	1,500,000	1,000,000
Wool production	Lower production[11]	4,480,000	2,700,000
	Total:	46,820,000*	24,287,000*

Source: After Maunder (1971b).
Notes:
* These costs may be compared with the total value of all exports from New Zealand in 1969/70 of $1164 million. The high cost therefore represents about 4% of the total value of exports.
1 Transfer of 1,000,000 ewes and lambs from Canterbury and North Otago to Southland. Maximum cost 1,000,000 at transfer cost of 50 cents; minimum cost 1,000,000 at 25 cents.
2 1969/70 wheat production 10.6 million bushels, or 6 million bushels less than 1968/69. Estimated import requirements for 1969/70, 5 million bushels. Maximum cost 5 million bushels at $1 a bushel; minimum cost 60% of this.
3 Subsidies for fertilizer price and transport. $13.5 million allocated in 1970 Budget. Probable maximum cost due to drought conditions 50% of this; minimum cost possibly 20% of the $13.5 million.
4 1970 Budget provisions $3.8 million. Minimum cost possibly 75% of this.
5 Budget provision to set up Special Agricultural Assistance Fund of $10.0 million. Budget of $5.0 million for 1970/71. Maximum cost due to drought conditions probably 60% of the $10,000,000 over 1970/71 and 1971/72. Minimum cost probably 20% of the $10,000,000.
6 Additional fuel used in thermal plants to overcome water deficiency in hydro stations. The cost in 1969/70 was $3.5 million more than cost in 1968/69. This probably represents maximum cost due to drought. Minimum cost probably 80% of $3.5 million.
7 1000 acres of experimental plantation lost owing to fire. Plantation was five years old and had an estimated value of $150/acre. Probable maximum cost 1000 acres × $150. Minimum cost probably 50% of this.
8 The 1969/70 expenditure on forest fire prevention was $240,000 compared with about $100,000 in a 'good' year. Probable maximum cost due to drought conditions $140,000; minimum cost about 80% of this.
9 The 1969/70 butterfat production was 50 million lbs less than in 1968/69 and 34 million lbs less than the 1964/65 to 1968/69 average. In 1966/67, the *net* return to farmers (after expenses) was about 16 cents per lb of butterfat and the gross return was about 35 cents per lb of butterfat. Probable maximum loss of 50 million × 30 cents, and minimum loss of 34 million × 20 cents.
10 Slaughterings of lamb in 1969/70 of approximately 27 million were up on 1969/70, but average weight was down 0.6 lbs per lamb. Maximum loss in production 27 × 0.6 million lbs of lamb at 14 cents/lb (minimum export price) = $2.0 million. Probable maximum loss to farmers $1.5 million, and minimum loss of $1,000,000.
11 In the Auckland, Wanganui, Wellington, Christchurch, Timaru and Dunedin wool centre hinterlands, the 1969/70 production was down 20 million lbs from 1968/69. At 28 cents/lb (approx. 1969/70 auction price), this equals $5.6 million. Probable maximum loss to farmers of 80% of this, and minimum loss of 50% of the $5.6 million.

> Restrictive policies in the countries of destination of our various products, combined with the threat of synthetic substitutes, have produced a considerable relative decline in the value of wool and dairy products. . . . these long-term effects have been accentuated this year by the disastrous drought which affected virtually the whole country.

These comments illustrate the importance to New Zealand of adverse climate conditions, and Maunder (1971b) attempted to measure in dollar terms the effect of this drought.

Initial calculations suggested that the total costs exceeded 3 per cent of the total value of exports from New Zealand. It was emphasized that any estimation of the true economic effect of this drought is beset with many difficulties. For example, no one organization or government department had (or indeed has) the responsibility of collating the various studies made by government departments and related producer and consumer groups on how the weather or climate affected their specific interests. Because of this deficiency, it was considered necessary to consider all the relevant 'reports' as were available, and from these piece together a comprehensive account of the drought.

In considering the economic effects it is appropriate to assess the drought impact according to items such as pasture, wool production, dairying, electric power, and external trade. In the Annexe to this chapter, p. 259, extracts taken directly from a sample of relevant annual reports or other documents are given. From these 'reports' the costs of the 1969/70 drought in New Zealand (Table VI.2) were estimated. Many assumptions had to be made; nevertheless, it is believed that the costs indicated are of the correct order of magnitude. The range of costs shown in Table VI.2 varied from $24 million to $47 million (1970 dollars) and appear to be comparable on a per capita basis with the costs of drought in Australia, where the total loss in farm income in the two years following a major drought were estimated to be about $400 million.

6. Drought in Canterbury, New Zealand: What the climate record revealed

(a) Introduction

A severe drought in the Canterbury area of New Zealand occurred in 1981/82. The following account was originally written in September 1982 while the drought was still in progress (see Maunder 1982b) and provides a useful example of the day-to-day information that was available, and the decision-making that was (and is) necessary in a time of drought. It is considered reasonably typical of the kind of information that would be available in many other drought areas, and it shows clearly that decisions are always much more difficult to make 'before the rains come' than after the grass or crops have become green again.

(b) THE SITUATION IN SEPTEMBER 1982

The current drought in Canterbury is* one of a large number of droughts that have affected this area since early times, but the significance of the current drought is its severity. In purely economic terms, droughts are often difficult to classify or rank, but if one accepts the definition that a drought is a period of 'severe moisture deficiency' the ranking becomes a little easier. Various definitions of droughts are available, and they all endeavour to describe or measure the lack of moisture. But whatever the kind of drought, there remains the problem of 'too little water' required by 'too many things'.

This 'too little water' concept is all too readily apparent in Canterbury at the date of writing (early September 1982) as is evident in Fig. VI.3 which shows the November 1981 to August 1982 rainfalls as a percentage of the

FIGURE VI.3 November 1981 to August 1982 (inclusive) rainfalls in Canterbury/North Otago expressed as a percentage of the 1941–70 normal for the November to August period

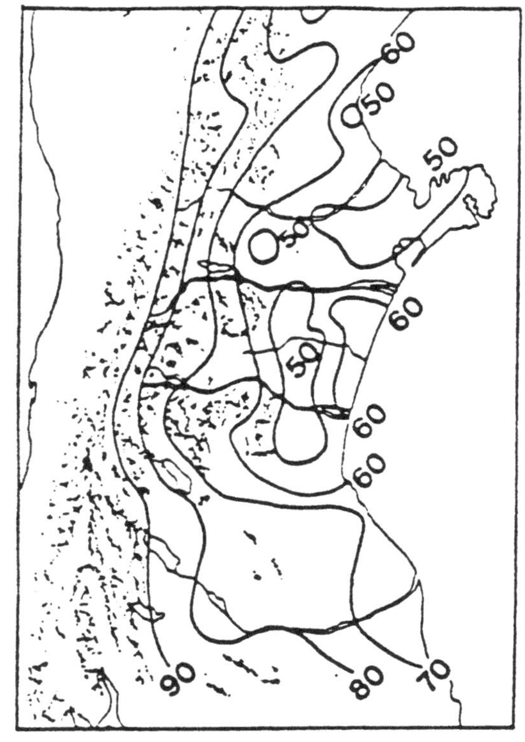

Source: After Maunder (1986a).

* Written as the drought was happening.

1941–70 normal. A brief summary of the severity of the drought in climatic terms, and a comparison with previous severe droughts that have occurred since the end of last century, are now given. In this analysis two periods will be considered – first, the 10-month period from November 1981 to August 1982 (inclusive), and second the 27-month period from July 1980 to August 1982. The reason for looking at these two periods is that the last significant rain in most parts of Canterbury was late in October 1981, and *prior* to that in early to mid-June 1980.

(c) PERIOD NOVEMBER 1981 – AUGUST 1982 AND PROSPECTS

At Christchurch (a typical station in the drought area) the total rainfall for the 10 months November 1981 to August 1982 (inclusive) was 308 mm or 54 per cent of the average for this 10-month period. Comparable records at this site are available since the late 1890s, and the 1981/82 total is the lowest recorded, being 6 mm less than the previous lowest November to August total of 314 mm recorded in 1896/97. The relevant data for the driest November to August periods is given in Table VI.3, other very dry seasons being 1931/32, 1968/69, 1921/22, 1970/71, 1932/33, and 1947/48. Also included in this table is the ranking of the 12-month rainfalls November to October inclusive; and this shows that the rankings of the lowest November to August rainfalls at Christchurch are very similar to the rankings of the November to October rainfalls. This indicates that very low rainfalls for the 10-month November to August period usually also persist as very low rainfalls for the 12-month November to October period. As can be seen in Table VI.3 the total rainfall for November to August 1981/82 is 308 mm, therefore for the 'season' *not* to remain the driest on record for the

Table VI.3 Christchurch 'seasonal' rainfall rankings: 1896–1982

Season[1]	*November to August*	*Season*[1]	*November to October*
1981/82	**308**[2] **(54)**[3]		
1896/97	314 (56)	1896/97	397 (59)
1931/32	334 (59)	1968/69	404 (61)
1968/69	337 (60)	1921/22	423 (64)
1921/22	347 (61)	1914/15	439 (67)
1970/71	362 (64)	1932/33	442 (67)
1932/33	364 (64)	1970/71	443 (67)
1947/48	380 (67)	1947/48	449 (68)

Source: After Maunder (1982b).

Notes:

[1] Ranked in order of rainfall totals.

[2] mm.

[3] Percentage of the 1941–70 normal rainfall.

November to October period any further rainfall needs to exceed 89 mm (to give a total of 397 mm, the November to October rainfall in 1896/97). Such a rainfall (that is 89 mm for the two months September and October) represents over 90 per cent of the average for these months.

A more detailed examination of the rainfall for Christchurch in September, October and November following an extremely dry November to August period is shown in Table VI.4. Specifically the average rainfall for the period November 1981 to August 1982 is 565 mm, and this may be compared with the record low of 308 mm in 1981/82, 314 mm in 1896/97, and so on. If September is added, the average is 611 mm, and this may be compared with 345 mm in 1896/97, and 360 mm in 1931/32. Thus if September 1982 has less than 37 mm (or 80 per cent of the average for September), 1981/82 will still be the driest November to September period on record. Similar information is also given in the table when the rainfalls for October and November are included. But the most significant feature of the data in Table IV.4 is that, if one considers the ten driest November to August periods, the total rainfall by the end of November in those seasons (that is November to November) was as low as 411 mm (58 per cent of average), and as high as 585 mm (83 per cent of average). Therefore, from a purely historical viewpoint, the prospects of 'reasonably substantial rainfalls' in the 13-month period November 1981 to November 1982* – which to the end of

Table VI.4 Christchurch 'seasonal' rainfall: 1896–1982

Season[1]	*Total rainfall: November to –*			
	August (mm)	*September (mm)*	*October (mm)*	*November (mm)*
1981/82	**308**	**354**[2]	**400**[2]	**444**[2]
1896/97	314	345	387	411
1931/32	334	360	486	512
1968/69	337	361	404	417
1921/22	347	411	423	505
1970/71	362	394	443	472
1932/33	364	400	442	483
1947/48	380	429	448	506
1906/07	389	494	536	561
1934/35	395	441	512	585
1914/15	396	400	440	473

Source: After Maunder (1982b).
Notes:
1 Ranked in order of November to August rainfall totals.
2 Assuming 'normal' rainfalls in September, October and November.

* Remember, this was written in early September 1981 while the drought was happening.

Table VI.5 Christchurch rainfall probabilities

September to February[1]		*Probable total*[3] *November 1981 to February 1983*	
Lowest	147[2]	455	53%[4]
10%	180	488	57
20%	211	519	61
30%	230	538	63
40%	249	557	65
50%	286	594	69
60%	308	618	72
70%	343	651	76
80%	371	679	79
90%	422	730	85
Highest	554	862	101

Source: After Maunder (1982b).
Notes:
[1] Probabilities based on historical record.
[2] mm.
[3] Actual rainfall November 1981 to August 1982 of 308 mm plus September to February probabilities.
[4] Percentage of 1941–70 'normal' rainfall for the 16 months November to February.

August 1981 has had the lowest rainfall on record – do not look very promising.

Finally, if we look a little further ahead to the end of February 1983 and examine rainfall probabilities from the historical record for the September 1981 to February 1983 period, it is seen from Table VI.5 that the lowest rainfall is likely to be about 455 mm (comprised of 308 mm, being the *actual* rainfall from November to August, plus 147 mm which is the lowest 16-month September to February rainfall on record). More realistically, the 50 percentile gives a rainfall of 286 mm, which if added to the 308 mm for the period November 1981 to August 1982 gives a 6-month total of 594 mm or 69 per cent of average, and the 70 percentile is only 651 mm or 76 per cent of average. Thus, from the historical record, there are only about three chances out of ten that the total rainfall from November 1981 to February 1983 will be more than about three-quarters of the average rainfall for this 16-month period.

(d) Period July 1980–August 1982 in an historical setting

During the 27-month period from 17 June 1980 to 2 September 1982, the rainfall at Christchurch was only 67 per cent of average. An examination of the historical rainfall record for Christchurch shows that the 26-month drought from July 1980 to August 1982 inclusive was (at the date of writing the original report) the second driest on record. Table VI.6 shows that the total rainfall of 1059 mm (74 per cent of average) is 19 mm more than the

Table VI.6 Christchurch 'two season' rainfalls: 1900–1982

Two seasons[1]	*Rainfall (mm)*	*% Normal*[2]
1931–32–33	1040	73
1980–81–82	**1059**	**74**
1968–69–70	1060	75
1930–31–32	1082	76
1969–70–71	1101	77
1958–59–60	1105	77
1914–15–16	1126	79
1947–48–49	1129	79
1971–72–73	1133	79
1970–71–72	1135	80
1929–30–31	1184	83
1913–14–15	1187	83
1920–21–22	1193	84

Source: After Maunder (1982b).
Notes:
1 26 months July to August.
2 Expressed as a percentage of the 1941–70 normal.

1040 mm recorded in the record driest 26-month July to August period which occurred in 1931–32–33. Other very dry 'two seasons in succession' occurred in 1968–69–70, 1930–31–32, 1969–70–71, and 1958–59–60.

(e) Advice for the decision-maker

Much more detailed information on the present and past droughts in the Canterbury area in New Zealand (or – for that matter – any other area of the world) *could* be given, including the all-important data on soil moisture and the number of days of soil water deficit (or 'days below wilting point'). However, sufficient historical climate data has been given in this case to show that the (then) current (September 1982) situation in Canterbury ranks amongst the driest of any of the 10-month periods from November to August, and also amongst the driest of any of the 27-month periods from July through two seasons to August. Analysis of the historical record also indicates that there is considerable persistence in the dry periods; that is, the very dry November to August periods in the past were also generally very dry when the months of September, October, and November were *added*.

In terms of soil moisture the 1981/82 season was extremely dry in the North Canterbury/Mid-Canterbury area with an estimated 79 'days of soil water deficit' (based on a Thornthwaite/weighted water balance computation; see Maunder 1980b). This is the highest value to date for this period since comparable computations were started in the 1949/50 season. If to this total is added the number of days of soil moisture deficit in the previous

1980/81 season, 68 days, then the two successive seasons' (1980/81 and 1981/82) combined total is 147 days. This may be compared with the previous highest two successive season totals in the area of 133 days in 1971/72/73, 128 days in 1970/71/72, and 106 days in 1954/55/56.

To the concerned farmer, as well as to the concerned community in which the farmer lives, two questions may well arise as a result of the mainly descriptive account given in this section. First, does long-range weather forecasting provide any answers; second, is weather modification (specifically rain-making) possible? Regrettably the answer to both questions is no, for although a considerable amount of work has been done on both subjects, there are few positive results which could lead to providing any real help to the farming community in this region of New Zealand. At the moment*, therefore, the most useful guidance that can be given is that, in terms of historical droughts, the current drought is among the most severe (at least in a climatological sense). In addition, the historical record indicates that no immediate change in the situation is likely, and that from a planning viewpoint, the wisest decision is to assume that the drought will continue.†

7. Drought in Africa: The Sahel experience

'Drought. Africa. These two words have been linked together in newspaper headlines around the world for more than a decade.' These words in the April 1985 issue of *Weatherwise* are used to introduce a paper by Derek Winstanley on 'Africa in drought: a change of climate?' and highlight the significance of drought in this part of the world. Clearly, the drought is far more important than the drought in the Canterbury area of New Zealand just discussed, and should warrant considerable space in any book on climatic uncertainties. However, as so much has been written on drought in Africa in recent years, it is considered that the following comments from a few of the relevant publications on this subject may serve to highlight the dilemma of drought in Africa, and not just repeat what has already been very well discussed in a variety of publications.

The paper already mentioned by Winstanley (1985) provides a very useful survey of the African drought situation, and he comments:

> Tropical Africa has suffered a decade or more of recurring droughts. This has been an important cause – but not the only cause – of famines, economic

* Again, it should be emphasized that this was originally written in September 1981 while the drought was happening.

† In retrospect, the actual rainfall at Christchurch for the September 1982 to March 1983 period was 301 mm or 86 per cent of the 1941–70 normal. The actual November 1981 to October 1982 rainfall was 410 mm or 62 per cent of the average (which meant that the 1981–82 rainfall for this period ranked third lowest), while the actual November 1981 to February 1983 rainfall was 574 mm or 67 per cent of the average.

problems, and social disruptions. Future development in many of the countries already affected by droughts will be dependent, in part, on rainfall conditions on the continent. This close association between rainfall and development raises many questions that have important strategic and policy implications, not only for the peoples of Africa but also for the international community.

Winstanley then goes on to discuss the climatological aspects of the African drought(s), and in a special section highlighting the climate impacts of the drought(s) he asks two questions:

1. What are the effects of repeated and persistent droughts in the sub-Sahara zone of the last decade?
2. What are the possible effects of a continuation in the trend toward lower rainfall?

The answers Winstanley provides in relationship to the impacts on rainfed food production, cash crops, inland fisheries, river basin development, economic development, and desertification are very pertinent. He states:

Rainfed Food Production
A slow rate of increase in food production, much less than that required to keep pace with the population growth rate of about 2.8 percent per year. Significance: food shortages.

Cash Crops
A slow rate of increase in the production of cash crops for export (e.g., peanuts, cotton). Significance: slow rate of growth in export revenue.

Inland Fisheries
A decrease in fresh water fish production particularly in the Senegal, Niger and Chad basins. Significance: reduction in food supply and export revenue.

River Basin Development
A decrease in the amount of water available in the major rivers for irrigated agriculture (food and cash crops) and hydroelectricity production. Significance: decrease in the production of food and cash crops, electricity shortages, reduced economic feasibility of constructing new dams.

Economic Development
A need for greater food imports and a reduction in export revenue. Significance: slower economic growth and reduced debt servicing capability.

Desertification
Reduction in biological productivity. Significance: acceleration of the process of desertification, especially when combined with increasing human pressure on the land.

The question of credible predictions of the climate is also examined by Winstanley (1985) – see also the paper on the value of a long-range weather forecast for the Sahel by Glantz (1977a) discussed on page 56 – and he states that in the absence of a credible climate prediction capability, our expectations of future climatic conditions must be based on analyses of the historical records. Such an analysis is made by Winstanley, and although not all will agree with his findings, his conclusions as to the future are relevant in any discussion of 'The Uncertainty Business'. He states:

> To assume that rainfall conditions over the next 30 years in tropical Africa will be similar to conditions over, say, the last 30 years must be regarded as wishful thinking. I don't know what the future will bring, but a continuation of the trends must be considered as a distinct possibility, indeed probability. The significance of a continuation of the trends to the welfare of the various African nations and peoples involved would seem to deserve very serious attention.

Many papers have been written on the very severe nature of the drought in the sub-Saharan Sahel – particularly from a 'climatic viewpoint'. One such paper is that by Hulme (1984) who, in analysing the exceptionally dry year of 1983 in Central Sudan, noted that most reports and studies of 'drought in Africa' tend to focus on the West African countries. In this particular study, Hulme analysed rainfall data from thirty rainfall stations in Central Sudan (nineteen having records beginning prior to 1920) and concluded that in north-central Sudan, 1983 was the driest year this century. Hulme further commented that the consequences of the 1983 'wet' season for the Sudanese people are (or were) particularly disturbing, in that field work he carried out in January 1984 in the Ed Dueim area revealed crop yields of only 10–20 per cent of that gained in a 'normal' year amongst cultivators, and livestock deaths of up to 50 per cent of herd size amongst herders.

Clearly drought as severe as has occurred in Africa since the late 1960s calls for action, and Tooze (1984), writing in *Nature* (9 February 1984), stated:

> A call for the formation of a Sahel fund to combat the area's drought and desertification was issued last week at the end of the sixth meeting of the Permanent Interstate Committee for Drought Control in the Sahel (CILLS), at Niamey, the capital of Niger. The committee was formed ten years ago by eight West African governments (Cape Verde Islands, Chad, Gambia, Mali, Mauritania, Niger, Senegal and Upper Volta) to co-ordinate action during severe droughts. The sixth meeting was attended by heads of governments and the director-general of the Food and Agriculture Organization.

Tooze comments that conditions in the Sahel are part of the climatic threat to the world's poorest people, and notes that according to the Club du Sahel, formed by the Organization for Economic Co-operation and Development, several West African governments, as well as various development agencies, aid to the area has doubled in ten years and now amounts to 15 per cent of the total gross national product in the Sahel states and more than 20 per cent of the gross national product in half of them. Tooze then points to the difficulties associated with aid assistance to areas suffering drought (which is essentially but not entirely an environmental problem) and comments as follows on what is becoming a very serious problem:

> So dependent on aid has the region become that some agencies now fear that long-term massive food aid is sapping the ability of the region to acquire the resources necessary to respond to normal climatic swings, let alone to cope with disaster.

Finally in this brief discussion of drought in the Sahel, reference should be made to the excellent booklet *Climate Variations, Drought and Desertification* written by F. K. Hare and published by the World Meteorological Organization (see Hare 1985b). Hare comments that around the world's deserts there is a wide extent of semi-arid and sub-humid land supporting a large human population, and that in spite of its hazards, the arid zone – deserts and surrounds alike – has offered a challenge to humanity that for centuries has been successfully answered. Hare observes, however, that today this zone is in many places the scene of acute distress and even of tragic famine, and that in spite of the long traditions of human adaptation, many nations now find themselves unable to wrest a reasonable living from the vanishing soils and natural resources.

Four questions are posed by Hare in relation to the above:

(1) What accounts for this unhappy situation?
(2) Is it human failure?
(3) Are the causes to be found in human interference? Or
(4) Is it a natural deterioration of climate?

The answers to these questions, according to Hare, involve the problem of desertification, the spreading of deserts into formerly productive land – or, more correctly, the degradation of that land until it can no longer adequately support living communities. But why have the deserts spread? Again Hare provides useful answers:

> Firstly, the human populations of the arid zone have grown enormously in recent decades and the need for food, fibre, and resources has grown accordingly. In many areas the demands made on soils, vegetation and climate now greatly exceed their capacity to yield. The carrying capacity of the land has been exceeded.
>
> Secondly, many recent years have seen protracted drought. Under natural conditions such failures of expected rainfall would have little effect. Traditional human societies were adept at conquering drought. But the rise in population and various forms of regulation have made this adaptation more difficult. Nomadic pastoralism, for example, has declined as a livelihood system, as has shifting cultivation. Drought has in some areas driven the pastoralists into the zone of cultivation, with resulting conflict.

Clearly these are only some of the answers, and the whole (but simple) concept of drought being present when there is too little water for too many people, animals, or crops is very relevant. Hare also discusses the climatic stresses that underline the tensions discussed in the previous paragraph, the major finding (according to Hare) being that 'climate alone will not destabilise the productive cloak of life that shelters the world and yields so much to human needs – but, nevertheless, it cannot be disregarded'. In the light of the title of this book, *The Uncertainty Business – Risks and Opportunities in Weather and Climate*, Hare's comments are particularly relevant. Indeed the whole spectrum of drought in Africa involves both risks and opportunities, and clearly people and nations ignore the climate factor at their peril.

B. TRANSPORT/WEATHER RELATIONSHIPS

1. Productivity and transport

Given sufficient lead-time, the value of accurate weather-based forecasts of a nation's production is wide-ranging, involving many activities 'beyond both the farm gate and the factory gate'. The key factor in optimizing the value of these forecasts is the need for a greater realization within industry that weather and climate variations can and indeed must be taken into account. In the United States, among other countries, this is now taking place; indeed, in 1977 the US Secretary of Agriculture in looking at the agricultural sector of the US economy issued a ruling* that 'never again do I want to see an economic report come out of the [US] Department of Agriculture that assumes average weather'.

This 'average weather' and 'average climate' misconception is regrettably still a feature of many aspects of the production industries – and by inference the associated transport industries. Indeed, as previously emphasized, many decision-makers within industry appear to be quite content to go from one month to the next and one year to the next with an almost complete disregard of the effects of weather variations on productivity, marketing, and transportation.

The above comments could apply to many industries and to many countries, and they certainly apply to the New Zealand pastoral production and associated transportation industry. But one exception, in New Zealand, to this disregard of weather and climate variations is the NZ Meat and Wool Boards' Economic Service, which is one of a select group of decision-makers which are aware of the importance of weather and climate variations. For example, Thompson and Taylor (1975) of that Service stated that climate variations affect both the seasonal volume of production and its movement to markets, with implications for transport and sales which must be coordinated worldwide. They further commented:

---

* As stated in the *Bulletin of the American Meteorological Society*, 58:534 (June 1977).

Unpredictability of supply at whatever level disrupts the smooth flow of products from farmgate to consumer, increasing costs to all involved. Weather effects on flow patterns of major products within seasons can be significant, with implications for the scheduling of auctions in the case of wool and for meat, and the arrangement of shipments to export markets. Apart from the marketing aspects, steady product flows enable better use of transport and storage facilities and contribute to reducing costs.

2. Transport costs beyond the farm gate: A New Zealand example

New Zealand's success as an exporter of meat, wool, and dairy products has been highly dependent upon her ability to produce high-quality products at relatively low cost. For example, Flemming (1982) states that with the blessing of favourable climate and efficient farming, New Zealand can produce meat at the farm gate for a fraction of the energy input used in Europe or the United States. However, he further comments:

> By the time that meat has been processed at a freezing works the energy content has more than doubled and a further doubling occurs between works and market. In energy terms we have lost much of our competitive advantage by the time the meat reaches the market. . . . It is astonishing to realise that the direct energy used in producing, processing and transporting a lamb carcass to the UK is equivalent to half its weight in oil.

An indication of the costs of transporting lamb from New Zealand to the United Kingdom in the three seasons 1974/75, 1975/76, and 1976/77 is given in Table VI.7, which indicates the relatively low proportion of the final

Table VI.7 Major components of United Kingdom lamb export price/cost structure*

	Season		
Component	*1974/75*	*1975/76*	*1976/77*
Farm gate	38	55	68
Transport to works	2	2	2
Schedule (season average)	41	57	70
Works to f.o.b. (mostly processing charges)	21	25	29
f.o.b.	66	85	101
Freight, insurance UK levies, landing cost	44	55	67
Ex-hooks, Smithfield	111	140	169
Total costs, farm gate to Smithfield	73	85	101
Additional costs as %	192	155	148

Source: After Maunder (1978b), from Reserve Bank of New Zealand (per M. J. Walsh).
Note: * Cents per kg of PM grade.

wholesale price (at the Smithfield market in London) that is related to the on-farm product. A similar 'beyond the farm gate' situation applies in the transportation of wool as shown in Table VI.8 for the same three seasons, 1974/75, 1975/76, and 1976/77. It is considered that the meteorological/climatological input into minimizing these energy costs, and particularly those involved with transport, could be a significant factor in maintaining and even increasing New Zealand's market competitiveness in meat and other pastoral products.

The fluctuations in New Zealand's pastoral production from season to season is illustrated in Fig. VI.4. This product is closely linked to its transportation, since most of it is exported; and in this regard Troughton (1977) made some pertinent comments.

> Farm production is inherently seasonal because of climatic and biological constraints on plant and animal growth, and this introduces a distinctive and specialised component into the management of our agricultural system. A major problem is to reconcile the highly seasonally-dependent production with processing and market demand. This approach to management introduces the need to consider the whole agricultural chain from the farm to the market place as one integral system.

The month-to-month variations in the meat, wool, and dairy industries are also extremely important. In the specific case of lamb slaughtering in

Table VI.8 Major components of wool export price/cost structure*

	Season		
Component	*1974/75*	*1975/76*	*1976/77*
Farm gate	84	146	206
Transport, insurance, brokers' charge, Wool Board levy	8	11	14
Auction price (season average)	92	157	220
Delivery, and pre-shipment charges, buyer's commission	4	6	9
Estimated f.o.b.	96	163	229
Ocean freight and marine insurance	11	18	19
Estimated c.i.f.	106	181	248
Port charges and transport to mill	2	3	3
Estimated landed cost overseas	108	184	252
Total costs, farm gate to overseas mill	24	38	46
Additional costs as %	22	21	18

Source: After Maunder (1978b). Data from Reserve Bank of New Zealand (per M. J. Walsh).
Note: * Cents per kg of greasy wool.

FIGURE VI.4 Fluctuations* in New Zealand meat and wool and dairy production: 1973/74–1984/85

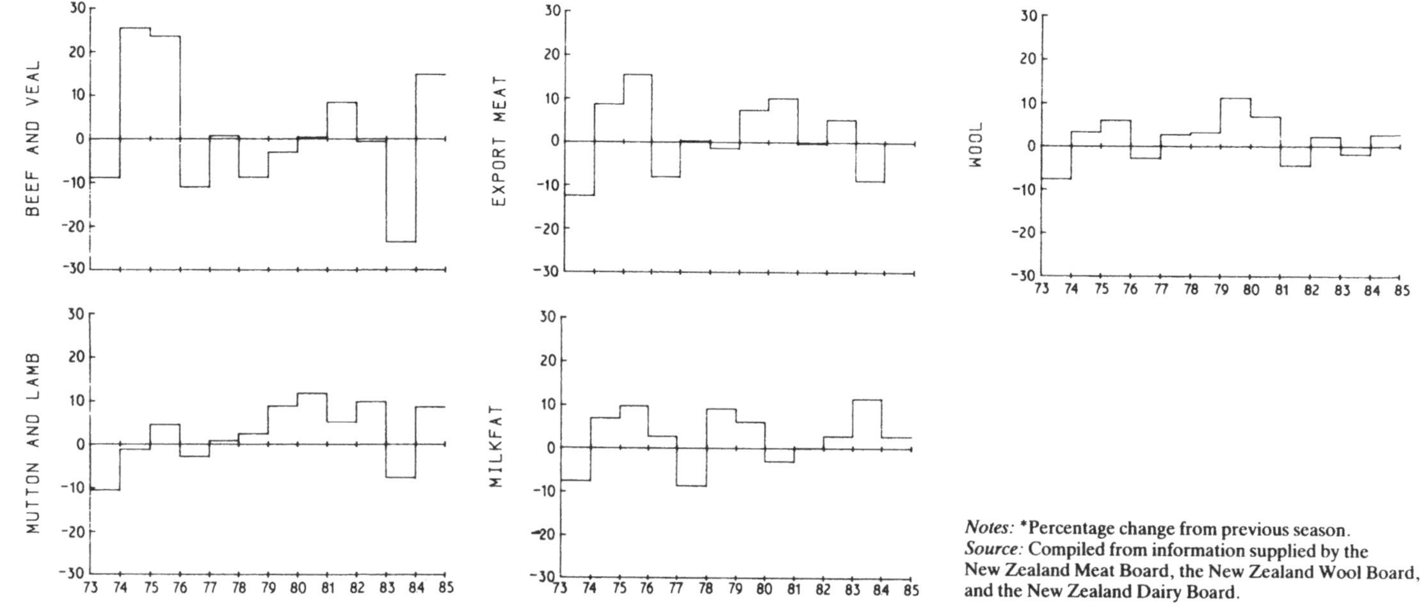

Notes: *Percentage change from previous season.
Source: Compiled from information supplied by the New Zealand Meat Board, the New Zealand Wool Board, and the New Zealand Dairy Board.

New Zealand (Table VI.9), Taylor (1977) noted that one of the main requirements for an efficient *production, processing*, and *marketing* chain is the accurate forward estimation of the volume of produce which the chain is required to handle. He further commented:

> Thus it becomes essential to establish a methodology for accurately estimating the volume of meat available for processing in advance, and adjusting this within the season as conditions change. . . . As a country so dependent on exports and hence transport can we afford to see our major export earner function at anything less than peak efficiency through the whole 'pasture to market' chain?

By way of example consider the 1977/78 season when the total value of pastoral-based exports from New Zealand was over $NZ2400 million. This sector represents about 70 per cent of the total value of New Zealand's exports and is therefore extremely important. A considerable proportion of the product is transported to customers around the world by container ships, the operating costs of which are very high (typical costs in 1978 were $15,000 a day while in port and $25,000 a day while at sea). Freight costs for container ships are based among other things on the number of days of voyage, the costs per day, and the load factor; therefore, any system which can increase the overall efficiency of the whole transportation industry of New Zealand pastoral production is worthwhile, not only to the farmer but also to the nation.

The weather factor in this efficiency aspect of the problem, particularly in

Table VI.9 Cumulative slaughterings of lambs for export (000) – New Zealand

To end of month	*Season 1975/76*	*Season 1976/77*	*Diff.**	*Season 1977/78*	*Diff.***
December	5868	6435	+567	4868	−1567
January	10579	10699	−120	9608	−1091
February	14181	14143	−38	14135	−8
March	17847	18419	+572	17908	−511
April	21166	21100	−66	20899	−101
May	23722	23721	−1	23458	−263
Average weight kg/head	13.7	13.3	−0.4	12.8	−0.5
Meat production*** (thousand tonnes)	325	315	−10	300	−15

Source: After Maunder (1978b), from information supplied by the New Zealand Meat and Wool Boards' Economic Service.

Notes:

* 1976/77–1975/76.

** 1977/78–1976/77.

*** 1 October to 31 May.

the effect of weather on seasonal production, is of course very significant. However, the real effects of weather variations tend to be hidden from the 'official' concern expressed by producer boards in New Zealand, as illustrated, for example in the NZ Meat Producers' Board 55th Annual Report (1977):

> With the rapidly-increasing container traffic, there is a need for more communication between the various parties to ensure maximum and efficient use of facilities and to avoid unnecessary delays. . . . The high daily running costs of container vessels make it essential that the quickest possible turnround be achieved if we are to avoid paying additional and unnecessary costs. This year our freight bill for meat alone is around $180 million.

C. Retail Trade and Weather Relationships

1. Some previous weather/retail trade analyses

In an early examination of weather and retail trade, Steele (1951) suggested that the weather might affect the sales of a retail store in four ways: first, the weather could be 'uncomfortable' to go shopping; second, the weather could produce situations which would physically prevent people from going to the store, as in the case of snow drifts or floods; third, the weather may have a psychological effect on people in that they may change their shopping habits; and fourth, some kinds of merchandise may be more desirable (and hence saleable) during a period in which certain types of weather conditions occur. In addition, weather conditions are often very important in the supply and quality of various commodities to retail outlets.

Many other factors also influence the demand for goods, but whether these are primarily economic, sociological, or psychological, the prevailing atmospheric conditions do have a considerable overriding influence. A basic factor is the 'urgency to shop', and this involves decisions related to both the actual and forecast weather, any alternative to shopping that a person might have, and the type of goods required. Some of these factors are taken into account by planners even before a store has emerged from the blueprint stage. However, the comfort of the shopper is only one facet of the complex problem of marketing, and it is evident that (irrespective of the type of shops or shopping centres available, or the means of transportation to and from such shops or shopping centres), weather conditions, as well as the seasonal climate, have a considerable influence on the type of goods sold and the amount of profit made by a retailer.

In some cases the astute retail manager utilizes such knowledge in planning day-to-day sales, for, by correctly assessing both the 24- to 36-hour and the 4- to 5-day weather forecasts, the day(s) on which there is likely to be the largest number of customers can be predicted. Such knowledge is also useful in determining those days on which maximum sales personnel should be available, and perhaps for deciding which days would be most suitable for holding unadvertised sales. Longer-range forecasts are also helpful, and any improvements in monthly and seasonal weather forecasts should enable those responsible for forward purchasing to become more efficient buyers, since such forecasts should be a guide to the kind and quantity of seasonable merchandise that is likely to be required by the customer next month or next season.

The specific effects of weather on the daily sales of three departments of Younker Bros. Inc., in Des Moines, Iowa, were the subject of the early study by Steele (1951). The period covered in this study was the seven weeks before Easter for the years 1940–48, and the results indicated that 88 per cent of the variance in the sales of the store were accounted for by the weather variables used in the analysis. In another early study, Zeisel (1950) examined the importance of weather on beer consumption in the state of Rhode Island. It was concluded that for every 1.0°F (0.5°C) temperature change, the consumption of beer changed 1.1 per cent above or below the expected level. Zeisel suggested that a similar analysis could also be made for carbonated beverages and iced tea sales, and from the preliminary data given in his paper it would appear that such sales, especially iced tea, would in fact show a much larger association with temperature changes than was the case with beer sales.

An important consideration in weather–retail trade relationships is (with few exceptions) the non-availability of data on sales of weather-sensitive commodities. Each category of weather-sensitive commodities probably has a unique set of weather responses. Some weather–retail trade studies are therefore of limited value, since they are concerned, through the deficiency of the data that must be used, with either *total* retail trade, or *sectors* of retail trade such as department stores or hardware stores. One exception to these studies is the paper by Linden (1962) concerning the sales of a single item of retail trade, namely women's winter coats in New York department stores. Linden found a strong association between the relative amount of business done in September and September's mean temperature. For example, the normal temperature in September in the New York area is about 67°F (19°C) and department stores realize about 18 per cent of their women's winter coat sales during this month. However, using data over a twelve-year period, Linden found that for each 1°F (0.5°C) deviation from the normal September temperature, there is an inverse deviation of about 0.6 per cent in September's contribution to the total women's winter coat sales in the season.

Several studies relating weather variations to retail trade in the United States have been compiled by the National Industrial Conference Board, and in a paper by Linden (1959) the effects of 'adverse' weather on customer traffic in department stores are discussed. In the analysis, weekly data for New York City for 1957 and 1958 were used. Each week's sales during the second year of the period were then compared with the sales in the related week of the preceding year, the difference being matched with the variation in weather conditions over the corresponding time. Linden noted that after adjusting for non-comparable weeks and for cases where the shopping weather was about the same, the data indicated that sales and weather moved in the 'same direction' four times more frequently than they moved in the 'opposite direction'.

The US Bureau of the Census has also studied the effect that weather has on retail trade, and some of the early exploratory work of the Bureau is detailed in a paper by Shor (1964). In this analysis monthly estimates of the irregular component of the seasonally adjusted retail sales series were developed for the eight-year period May 1953 to April 1961, for total United States retail trade and for several kind-of-business categories. The data applied to the United States as a whole and they were compared with individual monthly averages of selected weather factors. Regression analyses were carried out for the comprehensive 'all retail stores' group, and 'some' of the multiple correlations were significant at the 5 per cent level. Analyses of the various categories of retail trade were also made and Shor noted that in several autumn and winter months significant correlations occurred between weather and gasoline sales, and weather and drug store sales.

Some knowledge of the weather reflex of a particular product line is also extremely useful in improving marketing tactics. For example, Linden (1962) pointed out that sales, especially in the early period of a season such as winter or spring, may be compared with the weather record; such a comparison can well be a critical prerequisite for accurate short-term demand forecasting. Similarly salesmen's calls at retail outlets might be arranged to follow a period of weather which is especially favourable for a product, for it is likely that inventories would be lowest at that time, and that the merchant would be well disposed towards re-ordering.

There are of course many other ways in which weather knowledge might be used to increase sales. Much depends on the item, and in some instances promotions might specifically stress the particular buying impulses that are triggered by the climate. But as Linden (1962) stated: 'whatever tactics may be most appropriate for a particular product, it is a fact that weather has a powerful effect on demand. Moreover, with a little research and some imaginative application, marketers can make it a profitable ally.'

The effects of weather variations on the *total* retail trade of a national economy are also of considerable importance, and can pose difficult problems to economists trying to explain variations in certain key economic indicators. One major problem is 'explaining' the variations which take place from 'seasonally adjusted' data. For example, Petty (1963) commented on the movement in the Federal Reserve Bank of Chicago index of department stores sales, and the corresponding US Department of Commerce data for total US retail sales. As Petty aptly related in relation to the Chicago sales:

> We had to make some judgement as to how much of the poor showing to blame on (1) severe weather or (2) the consumer's persistent reluctance to buy. The relatively sharp drop-off in sales in areas hardest hit by heavy snows and extremely low temperatures suggested that abnormal weather was again exerting its influence on consumer buying. As someone has remarked: 'Most seasonal adjustments for last February "took out" only 2 in. of the 10 in. snowfall we had.'

Although it is evident that the weather does affect retail trade, the number of in-depth studies on the subject is relatively few; indeed the need to cite studies in this review from the 1950s and the 1960s reflects the sparse nature of published material linking weather and retail trade. One exception is the study by the author (see Maunder 1973a) which is discussed in the next section. Another exception is the most informative survey on 'Weather information – valuable economic tool in an era of low growth' published in the August 1985 issue of *Tokyo Newsletter* (see Gabe 1985).

The author of this survey, Masanobu Gabe, provides a valuable insight into how weather information *is* used in day-to-day decision-making, as the following extracts from his paper show:

> One leading brewery in Japan keeps data collected over the past 30 years showing the relationship between growth in beer consumption and the weather index. . . . The 'brewery's beer weather index' is compiled by recording daily maximum temperatures in 15 areas around the country where its branches are located. . . . When the maximum temperature registers one degree centigrade above normal on a fine day in July or August, beer sales that day increase by 2,470,000 large-size bottles. This is equivalent to an increase of about 8% over the average daily consumption.

But, as Gabe points out, it is also very difficult for this brewery (or possibly *any* brewery) to incorporate weather information into its planning process, and quoting from a member of the brewery's planning department, he states:

> Beer consumption is closely related to weather. And yet, it is not easy to plan beer production based on weather forecasts. Beer is not something that can be stocked for a long time; like fish, it must be supplied fresh. The important thing for us is to establish business operations that are not affected by the weather. This is why we try to develop various new products and new beverages and also suggest new ideas for living.

Gabe also notes that the demand for air conditioners is similarly affected by the weather. He says that the air conditioning and cooling industry in Japan compiles various statistical data for the October to September period (the 'refrigeration year') and he states:

> The data reveal that weather became a major determining factor in the demand for air conditioners around 1977–78. The intense summer heat in 1978 was responsible for the sale of 3,150,000 units, 450,000 more than 2,700,000 units anticipated. In 1982, which saw a cooler summer than usual, only 1,900,000 units were sold compared with expected sales of 2,400,000.

As with the consumption of beer, however, the interesting feature of the Japanese use of weather information in the planning process is to produce a product that is *less* sensitive to the weather and the climate. For example, Gabe gives the following comment from the Mitsubishi Electric Corporation.

> Planning production on the basis of the Meteorological Agency's long-range weather forecasts alone would be too risky. To minimize the risk, we have to give some allowance in the amount of production by assuming that the weather will be neither too warm nor too cold. The best strategy, however, is to develop products that can be used both in summer and in winter, regardless of the weather, such as heat pump-type air conditioners, which can both heat and cool. In so doing, we hope to stabilize demand from one year to the next.

2. A United States example

(a) The setting

In this example the formulation of weighted indices on a weekly basis specifically for the United States retail trade is discussed. The factors of 'large area' (i.e. the United States) and 'short time period' (i.e. a week) were chosen deliberately by the author (see Maunder 1973a) for two reasons: first, it is considered that nationwide (in this case, United States) economic activities are important to various 'high-level' decision-makers, and second, only weather over a short period (such as a week) has any real practical meaning to (in this case) the millions of 'low-level' decision-makers who each day go shopping in the United States. It could of course be argued that a weather index for a nation as large as the United States has little physical or practical meaning. However, it is strongly believed that a measure of nationwide weather can be computed (as discussed in detail in Chapter V), and that it can be of use to decision-makers. The alternative is that business, including business on a nationwide scale, will ignore an important aspect of the environment – the weather – an omission which can lead to incorrect decisions with unfavourable economic and social consequences to regions and nations.

A prime reason for this particular research into weather-economics was, therefore, to enable nationwide economic activities to be related to nationwide weather. Specific aspects of the United States nationwide weather are published weekly by NOAA in the *Weekly Weather and Crop Bulletin*, and the five-day nationwide weather forecasts issued by the National Weather Service complement this real-time weather information.

Various kinds of weekly economic indices are available for the United States; among these are those published by the United States Department of Commerce in their *Weekly Retail Sales*, and these data formed the basis of the 1973 investigation.* However it must be emphasized that these weekly data were available only on a nationwide basis. Consequently, irrespective of the value of such data on a regional basis, the fact that they are *not* available necessitates the calculation of *national* weather indices, if reasonably meaningful econoclimate studies on a weekly basis are to be made. It should also be noted that the publication *Weekly Retail Sales* is restricted to thirteen major stores types. Accordingly, no official weekly information was (or is) available relating to the weather-sensitive commodities such as ice cream, soft drinks, women's winter and summer clothes, refrigerators, air-conditioners, automobile tyres, beer, iced tea, paint, umbrellas, etc.

(b) FORMULATION OF A WEEKLY WEATHER INDEX

In order to use the available data in nationwide econoclimatic studies, it is therefore necessary to devise a weekly weather index for the United States as a whole. A problem that arises immediately is that people thinking of going shopping, or actually shopping, react to *weather* rather than to the individual *weather elements* of rain, snow, temperature, wind, humidity or sunshine. The combination of weather elements into a single useful retail-related weather index is, however, difficult. A further consideration is that the only suitable *published* weather data on a weekly basis for stations in the United States is that in the *Weekly Weather and Crop Bulletin*. It was therefore considered appropriate in the first instance to compute weekly weather indices in the form of weighted weekly precipitation indices and weighted weekly temperature indices, bearing in mind that an appropriate combination of wind, sunshine, humidity, temperature, and precipitation duration would presumably have been much more appropriate.

The nationwide precipitation and temperature indices were evaluated from weighted weekly temperature departures and weekly precipitation departures from the normal for 147 places across the United States. Most of these places each had metropolitan populations of 300,000 or more, and a combined population in excess of 120,000,000, this being equivalent to the then total urban population of the United States. A weighting was computed for each of the 147 places based on the buying power index published in the *Marketing Magazine*, as follows:

$$\text{Buying power index} = (5E + 3R + 2P)/10$$

where E = % of US 'effective buying income'
R = % of US retail sales
P = % of US population

* Publication of this series ceased in 1979 – accordingly research on weather and the national weekly retail trade in the United States is now made much more difficult.

Table VI.10 Weightings of selected metropolitan areas in the United States

Metropolitan area	*Population (1965)*	*% US retail sales (1968)*	*Buying power index (%) (1968)*
Atlanta	1,216,000	0.74	0.71
Boston	3,205,000	1.84	1.83
Chicago	6,689,000	4.11	4.22
Denver	1,073,000	0.66	0.62
Kansas City	1,183,000	0.94	0.75
Las Vegas	232,000	0.15	0.14
Los Angeles	6,765,000	4.25	4.19
Miami	1,061,000	0.77	0.63
New York†	15,821,000	8.68	9.27
Omaha	516,000	0.23	0.26
Phoenix	818,000	0.47	0.45
San Francisco	2,918,000	1.85	1.81
Washington, DC	2,408,000	1.63	1.67

Source: After Maunder (1973a).
Note: † Standard consolidated area.

A comparison of the total population, percentage of retail sales, and the buying power index for a selection of the 147 places is given in Table VI.10. The buying power indices* were next used to weight the differences from the normal weekly precipitation and normal mean temperature for the 147 stations. The weekly precipitation departures from the normal were obtained directly from the *Weekly Weather and Crop Bulletin*, and the weekly temperature departures from the normal were assessed using the 1931–60 normal temperatures for the specific week of the year.

The weighted temperature and precipitation departures for the 147 places were next combined into indices for the United States using the following expression:

* The 'Buying Power Index' used is as published in the '1969 Survey and Buying Power' of the issue of *Marketing Magazine* for 10 June 1969. The index includes the percentage of US retail sales, the percentage of US population, together with the percentage of the 'effective buying income' which can be defined as the 'income available to buy goods'. For example, in 1969 metropolitan Birmingham (Alabama) had 0.34 per cent of total US retail sales, 0.38 per cent of total US population, and 0.28 per cent of the total US effective buying income, giving a buying power index of 0.32 per cent, whereas the Standard Consolidated Area of New York had 8.7 per cent of US retail sales, 8.2 per cent of US population, and 10.1 per cent of the total US effective buying income, giving a buying power index of 9.3 per cent. Thus in comparison with the percentage of the total US population, New York had a higher than average buying power index, whereas Birmingham had a lower than average buying power index.

$$\text{United States Index I} = \frac{\Sigma W_i B_i}{\Sigma B_i}$$

where W = the rainfall or temperature departure from the average for station i (i = 1 . . . 147), and
B = the buying power index for station i (i = 1 . . . 147)

An abridged example of the calculation of the United States weighted temperature and precipitation departures from the normal for the week ending 11 February 1968 is given in Table VI.11. The period covered by this study extended over three years, nationwide precipitation and temperature indices for the United States being computed for each of the 154 weeks from April 1966 to March 1969. The degree of variability of these indices can be judged from the frequency distributions which are given in Table VI.12.

Some of the problems associated with the development of such weighted weather indices are discussed in Chapter V, but the main objection to

Table VI.11 Abridged example of the calculation of temperature* and precipitation** departures from the normal for the United States for the week ending 11 February 1968, weighted according to the buying power index of 147 localities

No.	*Station*	*Buying power index (US = 100)*	*Departure from normal*		*Weighted differences*	
			Temp. (deg. F)	*Precip. (in.)*	*Temp. (deg. F)*	*Precip. (in.)*
1	Birmingham, Alabama	0.32	−9	−1.3	−2.8	−0.42
15	Denver, Colorado	0.62	+4	−0.2	+2.4	−0.12
19	Washington, DC	1.67	−2	−0.5	−3.3	−0.84
22	Miami, Florida	0.63	−8	−0.2	−5.0	−0.13
32	Chicago, Illinois	4.22	−2	−0.3	−8.4	−1.23
50	New Orleans, Louisiana	0.50	−11	−1.0	−5.5	−0.50
65	Kansas City, Missouri	0.75	−3	−0.3	−2.3	+0.23
72	Las Vegas, Nevada	0.14	+3	+0.1	+0.4	+0.01
80	New York, NY	9.27	−4	−0.8	−37.1	−7.42
122	Dallas, Texas	0.79	−1	−0.6	−0.8	−0.47
132	Salt Lake City, Utah	0.26	+4	−0.3	+1.0	+0.08
147	Cheyenne, Wyoming	0.03	+7	−0.1	+0.2	+0.00
Total (147 stations)		63.12	–	–	−152.8	−35.35

Source: After Maunder (1973a).
Notes:
* Total weighted temperature departure = −152.8
Average departure = −152.8/63.12 = −2.4 deg. F (−1.3 deg. C)
** Total weighted precipitation departure = −35.35
Average departure = −35.35/63.12 = −0.56 in. (−14.2 mm)

Table VI.12 Frequency of the departure from the normal of weighted weekly temperature and precipitation indices for the United States: April 1966–March 1969

Temperature (deg. F)			*Precipitation (in.)*		
Departure from normal	*Number*	*%*	*Departure from normal*	*Number*	*%*
+8 or more	4	3	+0.8 or more	1	1
+5 or 6 or 7	7	4	+0.5 to 0.7	17	11
+3 or 4	15	9	+0.3 or 0.4	15	9
+2	16	10	+0.2	14	9
+1	18	12	+0.1	18	12
0	17	11	0	15	9
−1	19	13	−0.1	16	10
−2	17	11	−0.2	18	12
−3 or 4	23	15	−0.3 or 0.4	28	19
−5 or 6 or 7	13	9	−0.5 to 0.7	12	8
−8 or more	5	3	−0.8 or more	0	0
Total	154	100		154	100

Source: After Maunder (1973a).

assessing such indices appears to be their validity for large areas. For example, while no objection is usually made when a climate index is assessed from the several stations in a small area such as London, Rhode Island, or Malta, the assessment of a similar weather or climate index for the hundred or more stations in a large area such as Australia, Brazil, or the United States, usually raises objections as to whether such a composite weather or climate index is 'real'.

The same objections presumably apply, but do not appear to, to certain temperature indices used for research into climate change, in that temperature departures from normal are averaged over longitude and latitude zones, and in some cases for hemispheric zones. From a practical point of view, however, it is considered that weighted weather and climate indices for a country like the United States *do* have valuable uses and in most cases are much superior to the conventional approach of using only *point* climate and weather data. Specific examples assessing commodity-weighted weather and climate indices in a variety of situations using conventional as well as real-time weather and climate data are given in Chapter V.

(c) Details of a specific United States study

The weekly retail trade data for the United States published by the US Department of Commerce are *not* adjusted for seasonal or holiday variations; accordingly the 154 weeks' data for the three years used in the original study (April 1966 to March 1969) were first grouped into 11-week* overlapping periods, and the weather–retail trade relationships were assessed within these periods. The 11-week periods were centred at the mid-point of the months February to October inclusive in each of the three years. However, in considering the various 11-week periods, certain holiday weeks were excluded, these weeks being replaced by the nearest non-holiday week outside each period. In addition, because of Christmas, the analyses were not extended to include the 11-week periods centred on the mid-points of November, December, or January.

The association between retail trade sales and the weighted precipitation and temperature indices for the United States shown in Table VI.13 were assessed using the following regression equation:

$$x_1 = a_o + a_1x_2 + a_2x_2^2 + a_3x_3 + a_4x_4$$

where x_1 = retail sales for the specific kind of retail business for specific weeks in millions of dollars
x_2 = time factor (week 1, 2, . . . 11)
x_3 = 'buying power index' weighted precipitation departure for the United States for specific weeks (in inches)
x_4 = 'buying power index' weighted temperature departure for the United States for specific weeks (in °F)

The regression equation allowed an assessment of the relationship between variations in retail trade sales from a quadratic time trend over specific 11-week periods, and the variations in the weighted nationwide precipitation and temperature indices. For example, the computed regression equation for the 11-week period 13 January 1969 to 24 March 1969 for retail sales in the apparel group was:

$$x_1 = 357.3 - 26.2x_2 + 2.6x_2^2 - 60.1x_3 + 4.6x_4$$

The mean value of x_1 in this period was $313 million, the regression giving a standard error of the estimate of $11 million, an R^2 value of 0.94, and an F ratio of 22.2 indicating significance at the 0.1 per cent level. The interrelationships between, first, the weighted precipitation departures (x_3) and the retail sales (x_1), and, second, the weighted temperature departures (x_4) and the retail sales (x_1), were then examined using the relevant partial correlations. For example, in the equation given, the squared partial correlation of

* Eleven-week periods were considered as a compromise between a 'few' weeks, which would have provided too many variations, and a longer period which would have provided too few variations.

Table VI.13 Selected weekly retail trade sales* in the United States for the period January to March 1968 and associated precipitation and temperature indices weighted according to the buying power index of 147 localities

Period	*Retail business (millions of dollars)*					*Weighted indices***	
Week ending Saturday	*Total retail trade*	*Apparel group*	*Furniture and appliances*	*Lumber building hardware*	*Drug stores*	*Precip. (in.)*	*Temp. (deg. F)*
Jan. 13	5344	284	250	227	205	+0.10	−8.5
Jan 20	5562	296	281	255	215	+0.32	+1.9
Jan. 27	5581	273	276	264	208	−0.28	+1.8
Feb. 3	5706	277	295	275	203	+0.28	+6.4
Feb. 10	5720	292	287	286	210	−0.56	−2.4
Feb. 17	5772	264	278	297	227	−0.43	−5.8
Feb. 24	5778	284	302	287	200	−0.48	−6.3
Mar. 2	6049	275	295	315	207	−0.25	−2.8
Mar. 9	5957	306	291	316	214	−0.16	−1.7
Mar. 16	5967	306	271	330	199	+0.74	−1.6
Mar. 23	6121	327	279	342	206	+0.17	+0.8
Mar. 30	6548	376	304	376	196	−0.51	+7.5

Source: After Maunder (1973a). Data from Weekly Retail Sales Report – US Department of Commerce/ Bureau of the Census.

Notes:

* Weekly sales estimates are based on data from 2500 firms, covering approximately 48,000 retail stores in the United States.

** Weighted weather indices are for week ending midnight on the Sunday of the week indicated. The indices shown are departures from the normal for the specific week. Data are not adjusted for seasonal or holiday variations.

x_4 (temperature) on x_1 (retail sales) indicated that when the net quadratic influence of time (x_2) and precipitation (x_3) was removed, 49 per cent of the variance in the retail sales of the apparel group of stores in the United States during the 11-week period 13 January to 24 March 1969 was associated with variations in the United States weighted temperature index. The coefficients in the regression equation also indicate, in the example given, that drier than normal and warmer than normal conditions (for the specific weeks concerned) were associated with above-average retail sales in the apparel group during the late winter/early spring period of 1969. Similar regression analyses were made for the other 11-week periods in the three years April 1966 to March 1969 for all thirteen aspects of the retail trade, the computed partial correlations relating precipitation and temperature to retail trade sales being given in Tables VI.14 and VI.15.

Table VI.14 Associations† between weighted precipitation departures and total retail sales for various kinds of business – United States

Kind of business	*Feb.*	*Mar.*	*Apr.*	*May*	*Jun.*	*Jul.*	*Aug.*	*Sept.*	*Oct.*
Food group	+0.08	+0.08	+0.06	+0.29	+0.04	−0.21	−0.19	−0.08	−0.12
Grocery stores	−0.08	+0.08	+0.08	+0.31	+0.13	−0.21	−0.20	−0.11	−0.13
Chain grocery stores	+0.10	+0.05	+0.10	+0.42*	+0.05	−0.13	−0.13	0.00	−0.19
Eating and drinking places	+0.12	+0.24	+0.27	+0.04	+0.23	−0.15	+0.04	+0.02	−0.10
General merchandise, apparel, furniture	−0.35	+0.22	+0.25	+0.50**	+0.40*	+0.19	+0.05	−0.01	−0.01
General merchandise	−0.25	+0.30	+0.43*	+0.44*	+0.38*	+0.25	+0.09	+0.01	+0.04
Department stores	−0.22	+0.38*	+0.39	+0.45**	+0.34	+0.18	+0.03	−0.07	+0.04
Apparel group	−0.55**	+0.24	+0.06	+0.40*	+0.35	−0.01	−0.09	+0.04	+0.18
Furniture and appliance group	−0.06	−0.18	+0.05	+0.24	+0.23	+0.22	+0.25	−0.18	−0.45**
Lumber, building, hardware, farm equipment	+0.35	+0.58***	+0.07	−0.24	−0.08	−0.25	−0.03	−0.16	−0.38*
Automotive group	+0.15	+0.42*	−0.03	+0.05	+0.06	+0.31	+0.18	−0.15	−0.30
Gasoline service stations	−0.52**	+0.06	+0.28	+0.22	+0.13	−0.29	−0.28	−0.15	+0.03
Drug and proprietary stores	−0.37	−0.36	+0.18	+0.25	+0.35	+0.06	−0.16	−0.03	+0.20

Source: After Maunder (1973a).
Notes:
* Significant at the 10 per cent level.
** Significant at the 5 per cent level.
*** Significant at the 1 per cent level.
† Based on 11-week period centred on the months shown. Data applies to the 1966–9 period. The associations shown are partial correlations.

Table VI.15 Associations† between weighted temperature departures and total retail trade sales for various kinds of business – United States

Kind of business	*Feb.*	*Mar.*	*Apr.*	*May*	*Jun.*	*Jul.*	*Aug.*	*Sept.*	*Oct.*
Food group	+0.17	−0.07	−0.26	+0.25	+0.36	−0.07	−0.27	−0.08	−0.26
Grocery stores	+0.17	−0.06	−0.26	+0.21	+0.34	−0.04	−0.25	−0.05	−0.32
Chain grocery stores	+0.01	−0.10	−0.18	−0.29	+0.30	0.00	−0.10	−0.16	−0.32
Eating and drinking places	−0.07	+0.03	−0.14	+0.13	+0.51**	−0.23	−0.04	+0.11	+0.08
General merchandise apparel, furniture	+0.13	+0.48**	+0.27	+0.33	+0.42**	+0.25	−0.04	−0.14	+0.11
General merchandise	−0.13	+0.46**	+0.17	+0.33	+0.39*	+0.27	+0.05	−0.11	+0.09
Department stores	−0.01	+0.50**	+0.17	+0.36	+0.36	+0.23	+0.04	−0.10	+0.15
Apparel group	+0.36	+0.46**	+0.06	+0.24	+0.25	−0.08	−0.26	−0.20	0.00
Furniture and appliance group	+0.28	−0.01	+0.38*	+0.41*	+0.62***	+0.60***	+0.07	+0.04	+0.61***
Lumber, building, hardware, farm equipment	−0.08	+0.35	+0.20	+0.15	+0.01	−0.02	−0.14	−0.22	−0.14
Automotive group	+0.06	+0.45**	+0.56***	+0.24	+0.26	+0.28	+0.07	−0.10	−0.08
Gasoline service station	−0.15	−0.07	+0.01	−0.13	+0.12	−0.40*	−0.25	−0.11	+0.14
Drug and proprietary stores	+0.14	−0.22	−0.06	+0.10	+0.39*	+0.25	−0.14	−0.04	−0.11

Source: After Maunder (1973a).

Notes:

* Significant at the 10 per cent level.

** Significant at the 5 per cent level.

*** Significant at the 1 per cent level.

† Based on 11 week period centred on the months shown. Data applies to the 1966–9 period. The associations shown are partial correlations.

(d) RESULTANT WEATHER–RETAIL TRADE ASSOCIATIONS

An analysis of the results given in Tables VI.14 and VI.15 shows that in the case of the precipitation partial correlations for the period February to June (Table VI.14), 14 of the 65 partial correlations were significant at the 10 per cent level, compared with an expected 6.5. Similarly, in the case of the temperature partial correlations (Table IV.16) 15 of the 65 March to July correlations were significant at the 10 per cent level. Overall the analysis indicated the probable greater importance of temperature as a factor associated with retail trade sales, and also the relatively close association of late winter to early summer weather conditions with retail sales variations.

It is emphasized that the associations given are from only three years of overlapping 11-week periods, and that the results are purposely designed to apply to the United States as a whole. The published data on weekly retail trade also refer only to *store types* and not to weather-sensitive commodities. The associations obtained must therefore be treated as approximations only, but the results indicate several nationwide features: (1) drier conditions than normal (for the specific weeks concerned) appear to be generally associated with above-average retail sales (for the specific week concerned) in late winter and during the early autumn; (2) wetter conditions than normal appear to be generally associated with above-average retail sales in late spring and early summer; (3) colder conditions than normal appear to be generally associated with above-average retail sales in the autumn; and (4) warmer conditions than normal appear to be generally associated with above-average retail sales in spring and early summer.

As would be expected with weather and economic data averaged over such a large and heterogeneous area, the econoclimatic associations revealed by the partial correlations vary in their significance, and for some retail trade groups, and during some of the 11-week periods, the relationships have limited practical value. It may also be suggested that some or all of these features are already well known by marketing people, particularly in view of the fact that the analysis could be said to emphasize the generally held view that the variations in the retail trade in the United States are very often associated with the lateness or otherwise of the winter and summer seasons. However, most marketing people have only a very superficial knowledge of weather/sales relationships (particularly on a national scale); further, most economists and many meteorologists (including weather forecasters and climatologists) appear to be unaware of the quantitative relationships that exist between weather and retail trade activities. It is believed, therefore, that this study gives a valuable insight into the overall relationships between weekly weather conditions and retail sales on a national scale, particularly in regard to the seasonal sensitivity of various kinds of retail sales to precipitation and temperature variations from the normal.

(e) Further research

There are, of course, many other aspects of weather and retail trade. These include a question which is of considerable interest to economists, namely the permanency of lost sales. That is, how permanent is the loss of business experienced on bad shopping weather days, or 'is the sale that is lost today gained tomorrow?' According to Linden (1959), the evidence suggests that a specific item, not bought, is not always purchased when better shopping weather comes along. Perhaps even more important, however, is the fact that bad shopping weather means fewer trips to the store, a factor which is usually associated with lower total buying, since a single exposure of the buying impulse is usually less rewarding – from a retailer's point of view – than two exposures. As an example, Linden quoted the case of the near-record losses sustained by New York stores during a blizzard in mid-February 1958, which failed to produce any detectable swell in buying when the snows melted and the sun returned. For example, the Federal Reserve Bank index for this particular week was 25 per cent less than expected for the period, but in the week following the storm the index merely returned to normal, although the weather was fair and Easter was early.

Of course analyses of retail sales which consider only weather as a contributing factor omit the most important factors that determine the desire and ability of consumers to buy, namely the ability to pay, anticipated future income, price expectations, and present ownership of goods. Nevertheless, the weather can explain a substantial part of the short-run variability in sales, and the method of adjusting weather and climate data to fit various kinds of nationwide economic data as described in this section offers one solution to the data problem confronting studies in nationwide weather-economics.

As has been emphasized, it would be very desirable if weekly *regional* data on economically important activities such as retail trade were available for use in econoclimatic studies, but since in most cases this is not the situation, it is necessary to devise methods to incorporate *nationwide* economic data such as retail trade into econoclimatic studies. It is suggested that the model described here for the United States provides a useful method of translating weather information into meaningful indices, which may be used with profit by the decision-makers. It is further suggested that the application of these relationships could be a profitable venture for many businessmen in the retail trade. Of course some companies may already be using these relationships in their 'confidential' day-to-day and week-to-week activities. It is nevertheless reasonably clear that *much more* could be gained if there was a better appreciation of the atmospheric influences at work.

D. Manufacturing and Weather Relationships

1. An overview

The diverse impact of weather and climate on manufacturing may be broadly categorized into two broad divisions: those influencing the *location* of the industry, and those affecting the *operations* of the industrial plant once an industry has been established. Associated with these factors are the design of the plant and the planning of its operations. Initially, there is the importance of climate in relationship to planning, rather than weather in relationship to operational aspects. For example, the availability and flow of water, the need for heating, cooling, and dehumidifying the air, and the atmospheric stability are directly related to capital investments in manufacturing. Climatological and meteorological components which influence storage, the efficiency of workers, the weathering of raw and finished products, and the operation of transportation facilities are also important. Weather, as distinct from climate conditions, may also hamper essential outdoor activities, cause damage to goods and equipment, or interrupt power supplies. Wind and low relative humidities are also sometimes important in that these conditions may increase the fire hazard at manufacturing plants, especially if combustible or explosive materials are being processed or stored.

But the studies on the *specific* relationships between variations in the weather and variations in industrial output, or efficiency, are few. Nevertheless, in recent years an interesting development in what has been called 'industrial meteorology' has occurred, particularly in the United States. An early researcher, Boyer (1966), noted that the growth of industrial meteorology offered opportunities for meteorologists and for those businessmen who are alert to the advantages of being 'weatherwise'. It should be noted, however, that industrial meteorology in this context related to *all* commercial activities of meteorologists, and they are obviously *not* restricted to 'industrial' activities. A more detailed examination of the commercial and business aspects of national meteorological services and private meteorological (and climatological) companies has been discussed in Chapter III.

2. Location of manufacturing

Although many oppose any suggestion of climate control over the activities of people, it is nevertheless evident that the location of industry, and the migration of industry within a nation, in some countries, is partly related to the climate. For example, although the distribution of economic activity in many countries has been traditionally explained in terms of the interaction of four principal factors – access to capital, raw materials, labour supply, and markets – the final decisions on the actual site are now becoming increasingly based upon cultural and amenity factors. Among these is the decision of many people to move for other than economic reasons, including the desire to find a 'more pleasant place' to live (see Wilson 1966).

3. Perception of the effects of weather on manufacturing

The results of an early report on weather and manufacturing (US Department of Commerce, Weather Bureau, 1964) suggested that in the United States, manufacturing had the least economic benefit potential of any weather-user classification studied. Since in many countries, however, including the United States, an appreciable proportion of the national income is derived from manufacturing (e.g. Japan 36 per cent; Thailand 18 per cent; Australia 31 per cent; Iran 36 per cent), more comprehensive studies of the *specific* significance of weather and climate and of weather forecasts on manufacturing is perhaps long overdue. A few comments from the pioneer study by Bickert and Browne (1966) on the perception of the effects of weather on manufacturing firms in Colorado are however relevant. Two aspects of the problem were considered: first, the perceived and real effects of weather on various functional areas of manufacturing firms; second, the present and potential utilization of weather information by such firms.

Five manufacturing firms were selected for this study on the basis of their hypothesized sensitivity to weather in various aspects of their operations. First, a firm manufacturing precision mechanical components for the aerospace industry. In this firm production and quality control were the two operations most weather-affected. In particular, variations in temperature caused changes in dimensions and tolerances, and high humidity increased the cost of product maintenance and storage. Second, a brewery was considered. Here the chief weather effects were principally related to the source of its raw materials: barley, rice, and water. In addition, severe drought severely restricted the plant's water supply, and freezing temperatures in winter necessitated the use of insulated railroad cars for shipping. The third firm selected was involved in brick and ceramic product manufacturing. This plant was particularly weather sensitive; in particular, snow and cold presented a number of problems. For example, wet clay lowered the grinding efficiency, thus decreasing both output and quality. In addition, a drop in temperature below −12°C necessitated a change from natural gas to propane, this causing a substantial increased cost in the operation of the kilns. The ski apparel industry was also considered; this firm was sensitive to weather conditions, particularly as they affected marketing. Finally a firm involved in consumer and industrial durables was studied. Here a substantial proportion of the sales of this company was related to winter conditions. Consequently a lack of snow, ice, and freezing conditions during late autumn, when a considerable amount of advertising occurs, seriously inhibited sales.

In their conclusions, Bickert and Browne (1966) indicated that despite the fact that the *initial awareness* of the effects of the weather on the five companies was found to be minimal, the *actual* effects of the weather were found to be considerable, particularly in terms of costs to the companies concerned. It would appear, therefore, that in many cases managements' perception of the effect of weather is minimal. A programme to inform management that better decisions could be made if better use was made of weather and climate information and weather and climate forecasts, would therefore appear to be justified.

4. Impact of severe weather conditions on manufacturing activities

Few in-depth studies have been made on the impacts of severe weather conditions on manufacturing, but a recent investigation by Palutikof (1983a) of British industry is of interest. In this study a multiple regression analysis was used to evaluate the impact of a severe winter and a hot/dry summer on the 'performance' of industries in the United Kingdom. Unfortunately, the unavailability (or non-use) of 'industry- or commodity-weighted' weather information, as discussed in Chapter V, reduced the value of the results, but the survey did reveal some interesting contrasts. For example, the severe winter conditions of 1962/63 *increased* activity in the utility industry by 17 per cent over the previous winter, whereas activity in the brick/cement industry *decreased* by 14 per cent. The analysis also showed that a hot dry summer increased activities in the clothing and footwear, and the drinks and tobacco industries, but decreased activity in the pottery and glass industries, and in the utility industries.

E. CONSTRUCTION AND WEATHER RELATIONSHIPS

1. An overview

Three broad aspects of the association between weather and climate on the construction industry include their influence on structural and architectural design, the effect in economic terms of the impact of weather on the day-to-day operations of the industry, and the influence that weather and climate does, or more accurately should have, on urban, industrial, and transportation planning.

A brief review follows of the literature relating the benefits of weather forecasts and climatological services to the various weather- and climate-sensitive aspects of the building and construction industry. It is appropriate, however, that a brief comment should first be made on the relative importance of the various aspects of the building and construction industry. The example given is for New Zealand (see Maunder 1972b), but a similar analysis could be, or probably has been, prepared for other countries.

2. The building and construction industry in New Zealand

A recent census of the building and construction industry in New Zealand showed that 66 per cent of its expenditure was concerned with the construction, alteration, repair, maintenance and demolition of buildings, 17 per cent was associated with the construction of roads, railways, bridges, tunnels, dams, drains, wharves and jetties, and 17 per cent was associated with other building and construction activities. The total expenditure on building and construction can also usefully be subdivided into types of contractors, this showing the overall predominance in New Zealand of private companies (78 per cent of the total) and the importance of building contractors and civil engineering contractors within this sector.

The distribution of building permits issued by local administrative authorities is an additional measure of the importance of various aspects of the building industry in New Zealand, and a recent survey showed that 43 per cent was related to new houses and flats (apartments), 14 per cent to commercial buildings, 9 per cent to factories, 7 per cent to schools, and 7 per cent to alterations to houses and flats. It is believed that such an assessment of the relative importance of various aspects of the construction industry – or for that matter any weather-sensitive industry – is a necessary first step when considering the association between weather and climate conditions and the activities in the construction industry in any country.

3. Impact of weather on construction planning

The effects of weather and climate on the construction industry in many countries, especially those in which the winters are severe, include variations in the rate of employment of workers in the weather-sensitive industries, increased construction costs under adverse weather conditions, as well as the degree of risk and uncertainty in the planning of construction projects. The weather and the climate have in fact an economic impact at three levels – on the contractor, on the construction worker, and on the customer. Since no two jobs are exactly alike, and both construction practices and weather and climate conditions vary from region to region, and from country to country, it is not possible to evaluate the economic impact of weather and climate on the construction industry in any neat, closed form. However, its effect on a specific job can be estimated by the use of a simulation model, and such a model was presented at a conference in St Louis in 1971 by Benjamin and Davis (see Benjamin and Davis 1971).

The hypothetical job used in the study was a two-storey office building, the critical path network containing forty-one activities of different durations and different sensitivities to the various weather elements. In one specific case, a job was started on 1 April, with a level of reliability of weather predictions of 90 per cent. The total estimated job duration, with no allowance for weather-caused delays, was 101 days. However, because of adverse weather conditions both the total job duration and the total direct labour cost were greater than their estimates. Specifically, 21 out of the 321 decisions required in the job were incorrect because of the adverse weather, this resulting in twenty extra days' work being necessary. In addition, 'indirect' costs would be involved because of the delay in starting future jobs.

Many more questions were created during the development of this simulation model than were answered, and Benjamin and Davis commented that only after a more realistic model is developed will it be possible to determine what specific information the construction manager needs in order to make his daily decisions, and what it is worth to collect and disseminate this information to the construction industry.

4. The economic impact of weather on the construction industry of the United States

The United States Weather Bureau sponsored an early study (Russo 1966) to determine the nature and magnitude of losses due to the weather in the construction industry of the United States. In addition the potential capability of present and future weather forecast accuracy, as well as the use of meteorological and climatological services to reduce these losses, was examined. The resulting analysis indicated that the total annual loss due to weather varied from 3 to 10 per cent of the total potential volume of construction.

Among the conclusions reached by the study was that if specific weather and climate information had been made available to the United States construction industry and was appropriately used, then a potential annual saving of approximately 15 per cent of the estimated weather- and climate-caused loss was possible. In addition, the savings achievable if the weather forecast products for the shorter (0–24 hour) time periods were 100 per cent accurate were estimated to be about 30 per cent more than would have been achieved with the forecast accuracy then obtaining.

5. Applications of probability models in using weather forecasts to plan construction activities

The inherent variability of weather contributes substantially to the uncertainties in the planning of building projects, and thus their costs. This point is emphasized by Johnson and McQuigg (1972), who showed that through the use of probability models, historical weather data and information typically contained in construction 'logs' can be employed to obtain estimates of the probabilities that alternative activities can be undertaken. These probabilities are based upon the observed or forecast weather conditions.

Johnson and McQuigg pointed out that with the weather forecasts that are generally available, construction managers could use such models to make decisions regarding whom to call to work, or which construction activities could be active. It is clear that the cost of making decisions in this manner – which would include paying for the appropriate weather information – could be more than offset by savings in construction costs.

F. ROAD CONSTRUCTION AND WEATHER RELATIONSHIPS

1. The setting

A good knowledge of the expected working conditions is important for planning all phases of the road construction industry. For example, road construction firms need information on the number of days on which work is possible, both for bidding on road construction contracts and scheduling machinery and manpower, while contract-letting officials need such information properly to anticipate probable completion dates, and to schedule funds for payment to contracting firms as work progresses through the contract periods. Answers to such questions are not normally available from traditional climatological analyses, but in two papers by Maunder, Johnson, and McQuigg (1971a and b) a 'soil' moisture index was developed and applied to engineering data, to estimate conditions suitable for work in the road construction industry. It is considered appropriate that this important study should be given due space in this book; the original studies are edited to give prominence to these applied climatological questions.

Specifically, the analyses estimated (a) the hours of construction time available in various calendar periods; (b) the frequency of the different kinds of 'work-weeks'; (c) the relation between the types of work-weeks and the manpower and machinery requirements for specific road construction activities; and (d) the development and application of an index designed to estimate progress on construction projects on a regional basis. The analyses indicated that, given a reasonably precise translation of weather information into simulated operational values, the functional relationships needed to convert the simulated operational data into quantitative expressions in terms of costs could be developed.

2. The effect of weather on road construction: A simulation model

(a) INTRODUCTION

In a specific study (for complete details see Maunder, Johnson, and McQuigg 1971a) the effects of climate variables on the highway construction industry were estimated through their influence on working conditions during the main construction months. Data from two construction projects were combined with a soil moisture index to obtain conditions under which construction activities could proceed. This relationship was used to generate an experimental series of working conditions based on the available weather data. The resulting series of simulated working conditions were then assessed with regard to their potential as aids to planning and scheduling highway construction projects.

(b) Construction and meteorological data

Specific road construction operational data, for a road construction job carried out in Central Missouri, were obtained from two sources: the Missouri State Highway Commission and two private contracting companies. The operational data were kept primarily for accounting purposes, but they did contain reasonably good information on the quantity of material moved per machine and per operator per week during the four-year period in which the job was carried out. In addition, the operational records indicated the type of work, if any, performed each day. From these data, each day during the sample four-year period was classified into one of the following three categories by the resident engineer on the construction project: (1) a full workday, (2) a no-work day, or (3) a partial workday. Saturdays, Sundays, and holidays were excluded unless work happened to be done on those days.

Both of the road construction projects studied were located a few kilometres southwest of Jefferson City, Missouri. The nearest climatological station is located about 3 km from the construction site, at Lincoln University. Initially, daily precipitation data for the period from 1 January 1966 to 30 June 1969 were used because this covered the period for which the road construction data were available. Later, daily precipitation data from Jefferson City for the 50-year period beginning with 1 January 1918 were used in obtaining the experimental series based on the observed road construction data. Daily soil moisture measurements and climatological observations from the University of Missouri Atmospheric Science Department station near Columbia were also used in developing the 'soil' moisture index used in the analysis.

(c) Soil moisture index model

Development of the soil moisture index model to be applied to the road construction industry required consideration of two types of information. First, it was necessary to obtain information on daily soil moisture for a period comparable to the period over which the construction data were available. This information was obtained from measurements made with a neutron meter and a weighing lysimeter. Secondly, the moisture data was combined with information on the ability to operate heavy equipment in adverse soil moisture conditions, or the 'trafficability' of equipment. Ideally, this second type of information would have come from the company and the highway department records on the construction projects, but the precision of the records did not permit the refined estimates necessary for relating to trafficability. As an alternative, experimental information developed by the US Department of Agriculture and US Army Corps of Engineers (1959) for relating soil moisture and trafficability were employed. These data, modified according to the available information from the construction projects, are presented in Table VI.16.

Daily soil moisture index values were computed using the following relationship:

$$SM(n) = SM(n-1) + PRECIP(n) - LOSS(n) \quad (1)$$

with the constraint that $SM(n) \leq SM(max)$.

The soil moisture index for day n − 1 is SM(n − 1); SM(n) is the soil moisture index for day n; PRECIP(n) is the observed precipitation for day n; LOSS(n) is the index of soil moisture loss using the appropriate season and

Table VI.16 Index of daily soil moisture losses (in.)[a]

	Seasons								
Day of decline (n)	*Summer (June 11–September 30)*			*Transition (October 1–December 9, April 16–June 10)*			*Winter (December 10–April 15)*		
	A	*B*	*C*	*A*	*B*	*C*	*A*	*B*	*C*
1	0.20	0.11	0.06	0.16	0.11	0.06	0.12	0.06	0.03
2	.16	.09	.05	.13	.09	.05	.10	.05	.02
3	.13	.08	.04	.11	.07	.04	.08	.04	.01
4	.11	.07	.03	.09	.06	.03	.06	.03	.01
5	.09	.06	.02	.07	.05	.02	.05	.02	.01
6	.07	.05	.01	.06	.04	.01	.04	.01	.01
7	.06	.04	.01	.05	.03	.01	.03	.01	.01
8	.05	.03	.01	.04	.02	.01	.02	.01	.01
9	.04	.02	.01	.03	.01	.01	.01	.01	.01
10	.03	.01	.01	.02	.01	.01	.01	.01	.01
11	.02	.01	.01	.01	.01	.01	.01	.01	.01
≥12	.01	.01	.01	.01	.01	.01	.01	.01	.01

Source: After Maunder, Johnson, and McQuigg (1971a).
Notes:
[a] The values in this table are believed to be representative of soil moisture losses from a bare silt loam soil. Applications of this model to areas with different soil types would require revision of this table.
A If soil moisture index of previous day is >1.20
B If soil moisture index of previous day is 0.60–1.20
C If soil moisture index of previous day is <0.60

the previous day's soil moisture index (using Table VI.16); and SM(max) is the upper limit for the soil moisture index. In the case of the specific road construction project, SM(max) was set at 1.80 in. (46 mm), which is about the maximum available soil moisture in the top 12 in. (305 mm) of soil.

The selection of a soil moisture loss value to be taken from Table VI.16 proceeded as follows. On days when precipitation was greater than or equal to the maximum soil moisture loss for the particular column, the actual soil moisture loss for that day was considered to be the maximum for the appropriate column. On subsequent days, if the precipitation was less than this maximum amount, the soil moisture loss (from the table) was entered for n = 1, 2, . . . 12, depending on the number of days since the daily precipitation equalled or exceeded the maximum soil moisture loss.

Results of some typical calculations of the soil moisture index are given in Table VI.17. This shows, for example, that on June 2, 1966 the soil moisture index at the *end* of the day was 1.20 in. (30 mm), implying that the maximum soil moisture index loss on the *following* day would be 0.11 in. (2.8 mm) (see column B of the transition season, Table VI.16). Similarly, the loss on June 5 was 0.07 in. (1.8 mm) (the third day of decline in moisture column B), giving

Table VI.17 Examples of soil moisture index computations

1966 date	*Soil moisture index previous day (in.)*	*Maximum possible loss (in.)*	*Precipitation on day n (in.)*	*Is precipitation ≥ maximum loss?*	*Actual soil moisture loss (in.)*	*Soil moisture index on day n (in.)*
May 24	1.56	0.16	1.61	yes	0.16	1.80
May 25	1.80	.16	0.00	no	.13	1.67
May 26	1.67	.16	.00	no	.11	1.56
May 27	1.56	.16	.00	no	.09	1.47
May 28	1.47	.16	.00	no	.07	1.40
May 29	1.40	.16	.00	no	.06	1.34
May 30	1.34	.16	.00	no	.05	1.29
May 31	1.29	.16	.00	no	.04	1.25
June 1	1.25	.16	.00	no	.03	1.22
June 2	1.22	.16	.00	no	.02	1.20
June 3	1.20	.11	.17	yes	.11	1.26
June 4	1.26	.16	.00	no	.13	1.13
June 5	1.13	.11	.00	no	.07	1.06
June 6	1.06	.11	.00	no	.06	1.00
June 7	1.00	.11	.18	yes	.11	1.07

Source: After Maunder, Johnson, and McQuigg (1971a).

a soil moisture of 1.06 in. (26.9 mm) at the end of June 5. A similar day-to-day analysis of the soil moisture index for all days in the 50-year period from January 1918 to December 1967 was made.

(d) Concept of workdays

The series from the soil moisture index model were then combined with the data from the construction projects to produce a series of simulated working conditions. Four 'workday' categories were defined by the engineers of the Missouri State Highway Department, namely: a holiday, Saturday, or Sunday, when no work was to be done (symbolized by 0); a normal workday without any 'weather' restrictions (symbolized by 1); a no-work day due in the main to weather/soil moisture conditions (symbolized by 2); and a partial or restricted workday (symbolized by 3). Such a classification of days into 1, 2, and 3 depends on a number of things including the type of work being carried out. Nevertheless, the 'workday index' for the main road construction period of April to November appeared to be reasonably representative of working conditions as they applied to road construction.

To apply these classifications of workdays to the daily climate record, a preliminary analysis was made by comparing the daily precipitation and the computed soil moisture index for several hundred days in the 1965–8 period with the *actual* workday classification for the two road construction jobs. The analysis showed that in most cases it was possible to estimate the workday classification by considering the daily precipitation and the computed soil moisture index. In brief, workdays were defined in terms of the precipitation and soil moisture as follows:

(1) If soil moisture was 1.79 or greater, the workday was classified as 2 (no work).
(2) If the soil moisture was within the range 1.60 to 1.78, the workday was classified as 3 (partial work), unless the precipitation for the day was greater than twice the maximum soil moisture loss as given in the appropriate column of Table VI.16, in which case it was classified as 2 (no work).
(3) If the soil moisture was more than 1.40 and less than 1.60, the day was classified 2 (no work) if the precipitation for the day was greater than twice the maximum soil moisture loss as indicated by the appropriate column of Table VI.16; or 1 (full work) if the day was the third (or more) daily decline in the soil moisture index. If neither of these conditions applied, the day was classified 3 (partial work).
(4) If the soil moisture index was 1.40 or less, the day was classified as 1 (full work), unless the precipitation was greater than 0.4 times the maximum soil moisture loss for that particular moisture zone, in which case it was classified 3 (partial work).

Using this procedure the soil moisture series was translated into a series of values indicating working conditions. In Fig. VI.5 the soil moisture index is plotted along with the corresponding workday classification.

(e) Markov chain probability model

The output of the daily values of the work index for a 48-year period is voluminous in its raw form, but some of the patterns that emerge can in the first instance be portrayed in terms of monthly or seasonal averages. In addition, the information can be used in a summarized form in terms of expressions for initial and transitional probabilities of Markov chains. These estimates are particularly useful in viewing the persistence of sequences of favourable or unfavourable working conditions.

A first-order Markov chain probability model was used by Feyerherm and Bark (1965, 1967) to study the persistence of weather patterns. They defined two classes of days, 'wet' and 'dry', a day being classified as wet if the precipitation on that day exceeded the threshold value. This method was applied to the soil moisture/workday model by combining the classifications and defining an 'F' day as one which *would* allow full road construction work (type 1 day), and an 'N' day as one which *would not* allow full work (type 2 or

FIGURE VI.5 Graphical relationship between soil moisture index and work classification at Jefferson City, Missouri, USA, summer 1966

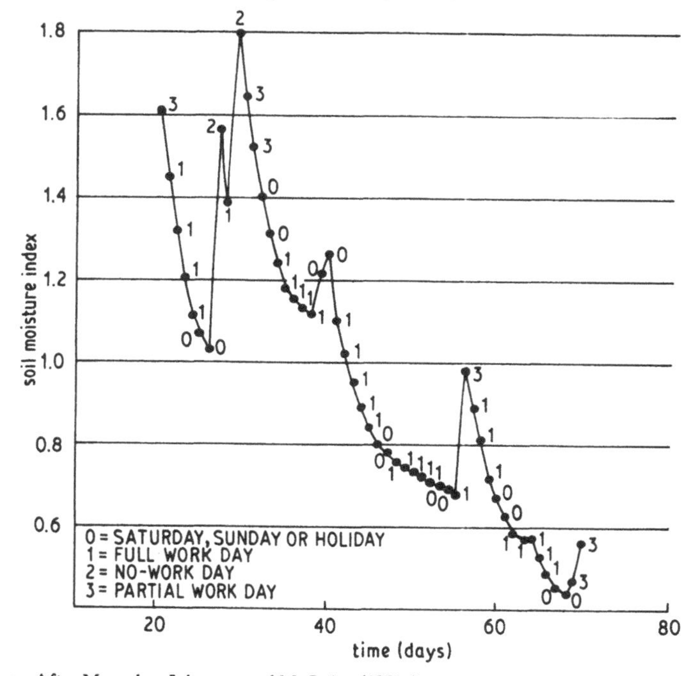

Source: After Maunder, Johnson, and McQuigg (1971a).

3 day). As discussed in the original paper (Maunder, Johnson, and McQuigg 1971a), various tests indicated that the probability of a sequence of full workdays and non full workdays expressed in terms relevant to the road construction project could be estimated from a Markov chain probability model of at least order two. Since a given day can be either wet or dry, there are thirty-two different possible workday sequences in a five-day work week. Assuming a second-order Markov process and using the computed probability values from the original paper, the probability of each of the thirty-two possible sequences were computed. The results (in the form of the probability that there will be at least n full workdays in a five-day work week for various periods) shown in Table VI.18 conform to the conventional ideas regarding the superiority of July, August, and September (in the United States) as months for road-building activity. It is considered that such information, in addition to its usefulness for planning in its present form, could be employed as input data for more sophisticated types of management decision models.

(f) Frequency of full and partial workdays and no-work days

The series of workdays obtained from the computed soil moisture index already described is now summarized. The data in Table VI.19 for the

Table VI.18 Probability that there will be at least *n* full workdays in a 5-day work-week chosen at random from the 25-day period shown in the first column

	Number of full workdays in 5*				
25-day period ending	*5*	*4*	*3*	*2*	*1*
March 21	0.25	0.44	0.58	0.71	0.83
April 15	.33	.47	.61	.73	.83
May 10	.16	.35	.52	.68	.80
June 4	.15	.25	.44	.64	.82
June 29	.23	.37	.53	.68	.83
July 24	.44	.65	.81	.91	.96
August 18	.40	.63	.79	.89	.95
September 12	.52	.69	.80	.89	.95
October 7	.30	.52	.66	.76	.85
November 1	.32	.61	.75	.84	.90
November 26	.38	.62	.79	.90	.96

Source: After Maunder, Johnson, and McQuigg (1971a).
Notes:
* For example, the probability of having at least 4 full workdays is computed as the summation of the following probabilities: $P(F,F,F,F,F) + P(N,F,F,F,F) + P(F,N,F,F,F) + P(F,F,N,F,F) + P(F,F,F,N,F) + P(F,F,F,F,N)$.

Table VI.19 Frequency of simulated operational data (excluding weekends) for Jefferson City, Missouri, USA, April–October, 1918–67

	Full workday		*No-work day*		*Partial workday*		
Month	*Number of days*	*% of days*	*Number of days*	*% of days*	*Number of days*	*% of days*	*Total number of days*
April	553	52	144	13	374	35	1071
May	587	53	159	14	361	33	1107
June	601	56	152	14	319	30	1072
July	800	72	87	8	219	20	1106
August	775	70	92	8	241	22	1108
September	720	67	116	11	236	22	1072
October	748	68	102	9	256	23	1106

Source: After Maunder, Johnson, and McQuigg (1971a).

months April to October are for all Monday to Friday work weeks in the 50-year period considered. The seasonal pattern is clear; there is a maximum number of full workdays in the period July to October and a minimum number in April and May. For example, in April, 52 per cent of the days were 'full work', compared with 13 per cent which were 'no work' and 35 per cent which were 'partial work'. These data may be contrasted with those for July, where the corresponding percentages are 72, 8, and 20. Moreover, the combined months of April and May in the 50-year period 1918–67 had only 1140 full workdays compared with 1575 such days in July and August. A similar analysis was made for all of the days in the period *including* Saturdays and Sundays, this analysis showing that there is little difference between the values which result from the two assumptions regarding possible work weeks.

(g) Applicability to an index of workability

One direct application of the information produced by the simulation model is the calculation of an 'index of workability', which is essentially an index of whether work can or cannot be done. The actual 'workability index' employed was calculated by computing the number of work hours and expressing this as a percentage of a total possible number of work hours. For this analysis a full workday was considered to be 8 hours, a no-work day 0 hours, and a partial workday 4 hours. These values are arbitrary and could be varied according to circumstances, but they are a reasonable approximation of what actually occurs.

The workability index was computed on a *daily* basis, considering all days including Saturdays and Sundays, for Jefferson City for the 50-year period

1918–67. The extremes, means, and standard deviations for the months April to October are given in Table VI.20. These data show that, on the average, 70–80 per cent of the total possible time could have been worked, the monthly average in the construction season varying from a low of 69 per cent in April and May to a high of 82 per cent in July. The highest percentages for the 50-year periods were in all cases above 90 per cent, and in seven months they were at least 97 per cent. By contrast, in April 1922, the index indicated that only 38 per cent of the possible work *could* have been done.

The actual workability index for two selected years (1936 and 1924) is shown graphically in Fig. VI.6, 1936 being a dry year for road construction, and 1924 a wet year. Similar graphical analyses were obtained for all years, and a graph showing the variation in the work index in May for the 41-year period 1918–58 is shown in Fig. VI.7. The graph shows the variation in the work that could have been done using all days including weekends, and the difference between a wet May and a dry May is quite evident.

The simulation model discussed in this paper is based on a comparatively short 4-year record for two road construction projects, and a comparatively long 48-year record of daily meteorological observations. The application of the model to generate 'experience' over the 48-year period appeared to provide useful information, and it seems reasonable to suggest that such an analysis is potentially more useful to managers of road construction projects than either a short period of operational records or a long period of weather records taken separately.

The potential application of this and similar simulated series of workdays

Table VI.20 Workability index for Jefferson City, Missouri, USA, 1918–67

Month	*Highest (year)*	*Mean (s.d.)*[a]	*Lowest (year)*	*Range*
April	0.92 (1956) (1959)	0.69 (0.12)	0.38 (1922)	0.54
May	0.94 (1934)	0.69 (0.12)	0.42 (1935) (1938)	0.52
June	1.00 (1936)	0.71 (0.15)	0.45 (1935)	0.55
July	0.97 (1940)	0.82 (0.09)	0.60 (1961)	0.37
August	0.97 (1960)	0.80 (0.09)	0.61 (1952)	0.36
September	0.98 (1939) (1940) (1950)	0.79 (0.12)	0.48 (1926)	0.50
October	0.98 (1964)	0.79 (0.11)	0.39 (1941)	0.59

Source: After Maunder, Johnson, and McQuigg (1971a).
Note: [a] Standard deviation.

FIGURE VI.6 Comparison of work index values for the wet year 1924 and the dry year 1936, April–October

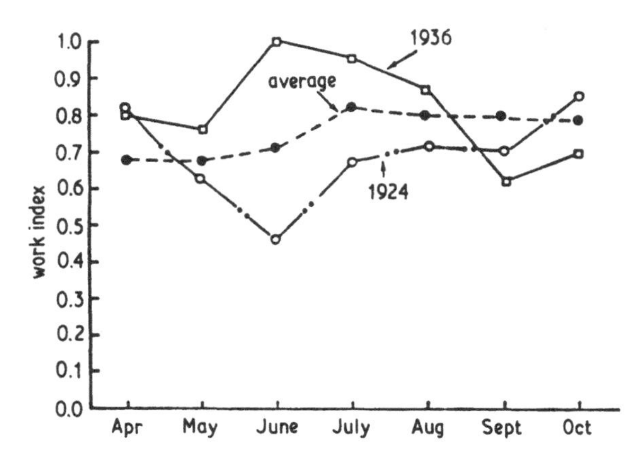

Source: After Maunder, Johnson, and McQuigg (1971a).

to economic problems resulting from road construction activities is apparent. In particular, with more complete operational data and with further refinements in the model, a long series of simulated workability index values could be used in long-term planning, bidding strategy, and contract negotiations.

FIGURE VI.7 Work index value variation in May at Jefferson City, Missouri, USA, May–July 1918–58, if weekends are included as workdays

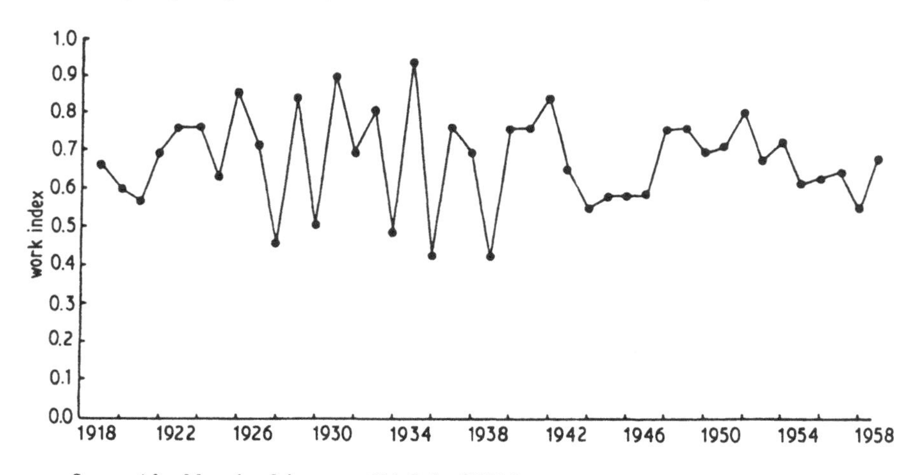

Source: After Maunder, Johnson, and McQuigg (1971a).

3. The effect of weather on road construction: Applications of the simulation model

(a) Introduction

The previous section discussed the application of a soil moisture index to engineering data to estimate conditions suitable for work in the road construction industry (see Maunder, Johnson, and McQuigg 1971a). In this section, the series of values generated by the model are examined for potential usefulness to the road construction industry (see Maunder, Johnson, and McQuigg 1971b).

Four specific applications of the experimental series are considered: (1) the hours of construction time available in various calendar periods; (2) the frequency of characteristic types of work weeks; (3) the relation between types of work weeks and the manpower and machinery requirements for specific road construction activities; and (4) the application of the index to the estimated progress on construction projects on a statewide basis. The purposes of these applications were to illustrate the advantages of simulation modelling techniques for providing useful information to industry and government. They also indicated the types of additional benefits that could be derived from such modelling procedures in the presence of more refined information on soil moisture, and on the standard physical construction processes of grading, scraping, excavation, paving, etc. associated with road building.

(b) Construction time available

Knowledge of working conditions is important for planning all phases of the road-building industry. For example, construction firms need information on the number of working days expected to be available, both for bidding on road construction contracts and for the scheduling of machinery and manpower. Contract-letting officials also need such information to anticipate properly the scheduling of funds for payment as work progresses. Questions for which the road construction industry needs answers therefore include:

(1) How many hours of work are available during the normal construction season, and what kind of variation can be expected?
(2) If the construction project is started at a particular time, and a certain number of hours or days is required for its completion, can the job be completed by a pre-specified date?
(3) If several jobs are initiated at different starting dates, and all require an equal amount of work, when will each one be completed?
(4) If a job is started on a particular date, and the succeeding period is 'unusually wet', what is the probability of being able to complete the job in the prescribed time?
(5) At what point in a 'wet' road-building season does it become virtually impossible to complete a specific construction job on schedule?

To provide information related to the answering of these questions, the daily series of working conditions was converted into an index expressed in hours of work-time. As noted in the previous section, the conversion was accomplished by weighting 'full workdays' by 8 hours, 'partial workdays' by 4 hours, and 'no-work days' by 0 hours. The resulting index appeared to be the most appropriate option available, given the data limitations. Other work-time indices could, of course, be developed using weights suggested by more precise construction project records.

The number of hours available for work during specific periods can be summarized in a number of ways. The options presented here were selected because of their direct bearing on the previously listed questions. Results of the first summarization are shown in Fig. VI.8. Beginning with day number 50 (that is 19 February) in each of the sample years 1918–65, the cumulative number of hours of working time were recorded for each day throughout the road construction year. The curve in Fig. VI.8 labelled 'mean' is the average value of cumulative work-time beginning on 19 February, and includes weekends and holidays as well as the regular work week.

FIGURE VI.8 Cumulative work hours (including weekends) for Jefferson City, Missouri, USA (1918–65)

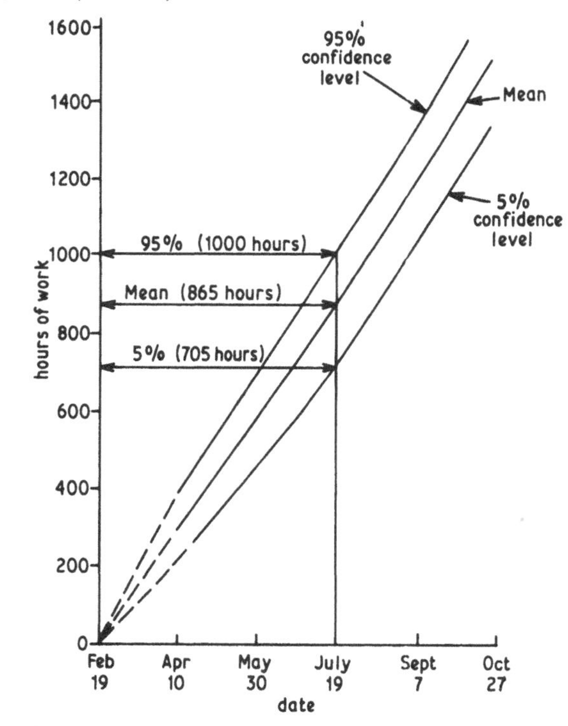

Source: After Maunder, Johnson, and McQuigg (1971b).

As an example of the applicability of the information presented in Fig. VI.8, consider the 200th day of the year (19 July). A work project started on the 50th day of the year (19 February) would have on *average* about 865 hours of working time logged by 19 July. In many instances, however, construction projects do not operate on a 7-day week or have options as to whether they may operate 5 or 7 days a week. Table VI.21 was produced, using the work-time index, to illustrate the influence of 5- and 7-day work weeks and the calendar date on available work-time. The data shown here for every fifth year of the 50-year period indicates the cumulative available work-time for time periods during the main road-construction period in Missouri, beginning on 10 April. The effect of having 7 days from which to select a work week of 40 hours is apparent from the information shown in the table, the 7-day work week providing a cumulative work-hour value which is about 45 per cent larger than the 5-day work week for the 10 April to 26 October period.

Another useful method of assessing the influence of working conditions on work-time is to assume a fixed amount of work-time (say 1000 hours) and further assume a variable starting time. In any construction period there will also be a 'run of days' that will be the shortest period of time during which the required work can be accomplished. In addition, there will be a 'run of days' that will be the most unfavourable for road building, and thus represent the longest period of time that would be required to complete the desired amount of work. Table VI.22 contains sample values of the type just described for every fifth year in the period studied. These data suggest that a road-building project in central Missouri requiring 1000 hours of work is

Table VI.21 Sample values of cumulative hours of worktime for road construction at Jefferson City, Missouri, USA. (Data in parentheses are for a Monday–Friday work-week: data not in parentheses are for a 7-day work-week)

Year	*April 10–May 29*	*April 10–July 18*	*April 10–September 6*	*April 10–October 26*
1920	232 (164)	568 (392)	884 (628)	1188 (852)
1925	300 (228)	580 (408)	904 (636)	1172 (840)
1930	352 (256)	700 (504)	1020 (712)	1328 (916)
1935	204 (124)	444 (300)	768 (536)	1148 (850)
1940	288 (208)	620 (436)	936 (668)	1312 (936)
1945	212 (148)	460 (324)	800 (534)	1052 (764)
1950	304 (232)	624 (472)	952 (696)	1320 (960)
1955	264 (185)	552 (376)	924 (620)	1244 (800)
1960	292 (212)	612 (487)	996 (720)	1304 (980)
1965	296 (204)	548 (372)	808 (584)	1124 (784)

Source: After Maunder, Johnson, and McQuigg (1971b).

Table VI.22 Sample values of the range of cumulative calendar days required to complete 1000 hrs of worktime (weekends included). These values are based on rainfall data for Jefferson City, Missouri, USA

	Minimum time		*Maximum time*	
Year	*Starting date*	*Calendar days to completion*	*Starting date*	*Calendar days to completion*
1920	July 5	157	March 26	176
1925	July 9	158	June 2	171
1930	February 27	141	July 14	155
1935	June 30	149	February 25	194
1940	June 30	140	February 27	168
1945	July 11	155	February 26	204
1950	June 13	142	March 11	166
1955	July 25	140	February 27	169
1960	June 14	138	February 20	164
1965	July 11	157	April 11	181
Sample average	June 20	148	March 30	174

Source: After Maunder, Johnson, and McQuigg (1971b).

most likely to be completed with the least delay if it is started in late June or early July. At the other extreme, the same job started in late February or early March is likely to experience the greatest delay. The table also shows an 'extreme' range in the number of cumulative calendar days required to complete 1000 hours of work-time from 138 days in 1960 with a 14 June start date, to 204 days in 1945 with a 26 February start date.

(c) Machine and labour requirements for construction activities

In addition to recording daily information regarding working conditions, resident engineers on highway projects in Missouri (where the original study was made) include in their reports descriptions of daily construction activities and completion estimates for common and rock excavation, finishing, and the various operations associated with paving. Common and rock excavation are items of major importance in highway construction projects. They are also the construction activities which are in process during a major proportion of the completion time. Because of the importance of the common and rock excavation activities, the length of period over which they are active, the varying weather conditions under which they are attempted, and the availability of data on which to base the estimates, these two activities were selected for an 'input-output' analysis of the machinery and labour requirements for 1000 cubic yards (765 cubic metres) of common and rock excavation.

The relevant 'input-output' values for common and rock excavation were calculated by combining completion estimates from the resident engineer's weekly reports with the foreman's weekly time sheets. The weekly records used for the calculations were selected on the basis of variability of weather-related working conditions and homogeneity with respect to the types of activities occurring within the week.

The first selection criterion presented no problems since the time period used in the study provided ample variations in weather conditions. The second selection criterion, however, provided some difficulty since construction activities *other* than common and rock excavation were in progress during the weeks for which information was obtained from the time sheets. This made it difficult to attain the desired amount of precision in assigning input values to the common and rock excavation activities. The weeks included in the final analysis were therefore those for which such errors appeared to be of least consequence.

A finer classification would probably have yielded input-output values with less variability. However, attempts at refining the classifications using available information on the resident engineer's reports were unsuccessful. The implication of this is that the values given, although useful in demonstrating the applicability of the workday model based on soil moisture conditions, should be regarded only as providing an indication of their relative magnitudes.

The computed 'input-output' values for the weeks used in the final analysis relating to the machinery and labour requirements for one of the categories, common excavation, is shown in Table VI.23. The data given are the values (in hours) of the various inputs required for each 1000 cubic yards (765 cubic metres) of rock excavation. Estimates of the amount of excavation per week are also included. Input categories are highly aggregated, as indicated by the items included in the related table notes.

Table VI.24 summarizes the information according to a work week classification, a Monday to Friday week with two or less working days being classified as a marginal work week, and weeks with more than two working days being classified as full work weeks. Ideally, the information should have been subdivided into 1, 2, 3, 4, and 5 working days per week, but this was not possible because of data limitations. A comparison of the 'input' categories in Table VI.24 for the two work week classifications shows that the number of working days, and presumably the various weather factors, do have an influence on the efficiency of construction operations (but note the large values of the variances). For example, an average of 5.8 hours of bulldozer time was required to move 1000 cubic yards (765 cubic metres) of common excavation during marginal work weeks, whereas an average of only 4.6

Table VI.23 Weekly machinery and labour requirements per 1000 cubic yards of common excavation

Amount excavated† *(10^3yd^3)*	*Common labour (hr)*	*Bulldozers*[a] *(hr)*	*Motor graders*[b] *(hr)*	*Tractor scrapers*[c] *(hr)*	*Other*[d] *(hr)*	*Full days worked*†
8.0	3.4	2.1	0.0	3.1	2.0	0*
4.2	2.2	8.2	2.5	9.0	0.0	0
3.5	0.5	3.1	2.7	9.7	4.2	1
3.2	0.9	3.1	3.1	11.0	1.8	1
1.2	20.8	11.2	6.4	14.4	0.0	2
22.0	1.2	7.4	0.0	4.1	3.1	2
28.2	1.0	4.5	0.8	3.3	1.0	3
38.0	1.6	6.7	2.0	5.0	1.0	4
29.7	10.2	7.0	1.2	4.8	0.0	4
28.0	0.3	2.7	0.0	6.7	2.2	4
15.5	2.8	2.9	2.2	5.2	0.0	4
18.5	1.0	2.3	1.7	6.2	0.0	5
8.4	8.3	7.0	3.3	12.8	0.7	5
26.0	6.6	4.2	2.3	5.6	1.1	5
11.0	2.3	4.1	4.0	8.0	1.8	5

Source: After Maunder, Johnson, and McQuigg (1971b).
Notes:
[a] Types D-6, D-7, D-9.
[b] No. 12.
[c] Models 631 and DW-21 (rubber-tyred).
[d] Euclid end dumps, 8-D and 95 N.W. shovel, ¾-yd cranes, loaders, 2-ton tandem trucks.
† As estimated by highway engineer.
* Less than 1 full workday.

hours was required to move the same amount of common excavation in full work weeks. Similar savings of inputs are indicated for the common labour, motor grader, and tractor scraper categories.

Differences in the average costs of common excavation under the two work week classifications can be computed by multiplying the differences in input values by the rental costs for machinery and the hourly wages of operators, and such differences are quite substantial. As previously noted, the variances for the common excavation input categories shown in Table VI.24 are in many cases quite large and are to some extent a result of the broad input classifications. Nevertheless, the conclusions drawn from the means are supported by the quantity of common excavation carried out, and they show clearly that 'better' weather conditions allow more common excavation to be carried out with resulting decreases in the input requirements (per 1000 cubic yards) of both machinery and labour. By contrast, the data for rock excavation (not shown here) showed little relationship to the two working classifications. This difference in the relationship of the com-

Table VI.24 Estimated means and variances of machinery and labour requirements per 1000 cubic yards common excavation[a]

Working	*Common labour*[b]		*Bulldozers*[b]		*Motor graders*[b]		*Tractor scrapers*[b]		*Other*[b]	
days/week	*mean*	*variance*	*mean*	*variance*	*mean*	*variance*	*mean*	*variance*	*mean*	*variance*
⩽2	4.8	61.9	5.8	12.9	2.4	5.6	8.5	18.2	1.8	2.8
>2	3.8	13.0	4.6	3.4	1.9	1.5	6.4	7.4	0.9	2.1

Source: After Maunder, Johnson, and McQuigg (1971b).
Notes:
[a] Calculated from Table VI.23.
[b] Hours per 1000 cubic yds.

mon and rock excavation to weather is partly attributed to the type of material excavated and also to the greater significance of 'soil' moisture conditions to common excavation.

(d) INPUT-OUTPUT VALUES AND THE WORK WEEK SERIES

Input-output values estimated from the weekly data series can be used to expand the simulation model discussed in the previous section. The use of these values gives an insight into the types of information that *could* be obtained from such an approach, using the available weather records and more refined data about the physical processes of road construction. The input-output values for common excavation were utilized because of the observed differences in the values for the two types of work week. The method used was as follows:

(1) Means and variances of the output for the two types of work weeks (full workdays ≤ 2 and full workdays >2) were calculated from column 1 of Table VI.23.
(2) The means and variances so calculated were used to define truncated normal distributions for the weekly amounts of common excavation completed for the two types of work weeks assumed.
(3) The two output distributions were then linked to the work week information series for the 48-year period used in the analysis to produce estimates of weekly and annual amounts of common excavation completed.
(4) Input coefficients were multiplied by weekly outputs as a basis for calculating annual average coefficients.

The resulting information (Table VI.25) is arrayed and summarized for the 10 years in which the most excavation occurred, and the 10 years in which the least excavation occurred. The difference in the average of the amounts of excavation for the two groups of years is quite large; for example, an average of 902,000 cubic yards (690,000 cubic metres) of common excavation was possible in the 10 most favourable years, compared with an average of only 702,000 cubic yards (537,000 cubic metres) in the 10 least favourable years in the 48-year series.

Similarly, the different requirements of the common labour, bulldozers, motor graders, and tractor scrapers input classifications are also quite significant. An individual comparison of years *within* the two groups reveals that the amounts of excavation are remarkably close, a factor that could be anticipated from the broad classifications of the types of work weeks. More refined classifications would be expected to produce wider dispersion in input requirements. The results reported may therefore be regarded as conservative estimates of the differences between the two sets of 10-year records presented.

Data for a Monday to Friday work week were also computed and this

Table VI.25 Common excavation and input requirements for the 10 highest and 10 lowest amounts of excavation completed during the May 1 to September 1 period using Monday–Sunday

Group	*Amount excavated (10^3yd^3)*	*Year*	*Work-week types and numbers*		*Average input requirements*			
			≤2	*>2*	*Common labour*	*Bulldozers*	*Motor graders*	*Tractor scrapers*
10 yrs in which most common excavation occurred	974.9	1930	4	36	95.1	115.6	49.0	158.8
	931.9	1940	4	36	90.3	109.8	46.6	152.3
	915.4	1964	7	33	89.7	109.1	46.2	151.7
	914.1	1932	3	37	87.4	107.1	45.5	148.3
	903.1	1952	8	32	88.0	107.0	45.4	148.5
	886.1	1960	4	36	85.9	104.5	44.3	145.0
	884.4	1949	9	31	87.1	105.9	44.9	147.4
	876.0	1955	8	32	86.3	104.8	44.4	146.0
	868.8	1935	11	29	86.4	105.0	44.4	146.5
	865.0	1950	6	34	83.7	101.8	42.5	141.1
Average	902.0		6.4	33.6	88.0	101.1	45.3	148.6
Standard deviation	33.6		2.6	2.6	3.2	3.7	1.8	4.6
10 yrs in which least common excavation occurred	642.5	1929	10	30	63.7	77.4	32.8	107.9
	674.2	1965	10	30	66.6	80.9	34.3	112.8
	681.4	1926	12	28	66.6	81.0	34.3	112.5
	684.0	1921	12	28	68.3	82.8	35.1	115.9
	704.1	1924	10	30	69.6	84.6	35.8	118.0
	718.1	1944	10	30	71.0	86.3	36.5	120.3
	719.5	1945	14	26	71.7	87.1	36.9	121.7
	730.1	1948	11	29	72.4	88.0	37.3	122.7
	733.1	1958	9	31	72.6	88.3	37.4	123.1
	737.8	1933	9	31	72.4	88.0	37.3	122.4
Average	702.5		10.7	29.3	69.5	84.4	35.8	117.7
Standard deviation	31.1		1.5	1.5	3.1	3.6	1.5	5.2

Source: After Maunder, Johnson, and McQuigg (1971b).

showed that an average of 823,000 cubic yards (630,000 cubic metres) of excavation was completed in the 'top' 10 years in the Monday to Friday series, compared with 902,000 cubic yards (690,000 cubic metres) for the 'top' 10 years in the Monday to Sunday series. The differences between the numbers of full and marginal work weeks in the 10-year averages account for the relatively large change in the amount of excavation completed. Clearly, there are 'gains' to both contractors and government authorities from using the simulation type analyses discussed in this section.

(e) Extension to a 'State Workability Index'

Many of the weather-sensitive components of the road-construction industry are associated with the management of funding of particular construction projects, and relevant data concerning costs and expenditures for road construction were available for Missouri, through the Missouri State Highway Commission. Using such data, and the previously discussed information, an attempt was made to calculate a 'Workability Index' for the state of Missouri. The method, described below, is essentially based on the 'weighting' of weather and climate information discussed in Chapter V.

Missouri is divided into ten districts by the Missouri State Highway Department, records of expenses and disbursements being maintained for each district for each month. It seemed appropriate therefore to compute a daily soil moisture index for each district, and then to convert these values into district workability indexes. Three weather stations were selected for each of the ten districts, and the average daily soil moisture (for the period January 1968 to July 1969) for each district was computed, using the method described on page 213. Typical values for one day (day 105) in the period showed that the soil moisture index varied from 1.80 for District 1 to 1.34 in District 9.

A 'district workability index' (DWI) was computed from these soil moisture values using the following equation:

$$\text{DWI} = (D_1 \times 0) + (D_2 \times 1) + (D_3 \times 2) + (D_4 \times 3) + (D_5 \times 4) + (D_6 \times 5) + (D_7 \times 6)$$

where D_1, D_2, . . . D_7 are the number of days with soil moisture 1.80, 1.70–1.79, . . . <1.30 respectively.

The 'weights' 0, 1, 2, 3, 4, 5, and 6 in the DWI equation were chosen as being a first approximation of the number of hours of road construction that could be expected to be logged when the district daily soil moisture index was within the respective intervals. The workability index values, for each of the ten districts for the 18-month period February 1968 to July 1969, are given in Table VI.26. The table shows a variation from a high of 186 in July, August, October 1968, and May 1969 for District 6, to a low of 47 in June 1969 for District 4. With 186 as the maximum possible value, the low value represents only 26 per cent of the maximum possible, representing a substantial decrease in the available work-time.

The data in Table VI.26 were then weighted according to the relative importance of the various districts to the total Missouri expenditure on road

Table VI.26 District workability indexes: February 1968–July 1969 (Missouri, USA)

Month	*District* 1	2	3	4	5	6	7	8	9	10
1968										
Feb.	158	138	118	121	125	127	110	98	104	108
Mar.	181	181	183	153	101	118	79	66	68	77
Apr.	97	74	75	98	106	102	105	98	94	74
May	174	142	111	144	148	139	144	95	100	110
June	145	127	123	154	113	159	133	165	162	130
July	139	114	131	132	142	186	165	148	158	184
Aug.	125	158	154	136	121	186	120	108	161	185
Sept.	155	129	101	136	125	131	137	131	139	139
Oct.	142	168	153	148	153	186	147	137	128	152
Nov.	175	143	131	140	90	180	93	83	86	161
Dec.	133	182	147	163	138	164	143	142	125	102
1969										
Jan.	84	141	89	110	74	133	101	80	88	60
Feb.	111	151	88	108	84	109	121	85	91	112
Mar.	150	168	117	92	82	138	145	129	91	148
Apr.	87	65	78	82	75	103	70	61	73	67
May	118	123	127	134	147	186	148	166	176	140
June	104	52	67	47	52	140	120	108	180	124
July	84	94	95	71	105	75	127	116	160	172

Source: After Maunder, Johnson, and McQuigg (1971b).

construction. As a first approximation the value of work per district was used. The total road-construction expenditure in Missouri in 1968 (when the study was made) was $111 million, of which 29 per cent was used in District 4 but only 3 per cent in District 3. Because monthly district data on expenditures were not available, these percentages were used for all months for which a workability index was calculated. This assumes, of course, that the relative importance of road construction in the Missouri districts during the 18-month period did not change. Using the dollar values available, a monthly State Workability Index was calculated as follows:

$$SWI = (DWI_1 \times W_1) + (DWI_2 \times W_2) + \ldots + (DWI_{10} \times W_{10})$$

where: SWI = State Workability Index
DWI = district workability index (1–10)
W_{1-10} = weights for the 10 districts ($W_1 = 0.09$, $W_2 = 0.04$, $W_3 = 0.03$, $W_4 = 0.29$, $W_5 = 0.07$, $W_6 = 0.21$, $W_7 = 0.04$, $W_8 = 0.10$, $W_9 = 0.06$, $W_{10} = 0.07$).

Table VI.27 Missouri workability index and value of road construction completed

	1968		*1969*	
Month	*State workability index*	*Amount paid to contractors**	*State workability index*	*Amount paid to contractors**
April	96.0	9.3	81.6	9.9
May	135.0	10.2	150.6	16.1
June	148.6	9.2	96.0	15.9
July	152.3	12.8	96.4	17.7
August	147.9	14.2		
September	134.6	12.9		
October	154.7	16.3		

Source: After Maunder, Johnson, and McQuigg (1971b).
Note: * Millions of dollars.

The State Workability Indexes so obtained are shown in Table VI.27 together with the amount of money paid to road-construction contractors for the corresponding months. These sample values give some indication of a relationship during the road-construction season between the amount paid to the contractors and the computed workability index. However, it is reasonable to suggest that a closer relationship would be apparent if the weights for the districts were based on monthly rather than annual dollar values, or if a larger sample of daily rainfall values were chosen for each district.

(f) FUTURE RESEARCH GUIDELINES

Many years of precipitation data are readily available from a reasonably dense network of climatological stations operated by national meteorological services. These comparatively large samples of observations can provide the basis for useful statistical estimates of seasonal variations and short-period probabilities of the occurrence of certain favourable and unfavourable sequences of working days.

The previous discussion has shown that given a reasonably precise translation of weather information into simulated 'operational values', the relationships needed to convert the simulated operational data into quantitative expressions in terms of costs can be developed. In particular, the applications of the work index concept are indicative of the types of information that such a marriage of weather and climate data and operational data from road construction can produce. Although the particular implications of the applications described are limited by some subjectivity in the operational data and various assumptions as to the construction processes, the general implication regarding the potential pay-off of increased efforts in this area for the road-construction industry is clear.

A logical conclusion is that all participants in the road-construction industry could benefit from some additional deliberate research on the effects of weather on their specific activities. But more carefully designed records, perhaps obtained by sampling design to reduce cost and inconvenience, describing specific progress on construction projects are needed. In addition, on-site measurement of weather events and further efforts on the part of road-construction contractors to record information, in addition to that required for the purpose of cost accounting, are among the initial steps that could be taken. This would clearly allow the industry to identify better and measure quantitatively the impact that weather has on the industry, as well as the value of weather information to the industry.

G. Electric Power and Weather Relationships

1. An overview of early studies

Several studies have been made on the relationship between weather conditions and power consumption. Davies (1960), in an early paper, commented on grid system operation in Britain, and noted the difficulty of making accurate load estimates because the demand is very sensitive to changes in the weather. At around freezing point, for example, Davies noted that the increase in demand per 1°C sustained fall of temperature was at the time of the investigation as much as 290 megawatts (for comparison, this is equivalent to about 10 per cent of the *peak* winter load in New Zealand in 1985), and at lower temperatures this can rise to 400 megawatts with no signs of demand saturation. Other weather elements such as wind, cloud, fog, and precipitation also cause considerable variations in electricity demand. For example, Davies found that near freezing point, a 25 knot (46 km/h) wind increased the then demand in Britain by about 700 megawatts as compared with that on a calm day, this being equivalent (in power demand) to a drop in temperature of 2° to 3°C. Cloud also has an appreciable effect on the lighting load and can cause a variation in Britain of more than 1200 megawatts. For example, dark clouds over London often increase the demand by 350 megawatts, after the effect of temperature has been accounted for.

Several additional studies have been published on the effect of weather on various aspects of electric power production, transmission, and consumption, including early studies by Dryar (1949), Stephens (1951), Nye (1965), and Harris (1964). These studies showed clearly the important effect that weather and climate conditions can have on electric power consumption, and most studies emphasize the effect of cold winds and cloud as factors in increasing consumption. However, an increasing summer demand for power is occurring in those parts of the world where air-conditioning is common. An interesting side-effect of this development was explored by Johnson, McQuigg, and Rothrock (1969) who considered the theoretical decrease of summer daytime temperatures over the mid-west of the United States through the creation of contrails, and its possible effects on electric power consumption and production costs. Further details of this study are given in Section 5 below.

Four additional contrasting reports on the effect of temperature on electric power consumption in the early 1970s are of interest. The first, by Hamilton (1970), stated:

> This could be a harsh winter. Even if the weather is no worse than usual, Britain's power supplies for the coming months have a distinctly unhealthy look about them, and it seems unlikely that we shall get by unscathed. The last mild winter was in 1966–67 and since then Britain has sniffled and slithered through three cold winters, with the customary expressions of pained surprise at the unusual conditions, and the customary grumbles about the lack of preparation for them. Yet colder winters are now the rule, rather than the exception. Britain appears to have passed through a peak of warmth somewhere around the mid 1940s, and since then has been getting colder.

In the United Kingdom, therefore, a then (1970) current problem was the relative 'severity' of winters with a consequent possible shortage of electric power at certain peak periods.

The second aspect, from the *Wall Street Journal* of 23 January 1970, stated that, responding to grim predictions of a 'power crisis' in New Jersey, the General Public Utilities Corporation said it would rush into construction three combustion turbine generating plants capable of producing a total of 400 megawatts of electricity. The 'power crisis' referred to was again weather related, namely that the growing number of air-conditioners and other electrical appliances in the United States were (and still are) consuming critical amounts of electric power during the very hot days of summer.

The third report associating temperature variations with electric power demand concerns New Zealand, as reported in the *Evening Post* of 14 August 1971:

> The New Zealand annual maximum demand for electricity increased by 6.1% for the year ended 31 March (1971), compared with the 8.2% forecast. Detailed analysis of the weather conditions at the time of national peaks suggested, however, that in a year of average temperature, the increase would have been 7.6%, states the Annual Report of the Committee to Review Power Requirements tabled in the House [of Representatives] this week.

The inference from this statement is clear, namely that the probability of various temperature trends should be considered in any long-term planning of New Zealand's power requirements. Indeed, as the report in the *Evening Post* noted: 'The economic planning of the generation, transmission, and distribution of electricity required careful forecasting of the future requirements. Serious effects could result from poor forecasting.'

In the fourth report, an analysis of the relationship between the *monthly* temperature departures for New Zealand (weighted according to the distribution of the human population) and the random oscillation of the total

FIGURE VI.9 Monthly New Zealand electricity generation and associated weighted temperature departures, April–September, 1961–70

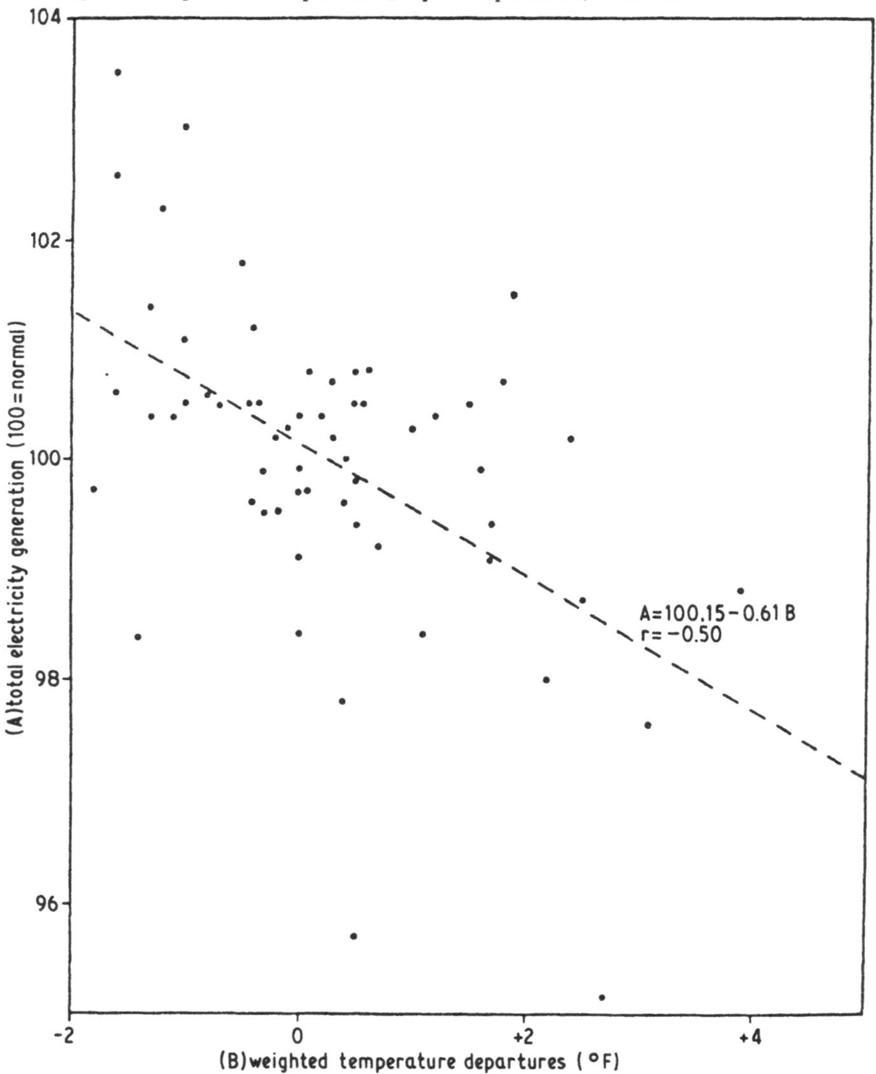

Source: After Maunder (1974a).

monthly electricity generation in New Zealand (as determined by the Department of Statistics, after allowance for seasonality, trend, and 'work' days) was assessed by the author (see Maunder 1974a). The analysis (see Fig. VI.9) covered the sixty April to September months in the 1961–70 decade, and indicated that 25 per cent ($r = -0.50$; significant at the 0.1 per cent level) of the variance in the random oscillation of New Zealand monthly electricity generation was associated with the departure of the weighted monthly New Zealand temperature indices from the 1931–60 normals. A similar analysis for other periods and for other similarly developed countries in middle or high latitudes would no doubt indicate a similar close relationship between winter temperatures and the demand for electric power during the winter period, and clearly if other weather parameters are included (such as wind, illumination, humidity) then the relationships would be even closer.

2. Electric power consumption and temperatures: A New Zealand example

(a) THE SETTING

In New Zealand, the natural resources of water and geothermal steam are harnessed to provide a substantial proportion of the country's electricity requirements. However, an increasingly important source of electricity is produced through coal, oil, and gas fired stations. For instance, in 1969/70, 1470 million kWh of electricity were generated by the oil, gas, and coal fired stations, compared with 1243 million kWh generated by geothermal steam, and 10,000 million kWh produced at the hydro stations. The year cited was extremely dry, and this necessitated using the non-hydro stations for long periods. More recently, the percentage of non-hydro electricity has increased, and in 1981/82, 18,829 million kWh of electricity was generated by hydro stations, 3093 million by geothermal steam stations, and 316 million by oil, coal, and gas stations. Thus in 1981/82 15.3 per cent of New Zealand electricity was produced by non-hydro means, and 1.4 per cent was produced by oil, coal, or gas stations.

The percentages of non-hydro and non-geothermal power are particularly important, because the cost of producing electricity at non-hydro stations is several times the cost of producing electricity at hydro stations. For example, data from the Electricity Division of the Ministry of Energy showed that the cost per kWh in 1968/69 was 0.32 cents for power from hydro stations and 0.95 cents for power from 'thermal' stations, whereas in 1981/82 the respective costs per kWh were 1.1 cents for power from hydro stations and 5.0 cents for power from 'thermal' stations. It is evident that the significance of these differences will become more important in the future as the proportion of oil, gas, coal and possibly nuclear power station generation capacities increase relative to the hydro station generation capacities.

The demand for power in New Zealand is related to a number of factors. One of these factors is temperature, and discussions with the Electricity Division of the Ministry of Energy have shown the importance of being able to predict 3 to 30 hours in advance the demand for electricity. Such predictions allow the phasing into the New Zealand power supply system of the more expensive non-hydro plants in the most economical manner. It therefore follows that, from an economic point of view, it is important to be able to predict what proportion of the national non-hydro capacity will be needed to meet the variation in the demand for power.

(b) ELECTRIC POWER CONSUMPTION IN NEW ZEALAND

Figures showing the consumption of electricity for each half-hour are available from local distributing authorities for specific areas in New Zealand, and from the Electricity Division of the Ministry of Energy for New Zealand as a whole. These can be directly compared with the hourly observations of temperature, wind, and sunshine, as well as hourly forecasts of these same parameters.

In an earlier study by the author (see Maunder 1971c) half-hourly electric power consumption values were obtained from the Municipal Electricity Department (MED) of Wellington City for the Wellington City area, and from the Electricity Division of the Ministry of Energy for New Zealand. Data are available for every day, but since all days are *not* equal from an electric power consumption viewpoint, it was considered to be more meaningful to compare the same day of each week, and Wednesday was taken to be the most 'normal' day of the week. Accordingly the investigation cited considered the half-hourly power loads for successive Wednesdays and assessed from these data the 'average' Wednesday consumption. The original study considered the nine Wednesdays in the mid-winter months of July and August 1969.

The data applying to the Wellington MED area consisted of the 24 half-hourly loads (e.g. 12.30 a.m. to 1 a.m., 1.30 a.m. to 2 a.m., etc.). In a few cases, some 'control' had been placed on the system (i.e. electric power to water heaters and storage heaters was cut by the supply authorities), and the data applicable to these times were 'adjusted' slightly in order to make the whole series of power loads compatible. The half-hourly electric power consumption for the nine Wednesdays was then averaged, the differences between the actual consumption and the average for specific half-hour periods exceeding in some cases 20 per cent of the average consumption. For example, on 2 July 1969 (the first Wednesday of the period), the half-hourly differences varied from 10 per cent above the average consumption in the early morning, to 18 per cent above the average between 8.30 a.m. and 9 a.m., and 21 per cent above the average between 4.30 p.m. and 5 p.m. In contrast, on Wednesday, 13 August 1969, from 4.30 p.m. to 5 p.m., the consumption was 23 per cent below average. The variation in electric power

Table VI.28 Electric power consumption on Wednesdays in New Zealand during July and August 1969. Differences (in megawatts) from the average Wednesday consumption during this period

		Successive Wednesdays								
Half-hour periods	*Average*	*1*	*2*	*3*	*4*	*5*	*6*	*7*	*8*	*9*
4.30a.m.–5a.m.	930	+30	−20	0	+80	+20	−30	−40	0	−30
7.30a.m.–8a.m.	2250	+180	+60	+40	+200	+80	−100	−80	−100	−260
9.30a.m.–10a.m.	2290	+180	+20	+50	+160	+90	−170	−210	−30	−50
1.30p.m.–2p.m.	1980	+240	+60	0	+220	+40	−220	−230	−130	+50
5.30p.m.–6p.m.	2520	+160	+50	+40	+110	+60	−100	−190	−90	−30
8.30p.m.–9p.m.	2290	+160	+30	0	+100	+30	−110	−120	−60	−40

Source: After Maunder (1971c).

demand during the peak late afternoon hours in the Wellington City area was therefore considerable, and is caused principally because of the large temperature-related domestic consumer demand for home heating.

A similar analysis of power consumption in New Zealand was made using the half-hourly power consumption data supplied by the Electricity Division of the Ministry of Energy. A selection of the actual consumptions expressed as differences from the average are shown in Table VI.28. As indicated, some of these differences are substantial. For example, the New Zealand demand from 1.30 p.m. to 2 p.m. on Wednesday, 2 July 1969, was 240 megawatts (or 12 per cent) above the average.

As previously noted, since the non-hydro generating stations (coal, oil, or gas) are used mainly to supply *peak* loads, a considerable proportion of these peak loads (up to 20 per cent on occasions) must be satisfied through the generation of the more expensive coal, oil, or gas produced electricity. Consequently, if the peak demands could be forecast more accurately (using, for example, detailed temperature, wind, and illumination forecasts), it could be expected that the ratio of non-hydro to hydro produced electricity would be minimized. The significance of temperatures and forecasts of temperatures on electric power demands in New Zealand is now discussed.

(c) Electric power consumption and temperature relationships

In order to assess the relationship between electric power consumption and temperatures, a comparison was first made for Wellington City of the differences in the half-hourly power consumption from the average and the temperature differences from the average for the same hours. This showed that in general the larger the negative departure of temperature, the larger the positive departure of electric power consumption. A summary of the results is given in Table VI.29, which shows that the average departure in electric power consumption is of the order of 16 megawatts over a half-hour period (compared with an average total consumption of about 100 to 140 megawatts), whenever the temperature departure for the hour is either 5 to 7°F (2 to 4°C) above average, or 4 to 6°F (2 to 3 or 4°C) below average.

The *cost* of these variations in power consumption to the Wellington MED is a little difficult to determine accurately because of the relatively complicated price structure that the New Zealand Electricity Department then used for assessing the charge to the distributing authorities.* (A slightly different price structure is now used, but it still principally relates to a 'base' price, plus a 'penalty' for usage during periods of high demand.) The then current price structure was as follows. *First*, the year was divided into two

Table VI.29 Relationship between electric power consumption and temperature departures at Wellington, based on hourly* data on Wednesdays in July and August 1969.

*Temperature departure from average (°F)***	*No. of occurrences*	*Average electric power consumption departure from average (megawatts)*
+7	2	−15.5
+6	4	−18.0
+5	11	−16.1
+4	8	−15.0
+3	4	−7.8
+2	12	−1.3
+1	16	−1.2
0	16	+0.4
−1	13	+3.2
−2	10	+4.9
−3	9	+11.4
−4	5	+17.4
−5	7	+18.7
−6	2	+21.0
+7 to +4	23	−16.3
−3 to −6	23	+15.8

Source: After Maunder (1971c).
Notes:
* Considering the half-hour periods 7.30–8 a.m., 8.30–9 a.m., to 7.30–8 p.m. (the average power consumption during these hours varies from about 100 to 140 megawatts).
** Original study in °F.

parts, the three months ending 30 June (mid-winter), and the nine months ending 31 March (mid-autumn). In both periods the three highest half-hourly peaks were taken (not more than two being counted on any one day) and these were averaged. This average peak demand was then (1971) assessed at $23.50 per kilowatt of demand. An additional *one* megawatt of demand therefore cost $23,500. Thus in the case of the Wellington MED with an average peak load of approximately 140 megawatts, the demand charge was $3,290,000. It follows, therefore, that it was (and is) highly desirable for distributing authorities to keep the peak demands as low as possible. The successful forecasting of very cold conditions – when peak demands are most likely to occur – would therefore have obvious economic advantages. The *second* charge made by the New Zealand Electricity Department is for the *actual* electric power consumption, which in 1970 was 0.27 cents per kWh. This charge is 'offset' by the charge to the consumer and in 1969 the average revenue per kWh to the Wellington MED was 1.43 cents. Thus the gross return per kWh was 1.16 cents. Accordingly, an additional 15 megawatts of demand in the Wellington area (such as would arise if the temperature was 2 to 4°C colder than normal) would, provided it was less than the three 'peak' demands used to determine the 'demand charge', have given a gross return to the Wellington MED of $174 per *hour* (15,000 kWh × 1.16 cents). Naturally, both these charges must be considered by the local distributing authorities, and in simple terms, the aim of the local controllers is to sell a high base demand but keep the peak demands as low as possible.

The analysis of the relationship between differences in hourly consumption and differences from the average hourly temperatures was also carried out for New Zealand in a similar way to that described for Wellington City, with the exception that the average hourly temperatures and temperature departures were *weighted* according to the consumer demand for electric power. The human population weighted temperature departures were next compared with the differences from the average half-hourly electric power consumption, for each half-hour period for the nine Wednesdays in July and August 1969. The data in Table VI.30 shows that in July and August 1969 temperature departures in the 3° to 7°F (2° to 4°C) range were associated on the average with departures in the electric power consumption in New Zealand of the order of 150 to 200 megawatts (from the 'expected' values) over a half-hour period (during the period 6 a.m.–11 p.m.). These departures were approximately 10 per cent of the *average* usage of electric power during these periods.

The value of knowing about these variations in advance – that is, knowing

---

* The pricing structure and the methods of administering such changes will vary between generating authorities in each country. The example for Wellington, New Zealand shows that *local* knowledge of the procedures used is essential if useful applications of weather information are to be made.

Table VI.30 Relationship between electric power consumption departure from average and weighted temperature departures in New Zealand on Wednesdays in July and August 1969*

*Temperature departure from average (°F)***	*No. of occurrences*	*Average electric power consumption departure from average (megawatts)*
+6° to +7°	4	−222
+5° to +6°	7	−253
+4° to +5°	3	−180
+3° to +4°	14	−140
+2° to +3°	18	−15
−3° to −2°	11	+90
−4° to −3°	3	+134
−5° to −4°	6	+170
−6° to −5°	9	+187
−7° to −6°	2	+140
+3° to +7°	28	−188
−3° to −7°	20	+169

Source: After Maunder (1971c).
Notes:
* For half-hour periods 6.30 a.m.–7 a.m., 7.30 a.m.–8 a.m., to 10.30 p.m.–11 p.m.
** Original study in whole °F.

at, say, 4 p.m. today that electric power consumption tomorrow will be about 10 per cent more than average, because of the colder conditions expected, is considerable, and as far as New Zealand is concerned, is principally related to the difference in the cost of producing electricity from the hydro and non-hydro stations. Specifically, if an *additional* 200 megawatts of power is required, and half of this power has to be provided by thermal means, the additional cost – over hydro-produced power – was at 1982 prices about $3900 per *hour* (100,000 × 3.9 cents/kWh where 3.9 cents is the estimated additional cost of non-hydro generated power). Accordingly, knowledge of tomorrow's temperatures and hence knowledge of tomorrow's electric power consumption should assist local and national decision-making. Specifically, such decision-making would lead to a reduction in the need to 'fire' thermal plants to produce the more expensive electricity, when such electricity is either not needed or when it could be supplied by hydro-electric means at a much lower cost.

(d) CONCLUSION
It is evident that weather conditions, and in particular temperatures, have an important impact on electric power consumption. Specifically, the analysis described shows that on 'cold' or 'warm' days in winter in New Zealand, the electric power consumption increases or decreases by a factor of 10 to 15 per cent. Such variations are economically significant.

Clearly, the decision-making involved in the economical generation and supply of electric power is considerable, and the analysis indicates that the 'correct' use of temperature forecasts could prevent the unnecessary generation of more expensive power. In addition, prior knowledge of very cold periods could assist local supply or distributing authorities in estimating the peak demands for power, and thus avoid the unnecessary cutting of water heaters, storage heaters, etc. It is of course difficult to forecast these very cold periods more than two or three days in advance; however, useful 12- to 36-hour forecasts of temperatures, wind, and other weather parameters can be made, and since 1983, forecasts of temperatures for thirty-five places in New Zealand have been made operationally. Such forecasts provide valuable decision-making information to the electrical supply industry of New Zealand.

3. Monitoring electric power consumption: The value of weather information to the New Zealand decision-maker

The value of relevant meteorological information to the decision-making involved in the correct monitoring, pre-scheduling and, therefore, generation of electric power is well established. For example, press statements such as that issued on 23 March 1976 by the New Zealand Electricity Department that 'Allowing for temperature corrections, electricity used last week was 2.9% below the national guidelines', and 'Oil generation rose last week, fuel costs amounting to $1.1 million', are now common, and have obvious meteorological implications. Indeed, the New Zealand Meteorological Service has for several years been supplying to the Electricity Department relevant meteorological information on a routine basis. This information comprises forecasts of weather conditions, especially temperatures, as well as detailed analyses of temperature departures from the average (weighted according to the distribution of the human population) for the immediate past.

The first type of information provided comprises special weather forecasts for the Electricity Division of the Ministry of Energy which are issued every morning, and comprise (1) general forecasts for the North and South Islands of New Zealand; (2) temperatures, wind, and 'illumination' forecasts for noon and 5 p.m. today, and 8 a.m., noon, and 5 p.m. tomorrow for eight urban areas; and (3) forecasts of rainfall in nine hydro-catchment areas. These forecasts are sent to various Island system controllers and are used in assessing the demand for electricity during the next twenty-four hours, and the likely storage available in the various hydro-catchment areas.

The research previously discussed has shown that a 1°C change in temperature causes a change in power demand of about 2 per cent of energy. If generated by oil or gas rather than water, this energy involves additional generating costs of the order of $2000 for every hour that the extra energy is required. Accordingly, the correct pre-scheduling of power generation (that is power generation at minimum cost) is closely related to the accuracy and availability of temperature forecasts. The additional information provided on the expected rainfall in the key catchment areas of New Zealand is also of significance to the system controllers in the Electricity Division in their day-to-day control and use of water storage for hydro-electric generation.

The second type of weather information supplied to the Electricity Division comprises a detailed analysis of the temperature departures from the average at twenty-four stations in New Zealand at 8 a.m., noon, and 5 p.m. for each day. The actual station-hour departures are weighted according to the distribution of the human population, for each Island and for New Zealand as a whole, and the resulting temperature indexes are used by the Electricity Division in 'monitoring' and 'explaining' the hour-to-hour demand for electricity. The information is also a vital component in assessing how closely the actual electricity demand, *adjusted for the temperature conditions*, meets any government 'guidelines' on national electricity consumption, as well as in correctly assessing any additional 'control' which may be required by the central authority on the sixty local supply authorities in New Zealand. The incorporation of wind and illumination factors into these temperature deviations such as provided in the daily forecasts, would provide even more consumer/producer-related information.

The *real* value of the weather information and weather forecasts supplied to the Electricity Division in New Zealand to minimize the cost of generating power is, however, difficult to assess. But it is evident the weather is a dominant variable affecting domestic power consumption in New Zealand, and that in the absence of the vital weather information cited above, the costs of power generation would be increased by at least $2 million a year (equivalent to about 12 per cent of the annual budget of the New Zealand Meteorological Service).

Any study relating weather conditions to electric power demand and the assessment of the value of weather information to the decision-maker is clearly related to the deviations from the 'expected' in both electric-power demand and the weather. But it is at times very difficult to obtain a true measure of these deviations. For example the importance of non-weather conditions such as short-term variations in the industrial demand for electricity, changing consumer patterns, the impact of holidays, weekends, sporting events, and even television programmes, are often very difficult to quantify. Similarly, with weather conditions, it is current practice to express temperatures as a departure from so-called 'normals'. However, temperatures (and other aspects of the weather) vary from decade to decade as well as from year to year; accordingly a temperature deviation from the 1951–80 'normal' may give a somewhat false impression of the 'real' warmth or cold of any period. More importantly, in most cases it may well be that it is *not* the temperature departure from some arbitrary mean or normal that is important, but rather the temperature difference from week to week, from month to month, from season to season, and from year to year. For example, the winters of both 1981 and 1982 in New Zealand were among the warmest winters since instrumental records began in the 1860s; hence the winter of 1983 which had 'normal' temperatures (based on the 1951–80 normal) was justifiably described by many as being rather cold, with important economic consequences to both the generating and distribution sectors of the New Zealand electricity industry.

4. Summer electric power consumption and temperatures: A United States example

(a) Population-weighted cooling degree days as economic indicators

In the United States, population-weighted cooling degree days are computed by the National Weather Service to assist decision-makers in the energy sector particularly during very hot conditions (see Le Comte and Warren 1981). It is suggested that such information could also be used to 'adjust' many indicators of United States economic activities, in order that a more 'correct' picture of the economy may be presented. Specifically, it was suggested by the author (see Maunder 1982b) that the 'true' indicators of the United States economic activity during the 1980 summer would have been somewhat lower than those published and, more importantly, would have given different week-to-week variations if the effect of temperature had been taken into account. How much lower the true economic activity would have been is a matter of analysis and opinion, but an inspection of the data in Table VI.31 is a useful starting point. This shows that the total United States electric power production in a 14-week period in 1980 of 673,239 million kWh was 6.3 per cent higher than the 633,430 million kWh produced in 1979. During the same period there was an increase in the population-weighted cooling degree days (such indices were computed by the then NOAA Center for Environmental Assessment Services (US Department of Commerce)) from 844 in 1979 to 1075 in 1980, giving an average *weekly* increase of 16.5 cooling degree days or 27.4 per cent. Disregarding purely 'economic' factors, it therefore appears that for the 14-week period, the average weekly increase of 16.5 population-weighted cooling degree days (equivalent to an average 27.4 per cent increase) was associated with a 6.3 per cent increase in

Table VI.31 United States electric power production and cooling degree days: summer 1980 and 1979 and differences

	Electric power production (millions kWh)			*Cooling degree days (population-weighted)*		
Week ending	*1980*	*1979*	*1980/1979 (%)*	*1980*	*1979*	*1980/1979 (%)*
June 28	46894	44256	106.0	83	46	180
July 5	45838	42332	108.3	77	48	160
July 12	49165	46206	106.4	90	78	115
July 19	52635	47691	110.4	110	75	147
July 26	49943	48066	103.9	84	87	97
Aug. 2	50126	49200	101.9	97	90	108
Aug. 9	51834	49516	104.7	105	85	124
Aug. 16	49507	43316	114.3	75	36	208
Aug. 23	47981	45679	105.0	73	64	114
Aug. 30	48138	46407	103.7	81	74	109
Sept. 6	46844	44980	104.1	75	67	112
Sept. 13	45549	43370	105.0	53	42	126
Sept. 20	44434	41239	107.8	39	25	156
Sept. 27	44351	41172	107.7	33	27	122
Total (14 weeks)	673239	633430	106.3	1075	844	127.4
Average (14 weeks)	48089	45245		76.8	60.3	

Source: After Maunder (1982a).

the total industrial, commercial, and domestic electric power consumption in the United States.

There were of course pure economic factors which *also* caused an increase in electric power consumption from 1979 to 1980. Indeed, an analysis of electric power consumption in the United States for the 4-week midsummer period from mid-July to mid-August for the twelve years prior to 1980, shows an overall *average* rate of increase of 2854 million kWh (per week) *per year*. Assuming that this *average* rate of increase also applied to the change from 1979 to 1980, the 'expected' average weekly power consumption during the 4-week period in 1980 would therefore have been about 50,626 million kWh compared with an 'expected' consumption of 48,789 million kWh for the same period in 1979, giving an 'expected' weekly increase from 1979 to 1980 of 3.8 per cent. In comparison the *actual* average weekly power production during the same 4-week period was 51,135 million kWh in 1980 and 48,618 million kWh in 1979, giving an actual increase of 5.2 per cent. The actual difference was therefore 1.4 per cent more than the 'expected' difference based on a (first approximation) linear trend over the period 1969–80. Alternatively, it may be more useful to suggest that the average weekly (for this 4-week period) 1980 production was considerably above the 'economic trend', whereas the 1979 production for the similar period was below the 'economic trend'. It is suggested that this difference of 1.4 per cent is associated in the main with a 'non-productive' demand for electric power as a result of the 1980 summer being very much warmer than the 1979 summer.

If it is further assumed that the analysis for the 4-week period over the twelve years (1969–80) gives a reasonably good estimate of the 'economic trend' for any of the fourteen 'summer' weeks (an analysis *could* have been made for each week, or for any combination of weeks but this was not done), it may be inferred that in the 14-week period in 1980 from late June to late September, the weather-related* increase in electric power consumption in the United States was 1.4 per cent. As previously noted the population-weighted cooling degree days for the United States for the same period in 1980 and 1979 showed an average weekly increase of 16.5 population-weighted cooling degree days, equivalent to an increase of 27.4 per cent. It is evident therefore that some (or even all) of this increase in cooling degree days was directly associated with the increase in the national electric power consumption of 1.4 per cent. It should of course be noted that the published electric power production includes industrial, commercial, and domestic consumption requirements. However, most, but not all, of the *weather-related* increase in demand is likely to have been in the domestic sector. On a unit base, therefore, it is suggested that a 1 per cent increase in the requirements for *cooling* (using as an index of cooling the national population-weighted cooling degree days) is associated with a weather impact on the United States total electric power consumption of 0.051 per cent, or alternatively that a weekly increase of *one* population-weighted cooling degree day is associated with a weather impact on the United States total electric power consumption of 0.085 per cent. It should be noted that while these percentages may appear to be small, their impact on local and regional electricity supply companies in times of high summer temperatures when the demand for electricity for air conditioning is at a premium, can be very substantial.

* Although the 'economic' trend has been taken into account it is clear that there may well have been other non-weather factors which were important. Nevertheless, it is considered that most of the remaining differences were weather related.

(b) ELECTRIC POWER CONSUMPTION AND ECONOMIC INDICATORS

A major component of the weekly indices of economic activity published in the United States, such as those in *Business Week* and other business journals, is the electric power production computed by the Edison Electric Institute. The relationship between these economic indicators and electric power consumption is discussed in detail in Chapter VII, but as indicated in the previous section an analysis of the electric power production in the United States for the summer weeks of 1980 and 1979 (see Table VI.31) showed very significant weather-related differences. Moreover, since the electric power production component in the *Business Week* economic index is weighted by a factor of 17.3 per cent, it is evident that an increase in electric power production (from one week to the next) will have a significant impact on the index of economic activity. It also follows that this will occur *irrespective of the reason for the increase* (or decrease) in electric power production, since the *Business Week Index* (and other similar business indicators in the United States) are based on the premise that an increase in the *production* of steel, wheat, *and* electric power, or an increase in the *consumption* of commodities – including electric power – automatically assumes that there has been an increase in 'economic activity'. But the question must be asked: are *all* increases in economic activity *real*, and more specifically can a business index which is not corrected for the weather be a reliable indicator of 'real' economic activity?

While such a premise is appropriate in most production and consumption activities, it should not automatically apply to electric power production and consumption. This is because a proportion of the variation in demand clearly reflects 'environmental' conditions (and in particular the weather) rather than economic 'strength'. For example, in New Zealand an increase in electric power consumption *as a result of colder than usual conditions* is directly associated with the need to use high-priced imported oil for generation, rather than low-priced domestically produced hydro power. That is, abnormally cold conditions are a direct *cost* to New Zealand, and they clearly *cannot* be regarded as an indicator of increased economic activity. Further, although it is evident that the economic structure of the United States is quite different from that in New Zealand, there appear to be quite strong reasons for compilers of economic indicators in the United States to depart from the normally accepted economic growth models in which increased electric power consumption (at least for domestic purposes) *from whatever cause* is automatically assumed to be related to economic strength.

The reason for this suggestion to depart from traditional economic thinking is that *part* of electric power consumption is weather related, and it seems important to separate 'real' economic growth from that associated with a 'temporary' weather/climate variation. It is therefore suggested that an increase in electric power consumption to cool or warm buildings and the people inside them – above that which would be considered average for a specific week of the year – should be considered 'unproductive' (except in the sense that it may allow people to be more comfortable and hence be more productive). Of course, an increase in electric power consumption means additional revenue for the private utility companies, and clearly in the United States where most electric power is produced by private companies, these companies will rightly state that an increase in electricity consumption from *whatever cause* is to *their* economic disadvantage. However, from a *national* viewpoint the question must be asked whether it is *always* an indicator of *national economic prosperity*? It is therefore suggested that an increase in electric power production as a result of higher summer temperatures or colder winter temperatures should, in some cases, be considered a *negative* contributor to indicators of national economic activity, and not necessarily a positive contributor, which is the current situation in many countries.

5. Application of weather modification to the electric power industry: Two case studies

(a) Contrail formation and air conditioning

An increasing demand for power is occurring in those parts of the world where air-conditioning equipment is commonplace. One interesting aspect of this was explored in a study by Johnson, McQuigg, and Rothrock (1969) who considered the modification of the 'surface' temperatures by clouds formed as a result of contrails from jet aircraft, and the possible effects of such modification on electric power consumption and production costs in the mid-west of the United States.

In the analysis, electric power loads were related to temperatures by standard regression techniques, and then historical and experimentally generated series of average daily temperatures were subjected to a modification scheme based upon the formation of contrail clouds. The costs of generating electric power under modified and unmodified temperature conditions were further used as a basis for providing some insights into the implications of temperature modification for the electric power industry.

A series of 'modified' temperatures were then produced using a simulation model based on the reduction of solar radiation arriving at the surface of the earth through the 'creation' of contrail cirrus clouds. An important feature of the model is its ability to create a time series of daily temperatures that are reasonably similar to what might be produced by a deliberate attempt to modify summertime temperatures in the central part of the United States. For example, decreases in afternoon maximum temperatures were generated in the simulation model on about 20 of the 53 possible days during the months of July and August 1963, the 'resulting' change in surface temperatures being from 2°C to 4°C on most of the 'modified days'.

An analysis is also given in the original paper for the costs of producing power using an experimentally generated and modified 80°F (27°C) and 85°F (29°C) series, some of the results using the 85°F (29°C) series being shown in Table VI.32. This shows that the hourly cost differences at that time varied from $2027 to $50,425, with an average of $22,540.

In their summary, Johnson, McQuigg, and Rothrock (1969) indicated that the implications for the electric power industry were clear, in that the sensitivity of power demands to high summer temperatures can be quantified. Furthermore, the results of the model suggested that benefits from modification on high average temperature days can be a substantial factor, which may well prove highly beneficial to power-producing companies in the future *if* weather modification in the form of contrail formation on a regional scale is both possible and desirable. In this connection it should be noted that in the twenty years since this study was completed no known deliberate attempt has been made to create contrail clouds on any large scale. Never-

Table VI.32 Costs* of non-pooled power production in the United States mid-west for hourly loads based on an experimentally generated and modified 85°F series‡

	Cost ($)			*Temperature (°F)***	
Date†	*Actual cost*	*Modified cost*	*Difference in cost*	*Average actual*	*Average modified*
530608	106,876	104,849	2,027	76.7	75.3
610730	96,940	91,865	5,075	82.1	81.8
530629	123,640	114,043	9,597	81.3	79.5
550820	141,910	127,563	14,347	85.1	82.2
620820	136,583	121,161	15,422	79.8	77.1
530830	132,214	111,569	20,645	83.9	80.6
540828	131,074	109,248	21,826	82.2	79.4
540730	137,601	114,747	22,854	82.8	80.3
540815	108,875	85,095	23,780	82.7	77.4
550725	118,272	91,301	26,971	80.5	74.9
540711	163,198	131,527	31,671	85.9	81.3
630701	143,191	107,044	36,147	84.2	77.8
540827	143,910	103,562	40,348	79.7	75.2
540706	106,682	56,257	50,425	83.1	75.1

Source: After Maunder (1971a), from McQuigg, Johnson, and Tudor (1972).
Notes:
* In order of difference in cost.
† Year, month, day.
‡ Transmission costs between all 14 electric power production and distribution areas set equal to the highest per unit cost of power production.
** Averages of average modified and observed daily temperature at the 14 distribution areas.

theless, the possibility clearly exists, if a centrally controlled government considered that such intervention was economically desirable.

(b) CLOUD SEEDING AND HYDRO-ELECTRIC POWER

Perhaps the most practical use of weather modification in the utility field lies in hydro-electric power generation, and Eberly (1966), in an early but highly relevant paper, discussed in some detail weather modification and the operations of an electric power utility. In particular, he noted that weather modification in the form of increased precipitation offers one means of improving the efficiency of the hydro-electric portion of a power system.

A test programme of the Pacific Gas and Electric Company, the largest investor-owned power utility in the western United States, operating (when the study was made) sixty-seven hydro-electric plants with an installed capacity of 2226 MW, is described by Eberly, who outlined some of the factors that would need to be taken into account in appraising the economic value of weather modification against the alternatives. He noted that *if* an increase of precipitation from cloud seeding could be depended upon, it might be possible to introduce some efficiencies in the operations of a hydro-electric system, since water levels in reservoirs at the end of the dry season are usually maintained at a level which will provide a reasonable assurance that the lakes will be full by the end of the following wet season. Thus, if the company did not have to guard against dry years being so dry, the water levels in the reservoirs could be reduced, which would have the added effect of having more storage space available when the wet years came. In addition, an increase in runoff from cloud seeding could provide additional energy from a hydro-generating system, the increased generation possibly satisfying some of the growth requirements of the company, which could delay capital expenditure for new thermal plants.

The investigation of the potential benefits from cloud seeding for several different watersheds is reviewed by Eberly, who assumed that the 'increased precipitation' from seeding resulted in a 10 per cent increase in runoff, that the increased runoff resulted in a 10 per cent incremental increase in daily runoff, and that the value of the increased runoff is in the fuel saved by not using thermal generation. Using these and other assumptions the benefit/cost ratios for weather modification in three watershed types was assessed by Eberly, who indicated that if a 10 per cent increase from cloud seeding could be achieved, the benefit/cost ratios would vary from a low of 0.4:1 to as much as 14:1, depending on the climate conditions (i.e. dry, normal, or wet year) and the kind of watershed.

H. PASTORAL PRODUCTION AND WEATHER RELATIONSHIPS

1. The world scene from a New Zealand viewpoint*

The products of a country depend as much on what its people want to produce as on its natural advantages. Such a statement could be applied to most countries, but as an example of the sensitivity to weather and climate of *pastoral* production in various parts of the world, New Zealand has a unique situation. Indeed it is considered by many observers of the international pastoral product market that New Zealand has a significant climatic advantage for grassland agriculture, this advantage being translated into the development of New Zealand over the past century into the world's leading exporter of grassland products (Table VI.33).

Table VI.33 New Zealand and world food production and food exports compared*

Product	*New Zealand Exports as percentage of World Exports*	*New Zealand Exports as percentage of New Zealand Production*	*World Exports as percentage of World Production*	*New Zealand Production as percentage of World Production*
Lamb and mutton	50.0	80.9	14.4	8.9
Beef and veal	8.9	46.7	6.3	1.2
Butter	20.8	76.8	15.0	4.1
Cheese	6.6	97.5	11.7	0.8
Dried milk	10.6	85.4	39.5	4.9

Source: After Maunder (1980a). Data from New Zealand Monthly Abstracts of Statistics, FAO Production/Trade Yearbooks. Original table was published in *New Zealand in the Future World*, a booklet on 'sustainability' prepared by Diane Hunt for the NZ Commission for the Future, Government Printer, Wellington, 1979.
Note: * Based on 1977 production.

*Clearly, the viewpoints from other countries or for other 'crops' would be different. For example, *The Economist* (7 December 1985) in discussing the farming scene in the United Kingdom stated:

> Overall, farm net incomes for 1985 are expected to be no more than £1.2 billion. That will be not only a third down on 1984 but possibly the lowest real income since 1945. The main reason is the failure of the 1985 harvest. Particularly disappointing has been the wheat harvest. Overall yields look like being down some 20%. Quantity is down. So is quality – and price. The typical feed wheat grower could have expected to get £121 a ton for the 1983 harvest. . . . This year, the equivalent price is down to £106 a ton, and for many the cost of harvesting has been raised by heavy fuel costs for drying grain during the wet harvest.

Table VI.34 Dairy products – selected world climatic data

Stations	*A**	*B†*
New Zealand		
Glenbervie	12	12
Te Aroha	12	12
Tauranga	12	12
New Plymouth	12	12
Palmerston North	12	10
Christchurch	12	9
Australia		
Sydney	12	12
Gabo Island	12	12
Hobart	12	10
Canada		
Vancouver	11	7
Toronto	12	6
Denmark		
Odense	10	5
Studsgaard	12	5
France		
Caen	12	7
Brest	12	7
Netherlands		
Den Helder	11	6
Groningen	12	6
Switzerland		
Berne	12	7
USA		
Madison	12	6
Albany	12	7
Los Angeles	4	12
USSR		
Tartu	8	5
Moscow	10	5
Kazan	5	5
West Germany		
Munich	11	5
Munster	12	7

Source: After Maunder (1977d) from Maunder (1963).
Notes:
* Number of months with average rainfall of 38 mm or more (an index of 'moist' months).
† Number of months with average daily maximum temperature of 13°C or more (an index of 'growing' months).

To set the world pastoral production scene, and in particular the significance of New Zealand in this scene, it is useful to compare the New Zealand 'pastoral' climate with the climates of the other main pastoral production areas of the world, and to assess what climatic advantage New Zealand has over her agricultural competitors. (A similar comparison could equally well be presented comparing, say, the Indonesian climate with the climate of the chief rice production areas of the world and then to assess the climate advantages which, say, Indonesia has over her agricultural competitors.) For this purpose, climate data for stations located in or near the principal areas of each of the major pastoral producing countries of the world have been analysed and compared with relevant climate data for New Zealand stations. Unfortunately, only temperature and rainfall data are readily available for most significant pastoral areas, but as a first approximation such data may be translated into 'pasture growth potential'. A summary in the form of the number of 'moist' months and 'growing' months is given in Tables VI.34, VI.35, VI.36, and VI.37.

Table VI.35 Mutton and lamb products – selected world climatic data

Stations	*A**	*B†*
New Zealand		
Ruakura	12	12
Waipukurau	12	11
Bulls	12	11
Ashburton	12	10
East Gore	12	8
Argentina		
Buenos Aires	12	12
Bahia Blanca	8	12
Australia		
Dubbo	12	12
Hay	1	12
USA		
Abilene	7	12
Louisville	12	9
Salt Lake City	3	7
USSR		
Riga	7	5
Kazan	5	5

Source: After Maunder (1977d) from Maunder (1963).
Notes:
* Number of months with average rainfall of 38 mm or more (an index of 'moist' months).
† Number of months with average daily maximum temperature of 13°C or more (an index of 'growing' months).

Dairying areas in New Zealand have generally more rain than other major dairying areas of the world, with no 'dry' months (see Table VI.34). Mid-summer temperatures in most world dairy-producing areas are fairly similar, but in mid-winter there is a wide variation. For example, the New Zealand districts are a little cooler than areas in southeast Australia, but in general mid-winter temperatures in the major dairy areas of the world are colder, and in some cases more than 20°C colder than those experienced on the grasslands of New Zealand.

Mutton and lamb producing areas of the world have annual rainfalls of

Table VI.36 Beef and veal products – selected world climatic data

Stations	*A**	*B†*
New Zealand		
Dargaville	12	12
Rotorua	12	11
Taumarunui	12	12
Molesworth	12	7
Haast	12	9
Argentina		
Rosario	10	12
Buenos Aires	12	12
Australia		
Brisbane	12	12
Rockhampton	10	12
Cloncurry	4	12
France		
Clermont-Ferrand	9	7
Limoges	12	7
Ireland		
Mallow	11	7
USA		
Lincoln	7	7
Houston	12	12
USSR		
Kiev	10	5
Kaunas	10	5
Uruguay		
Montevideo	12	12

Source: After Maunder (1977d) from Maunder (1963).
Notes:
* Number of months with average rainfall of 38 mm or more (an index of 'moist' months).
† Number of months with average daily maximum temperature of 13°C or more (an index of 'growing' months).

Table VI.37 Wool products – selected world climate data

Stations	*A**	*B†*
New Zealand		
Whatawhata	12	12
Kuripapanga	12	9
Masterton	12	10
Hanmer	12	9
Lake Coleridge	12	9
Omarama	11	8
Mid Dome	12	8
Argentina		
Buenos Aires	12	12
Bahia Blanca	8	12
Santa Cruz	0	7
Australia		
Hay	1	12
Dubbo	12	12
Bourke	1	12
Port Augusta	0	12
China		
Hanchow	3	7
T'ai-Yuan	4	7
South Africa		
Grootfontein	5	12
Victoria West	2	12
USA		
Abilene	7	12
Austin	12	12
Cheyenne	5	6
USSR		
Ordzhonkikdze	7	6
Astrakhan	0	7
Semipalatinsk	1	5
Chernovtsy	6	7
Irkutsk	4	5

Source: After Maunder (1977d) from Maunder (1963).
Notes:
* Number of months with average rainfall of 38 mm or more (an index of 'moist' months).
† Number of months with average daily maximum temperature of 13°C or more (an index of 'growing' months).

between 380 and 1100 mm, the annual rainfall in the New Zealand areas being generally in the 750 to 1100 mm range. At many of the world stations at least four months are 'dry' (see Table VI.35). Temperatures in mid-summer in the New Zealand mutton and lamb producing areas are, however, lower than in all other major producing areas in the world. For example, average daily maximum temperatures in mid-summer are 10°C higher in the Australian and United States areas than in New Zealand. However, mid-winter temperatures in New Zealand are only a little cooler than similar mutton and lamb producing areas of Argentina, Australia, and Texas, and generally much warmer than in other major producing areas.

Beef and veal producing areas of the world have generally much warmer mid-summer temperatures than are experienced in New Zealand (see Table VI.36), the typical New Zealand producing areas having average temperatures in mid-summer of about 16°C, compared with 20 to 27°C elsewhere. In mid-winter, however, over half the climate stations in the beef producing areas of the world outside of New Zealand have colder temperatures than those recorded at Dargaville in northern New Zealand. All New Zealand beef and veal areas generally have sufficient rain, but in several producing areas outside of New Zealand 'dry' months are a common occurrence.

Wool producing areas in New Zealand are favoured with more rain than other world wool areas (see Table VI.37). For example, Omarama, in one of New Zealand's driest areas, has only one month with an average rainfall less than 38 mm, and this is a shorter 'dry' season than is experienced at 16 of the 19 world stations. Mid-summer average temperature in New Zealand lie at the cool end of a 15 to 30°C temperature scale which is characteristic of world wool areas, average daily maximum temperatures in New Zealand being in the 21 to 23°C range, compared with 24 to 34°C at most other stations. In mid-winter, average daily minimum temperatures at the New Zealand stations are nearly as high as the milder wool-producing areas of northern Argentina and Australia, but in all other major wool-producing areas of the world night-time mid-winter temperatures are from 4 to 22°C colder than those experienced at Omarama which is one of New Zealand's colder wool growing areas.

More detailed information on some of these relationships is given in Maunder (1963), and early overviews of the relationship between climate variations and pastoral agriculture in New Zealand are given in Maunder (1966a) for dairy production, in Maunder (1967a) for wool production, and in Maunder (1967b) for meat production. Other relevant studies on the significance of climate factors on agricultural incomes in New Zealand (Maunder 1968a), and on the seasonal distribution of pasture production in New Zealand (Radcliffe 1974), should be noted. In addition, the high cost of housing animals during the winter in Europe and many parts of North America must be considered, which is in contrast with the situation in New Zealand where animal housing is not needed.

2. Livestock investment response: A New Zealand example

In all countries, production – whether agriculturally based or not – has considerable variability. This variability is caused by a number of factors, economic, cultural, political, and environmental. Generally, if production is not agriculturally based, the variations are economically manageable, but where production is agriculturally based, the variations can assume considerable economic and political significance. New Zealand is one country where the agricultural sector is economically very significant; the following example of productivity variations relates to that country, but an equally useful example could well be made for other countries.

Of New Zealand's total gross domestic product in 1984/85, 11.4 per cent was from the agricultural production group, and 6.1 per cent from the food production group. Moreover, nearly 70 per cent of New Zealand's exports are agriculturally based, and on a world scale New Zealand is a leading producer of pastoral-based products. However, despite the 'well-watered' image that New Zealand appears to have, there are quite significant fluctuations from season to season. Most agrometeorological research has traditionally focused attention on on-farm and off-farm activities such as marketing, production, animal health, crop protection, and crop quality. An additional factor is the farmers' investment response, and this aspect has been examined in detail by the Reserve Bank of New Zealand (see Walsh 1981). In their analysis, they considered that the main factors influencing farmers' livestock investment response in New Zealand are, in order of importance: (1) monetary terms of exchange; (2) climate conditions and biological factors; (3) technological change and innovation; and (4) the availability of resources of land, labour, and credit.

In this context, the Reserve Bank considered that climate conditions operate as a major constraint on growth, in that livestock feed is largely grown on the farm and it is not feasible or economic (except in exceptional circumstances) to import large volumes of supplementary feed in times of widespread drought. The effect of climate as a constraint on livestock investment by farmers is largely, therefore, that on pasture production and consequently on carrying capacity, especially during the winter period. The effect on reproductive rates is also significant, since feed intake just prior to conception and birth has a strong positive influence on lambing and calving percentages. Aberrations from 'normal' climate conditions, especially those of drought, also have an obvious effect on the current period carrying capacity and performance, but they can also have a significant lagged effect through to the following season and in some cases the following *seasons*.

Data for the twenty years to 1979/80 for New Zealand sheep and beef productivity measured in terms of stock units, as well as some of the main factors influencing investment in livestock, notably the terms of exchange index, real expenditure per stock unit, and the climate-variable 'days of soil water deficit' are shown in Table VI.38. The year-by-year changes shown in this table give some indication of the nature of livestock investment responses. However, a somewhat clearer picture can be obtained by aggregating the data into periods which have similar characteristics in terms of exchange and input volumes, and by examining the growth rates of livestock numbers which have resulted from the various influences. This simple procedure overcomes to some extent the problems of accounting for the effects of lags in the livestock investment response.

For the purposes of this analysis it is convenient to look at three five-year periods: (1) 1962/63–1966/67; (2) 1967/68–1971/72; and (3) 1972/73–1976/

Table VI.38 Agro-economic and climatic factors: The New Zealand sheep/beef farm

Season	*Sheep and beef stock units (million)* *(1)*	*Terms of exchange (1975/76 = 1000)* *(2)*	*Real expenditure per stock unit (in 1975/76 $)* *(3)*	*Days of soil water deficit* *(4)*
1960/61	60.1	1079	7.77	18.9
1961/62	61.0	940	7.54	34.5
1962/63	62.8	1049	7.90	29.4
1963/64	63.8	1214	8.94	40.5
1964/65	66.3	1064	9.27	17.5
1965/66	70.7	998	9.46	18.1
1966/67	74.8	918	8.80	33.9
1967/68	76.8	899	7.83	33.0
1968/69	77.5	951	8.32	30.4
1969/70	78.8	1075	8.54	34.2
1970/71	78.6	963	8.23	35.5
1971/72	79.2	884	8.29	28.7
1972/73	78.4	1362	9.68	57.1
1973/74	79.5	1194	10.04	39.4
1974/75	80.8	750	8.10	28.5
1975/76	82.6	1000	9.10	32.8
1976/77	83.9	1063	8.92	25.7
1977/78	85.9	877	8.48	54.1
1978/79	84.4	993	9.06	27.3
1979/80	89.2	964	9.16	9.9

Source: After Maunder (1986b), from Walsh (1981).
Notes:
(1) Estimated from national census figures, sheep at 30 June of second year, cattle at 31 January.
(2) N.Z. Meat and Wool Boards' Economic Service data.
(3) Derived from Economic Service Survey (see (2) above).
(4) Weighted average of the New Zealand Meteorological Service indices for sheep and beef cattle populations.

Table VI.39 Agro-economic and climatic factors: The New Zealand sheep/beef farm – selected periods

Annual average	*Period 1 1962/63–1966/67*	*Period 2 1967/68–1971/72*	*Period 3 1972/73–1976/77*
Sheep and beef stock units (Average annual % change)	+4.1	+0.4	+1.4
Terms of exchange index	1049	954	1074
Real expenditure per stock unit (1975/76 $)	$8.87	$8.24	$9.17
Weighted no. of days of soil water deficit	28	32	37

Source: After Maunder (1986b), from Walsh (1981).

77, as shown in Table VI.39. Walsh (1981), in his Reserve Bank of New Zealand paper, noted that the early periods correspond to conventional economic notions about the effects of relative prices on investment. For example, in the period 1962/63–1966/67, the growth of livestock numbers (approximately 4 per cent per annum) was associated with relatively high and stable levels of both monetary terms of exchange and expenditure per stock unit, as well as generally favourable climate conditions. There was some carry-over of the higher growth rate of stock numbers into the period 1967/68–1971/72, but generally this was a period of slow growth associated with significantly lower average levels of terms of exchange and real expenditure, while climate conditions remained fairly favourable but perhaps slightly less so than in the period 1962/63–1966/67.

In the period 1972/73–1976/77, however, there was a definite recovery in the rate of growth of stock numbers, but the growth was much less than might be indicated by the high average levels of relative prices and input volumes. For example, both of these measures on average were as high as or higher than in former periods of rapid growth of stock numbers. The reasons for this lower 'productivity' of expenditure included: (1) climate constraints in the form of drought, especially in 1972/73, but serious also in some regions in 1973/74, both of which affected stock-carrying capacity and reproductive rates in both these and subsequent seasons; and (2) relatively low levels of expenditure in the period 1967/68–1971/72 which had a damaging effect on livestock physiology and therefore also on reproductive rates, which carried over into the period 1971/72–1976/77. In addition, it also appears that growth requires stable as well as favourable terms of exchange and real expenditure.

In general, it could be said that the overall weather and climate conditions in New Zealand during the 1970–80 decade were not conducive to creating confidence in the farming community. For example, reference to Table VI.38 shows very high numbers of days of soil water deficit during the seasons 1972/73, 1973/74, 1977/78. Clearly, the impact of these droughts, coupled with the rapidly increasing costs of farming, both on and off the farm, need to be taken into account if livestock farming in New Zealand is to continue to prosper.

ANNEXE
Drought in New Zealand

The following are extracts from various publications of New Zealand Government Departments and Producer Boards on the effect of the 1969/70 drought in New Zealand. It should be noted that most 'annual reports' cover the financial year ending on 31 March; they may not therefore include all the impacts of the drought, but only those which occurred in the 1969/70 'season' which in many cases ended in June or July 1970.

WEATHER

NZ Meteorological Service: 1969/70
A local but severe drought developed in coastal areas between Oamaru and Timaru where the rainfall was persistently below normal each month from January 1969. By the end of October the total was a record low value of less than 250 mm for the 10 month period. In the North Island a long spell of warm, sunny weather began after Christmas and by the end of February drought conditions had developed from the Manawatu northwards through Taranaki to the Waikato. The effects were generally most severe in Manawatu where rainfall had been persistently low during the previous spring.

Note: Much more relevant commodity-weighted information is now available on this and subsequent droughts in New Zealand, but at the time this report was prepared only descriptive accounts such as that given above were published.

FARMING – GENERAL

Annual Review of the Sheep Industry
The widespread drought is another recent event with implications. It has been a hard taskmaster but has highlighted the need for more attention to fodder conservation. Drought, its impacts and after-effects, is a farming hazard not generally so drastic and widespread as occurs from time to time in Australia, but when it does strike a country with normally high and well distributed rainfall and relatively porous soils, the effects are very serious indeed. The seriousness of a drought in New Zealand is greatly compounded by the density of stock population, especially in view of the fact that this population has almost doubled in the last twenty years. The drought impact with its drain on feed reserves and the increasing need to build up winter feed

supplies, has clearly indicated the need for a better balance between summer and winter feed provision.

Lands and Survey Department
The seasonal conditions experienced throughout the year in most districts were not easy from a farming point of view, the most serious aspect being the widespread drought conditions prevailing over most of the country from late spring through to February–March. This resulted in some cases in forced sales of stock because of shortage of feed.

LIVESTOCK

Department of Agriculture
Conditions were generally favourable during the winter and spring of 1969, with lambing percentages higher throughout the North Island though in the South Island they were somewhat less satisfactory. The situation has been radically altered, however, because of the prolonged dry weather, and the effects of this are likely to be felt for some time. The most noteworthy event was the transfer during the early summer of about 1,000,000 ewes and lambs from Canterbury and North Otago to Southland.

Annual Review of the Sheep Industry
By early summer the drought situation in South Canterbury and North Otago became desperate and many more lambs went south for grazing. The widespread drought hit dairying and intensively stocked areas very hard and in many Canterbury and North Otago areas it was a real disaster following on, as it did, the persistent drought conditions of many years' standing.

MEAT PRODUCTION

Economic Review
Meat production in 1969/70 is expected to be about the same as last season, with an increase in beef production being offset by a reduction in lamb. Early expectations this season had been for a record lamb kill but delays at freezing works and the severe drought, followed by favourable autumn conditions, appear to have influenced farmers to retain a greater number of stock. As a result, sheep numbers could increase by about 3.5% this season.

DAIRYING

Economic Review
Despite the continued increase in dairy cow numbers, the amount of butterfat processed by dairy factories was about 9% lower than in 1968/69.

Department of Agriculture
Dry conditions in November caused some concern, though this did not markedly affect production. However, the prolonged dry weather in the main dairying districts from mid-January resulted in the most rapid fall in production for many years, large

numbers of herds being dried off before the end of March. The major problem confronting dairy farmers has been to conserve sufficient winter feed so that their cows will calve in good condition in the spring.

WOOL PRODUCTION

Annual Review of the Sheep Industry

Wool production in 1969/70 fell by 1.2%, but at 328.7 million kg greasy it was only 3.9 million kg lower than the record of 332.6 million kg in 1968/69. However, changes in total production (including sheep and lambs wool), and changes in production per sheep by regions, based on estimates of sheep wintered in the hinterlands of the eight wool selling centres were considerable.

CEREAL AND OTHER CROPS

Department of Agriculture

The area in *wheat* was lower in all districts as compared with the previous season. This, and the drastic fall in yield – the lowest for many years – means that the total harvest will be much lower and that some imports will be required. The area in *barley* was also smaller and the dry spring in the South Island reduced yields. Demand was brisk, and the widespread use of feed barley as an emergency measure may encourage more farmers to use it as a supplement under normal conditions. A larger area was devoted to milling *oats* in the South Island, and yields were reasonably good, except in North Otago. The area in *potatoes* was slightly smaller this season, and yields will probably be affected by the dry weather. Yields of *peas* in Canterbury fell by up to 45%, and they were also low in Marlborough, except from irrigated crops.

FODDER AND SEED CROPS

Department of Agriculture

Though in most districts the area in *fodder crops* was about the same as last season, the dry weather reduced the yield from turnips, swedes, and chou moellier. In North Otago, where the spring drought was very severe, all these crops failed. On many properties the widespread drought has unfortunately caused an invasion of weeds, which could prove expensive to control. An indifferent crop of *small seeds* is expected this season; in addition, farmers were frequently forced to graze areas normally shut for seed.

FERTILIZER

Budget*

It is pleasing to note that fertiliser usage has continued at a high level in spite of the drought earlier this year. In view of the effect which the drought and rising costs have

* The budget is the official Financial Statement of the New Zealand Government.

had on farm incomes and liquidity, the Government has decided to give an additional incentive to fertiliser usage. This incentive will take the form of a uniform subsidy of $5 per ton (1970 prices) on all fertiliser. Provision has been made in the (Government) Estimates for the expenditure of $13.5 million on the fertiliser price and transport subsidies.

AERIAL TOPDRESSING

Ministry of Transport
The upward trend in aerial topdressing continued until November when there was a decline from the previous year's figures. The fall-off in demand in the period November–March was probably attributable to the drought conditions prevailing at that time.

DROUGHT RELIEF

Budget
The Government announced earlier this year special relief measures to assist farmers running short of stock feed because of the extraordinary nature of the drought. This included assistance on the carting of stock for emergency grazing over a wider area than normally assisted, an allowance of up to $10 a ton (1970 prices) on the cartage of many types of fodder, and a refund of one half the on-farm costs of grain and stock-feed meals, purchased as emergency feed for sheep and cattle because of the drought. This assistance is being administered by local drought relief committees and provision of $3.8 million is made in the Estimates.

RURAL LENDING

Budget
To assist farmers, sharemilkers, and others engaged in agricultural and horticultural production who have encountered serious financial difficulties in their farming ventures for reasons beyond their control, (i.e. the drought) it has been decided to establish a Special Agricultural Assistance Fund of $10 million, by appropriating funds in the Vote [of the Department of Agriculture]. $5 million have been provided in the Estimates to meet expenditure over the balance of this financial year.

State Advances Corporation
As a result of the drought and other difficulties it has been necessary to continue to assist some mortgagors by either reducing or deferring their payments. Assistance of this nature was given to 939 farmers.

CATCHMENT AUTHORITIES

Soil Conservation and Rivers Control Council
The severe droughts in the western part of the Manawatu Catchment Board's area caused a high death rate in small plant-materials. In Marlborough the year was a record drought year and greatly affected farming incomes and has considerably slowed most conservation activity and applications for new work.

RIVER FLOWS AND HYDRO ELECTRIC CAPACITY

NZ Electricity Department
From April to early September, inflows to all catchments were 74% of average – only slightly above the lowest values previously recorded over the last 35 years. Inflows to the major North Island catchments were 75% of average for the year compared with a previous record minimum of 82%, while inflows to Lake Taupo were particularly low, being 67% of average for the year compared with the previous minimum over the last 66 years of 78%. Fortunately, unusually high inflows in the South Island catchments in early September allowed the drawdown of North Island lakes to be arrested by the high transfer of surplus energy over the inter-island transmission link.

ELECTRIC POWER

NZ Electricity Department
The abnormally dry conditions prevailing in many North Island areas during the year were reflected in the pattern of power system operation. Thermal generation was necessary to a much greater extent than in the previous year, and during the year fuel costing over $NZ6 million* was used compared with less than $NZ2.5 million worth of fuel used in the 1968/69 year. At Meremere extended running was necessary because of low water flows through the Waikato River Hydro Stations, and over 1,000 million units (kilowatt hours) were generated. Extended running of the Marsden plant was also required.

IRRIGATION

Ministry of Works
Water usage continued at a high level in Canterbury because of the persistent drought conditions in the area.

CONSERVATION

Soil Conservation and Rivers Control Council
The past year was notable for drought conditions in the Manawatu, Wanganui, and Taranaki districts. The western sand-country in these districts is showing considerable deterioration and an increase in the number of applications for financial assistance for conservation works is anticipated.

FOREST FIRES

NZ Forest Service
The winter drought, followed by above-average temperatures and drought conditions in many districts during the summer, gave rise to the most serious fire hazard since 1949. On exposed faces native scrub species have wilted and died, creating an explosive fuel in cover that is normally fire-safe. This has developed serious fire problems in urban surrounds as well as in forest and rural areas. Losses were much

---

* In 1985 the comparable costs were in excess of $NZ150 million.

greater than normal and 500 hectares of plantation in public and private forests was destroyed.

ROADS

National Roads Board

With the long dry spell, conditions for maintenance and construction have been generally good in the Waikato district with little flood damage in the winter months. In Hauraki, better than average weather conditions during the winter months followed by the driest construction season for many years have permitted good progress to be made on most works. South Canterbury has experienced an exceptionally dry season resulting in severe drought conditions over a large section of the district. This has resulted in heavy stock movements to outside districts, and the drought conditions generated considerable stock and feed movement to and from Southland and South Otago, giving rise to considerable additional maintenance.

EXTERNAL TRADE

Budget

Restrictive policies in the countries of destination of our various products, combined with the threat of synthetic substitutes have produced a considerable relative decline in the value of wool and dairy products. These long-term effects (associated with rapid price increases) have been accentuated this year by the disastrous drought which affected virtually the whole country.

ECONOMIC ASPECTS – GENERAL

Reserve Bank of New Zealand

Internally the drought of the early months of 1970, while now ended, will leave an impact on farming through reduced income, low stocks and winter feed, and deterioration in the health of livestock. It will take a long time for the farms most affected to recover.

Lands and Survey Department

Representations have been made to Treasury to find some way by which additional stock can be purchased, or prime stock replaced by store stock in times of unforeseen surplus growth of feed. It is wasteful farming practice not to use the extra feed completely or to continue to retain prime stock to control pasture growth. However, the traditional system by which Parliament controls expenditure where the funds are voted in the early stages of the financial (and the farming) year, makes it difficult to meet such unforeseeable events.

Economic Review

It is too soon to assess accurately the impact of the drought on total output, but certainly previous expectations of a substantial increase have not been realised. The increases in prices received for meat – and to a lesser extent for some dairy products – may not have been sufficient to offset the depressing effect of lower wool prices and the drought on total farm income.

VII Weather- and climate-based forecasts of economic activities

A. THE ASSESSMENT ROLE

Irrespective of the weather and climate measurements made, or the econo-climatic analyses done, the ultimate question to be answered is 'what does it all mean?' To answer this question, an assessment must be made using all available meteorological and non-meteorological information, and to do this skilled assessors are needed with vision, knowledge, and an apprecia-tion of *real* problems.

Such assessors (see, for example, Center for Environmental Assessment Services 1980, 1981, and Eddy *et al.* 1980) working in the agro-economic and meteorological fields are active in a number of countries, and a noteworthy aspect of the work of these groups is the very important (and different) concept that although last week's or last month's weather *is* history (from a traditional climatological and economic viewpoint), the assessment and measurement of its economic impact *is not* history. For example, economic 'indicators' are usually not available until several weeks after the fact, while agro-economic 'statistics' of events, production, consumption, etc. take months and in some cases years to compile and publish. However, weather and other environmental satellites, combined with the traditional meteoro-logical data collection systems, provide worldwide information on the state of the atmospheric environment within *minutes* of the 'events' taking place. Further, providing appropriate research analyses have been made, these atmospheric data can be converted almost immediately into agro-economic indicators, and can therefore become *first* estimates of that part of the agro-economy that is directly associated with weather and climate factors.

It therefore follows that a *forecast* of agro-economic data such as produc-tion or prices – *which at this moment do not exist* (or more correctly have not been collated or forwarded to any data collection system), can be made in certain circumstances using real-time weather and climate information. Further, if such agro-economic data are useful even several weeks or months *after* the event as is presumably the case, since a lot of such information is collected and published, then having weather-based 'estimates' of such indicators 'instantly' should make such data even more valuable. In addi-tion, because weather and climate factors are generally considered to be 'outside of the economic system' their direct effects are easier to measure than the effects of other economic variables.

Finally, it should be emphasized that since weather- and climate-induced trends abound in agro-economic data, it is evident that if they are *not* explicitly explained they may well mask the economists', agriculturalists', and politicians' understanding of a system or process and so hinder correct decisions being made. Any assessment using real-time weather and climate information must therefore provide the key to a fuller understanding of many agro-economic problems.

B. Explanatory and Empirical Models

1. Explanatory climate models

Climate models fall into two broad and overlapping categories with regard to the degree to which the models consider physical processes; they may be either 'empirical' or 'explanatory'. Models that do not embody all the salient physical-biotic mechanisms (such as those used in weather/production studies (e.g. Maunder 1980a)) may be called 'empirical' in that they employ a convenient but logical framework in order to correlate predictor and predicted variables. As explained in the next section, such a framework may *not*, therefore, be based on economic-physical-biotic mechanisms, because in many cases those mechanisms are either not known at all, or not known in sufficient detail to be used. In contrast, 'explanatory' models endeavour to specify accurately and explicitly the relationships between climate components and processes.

For various reasons, mainly related to the availability of relevant data, explanatory models have been much less developed by applied climatologists, although their utility is *potentially* much greater than their empirical counterparts. The purpose of an explanatory climatic model is to specify accurately and explicitly cause-and-effect or stimulus-and-response relationships between climate components and processes in such a way that the model quantitatively describes the fluxes of energy, mass, and momentum. Appropriately developed explanatory models can therefore be used to predict not only climate events but also variables in past or present atmospheres for which observations are not available. Examples of explanatory models are the current studies of the consequences of increased 'greenhouse' gases on atmospheric circulation and temperature through the statistical 'manipulation' of digital general-circulation models. This type of model also provides a means to predict the climate impacts of human activity, such as the effect of large-scale irrigation on latent heat flux and circulation. However, as pointed out by Tapp (1984) in a paper 'Weather, climate and predictability', it must be emphasized that the inherent variability of the atmosphere is *very much* larger than the capability of numerical models to predict it; consequently the 'explanatory' power of many explanatory models may be severely limited.

2. Empirical economic/weather models

Empirical models, as noted above, do not embody the salient physical–biotic mechanisms but they employ a 'framework' in order to correlate predictor and predicted variables. Interpolating or extrapolating climate information from such models is very important, some of the current models being used for (a) obtaining reliable relationships between radiation variables and weather data for differing time scales; (b) interpolating climate variables from sparse data-collection networks; and (c) estimating sensible and latent heat exchanges between the atmosphere and surface. In addition, they help to develop more useful space/time-orientated agrometeorological models which can be used for enhancing the real-time agricultural production forecasting work of Baier (1977), Sakamoto, Strommen, and Yao (1979), Maunder (1986b), and others. Many other more subtle relations between climate conditions and human activity await identification and description. Meteorologists often view the empirical model type of research and its results as only 'first' estimates of reality because most results are time and space specific. Clearly the models do not 'explain' environmental processes, but the statistical relationships resulting from such models very often condense complex cause-and-effect computations into simpler computations that reduce the cost of evaluating the model. In addition, in many cases – particularly in the area of agricultural production weather relationships for 'large' regions – such empirical models are the *only* models which can, for various reasons, be used operationally.

C. WEATHER- AND CLIMATE-BASED FORECASTS OF AGRICULTURAL PRODUCTION

1. An overview

The specific relationship between agricultural production and climate is usually assessed through the development of an agroclimatological model. Some agroclimatological models attempt to mirror some of the physical processes that cause variations in agricultural production (such as soil water availability), but in many cases a statistical relationship is established linking, say, variations in rainfall or temperature with variations in wheat or rice production. These statistical models do not therefore 'explain' environmental processes, but the relationships obtained from such models may condense complex cause-and-effect linkages into associations which are much more readily computed. For example, although radiation and soil water may in fact be the major physical linkages between the climate and, say, tomato or rice production, the common unavailability of such climate data usually means that for practical purposes the more readily obtainable data on rainfall, temperature, or sunshine are used.

In general, climate and agricultural production relationship models emphasize the importance of climate variability, and either the productivity of 'human-altered environments' such as the Iowa corn field, the Korean rice paddy, or the New Zealand dairy farm, or the productivity of 'natural' environments such as the grasslands of the Australian merino sheep or the Wyoming range cattle. The economic and political importance of larger-scale weather and climate fluctuations must also be considered, as they contribute to variability in global agriculture production. Specific examples of weather- and climate-based systems of forecasting agricultural production in New Zealand are now discussed.

2. Dairy production forecasts: A New Zealand example

(a) An overview

New Zealand is the world's largest *exporter* of dairy products, marketing and selling its products in more than 100 countries. For this reason, quality, regular supply, and price are very important to the New Zealand dairy industry, and of prime importance is the guarantee of a regular supply of specialized dairy products to an increasingly food-conscious world. This regular supply is dependent to a very large extent on the production from the dairy farm. It follows that an assessment of the expected production is of paramount importance to decision-makers in the New Zealand dairy industry.

(b) Dairy production variations

Milkfat processed by dairy factories in New Zealand varies from year to year (e.g. 275 million kg in 1976/77, 251 million kg in 1977/78, 274 million kg in 1978/79, and a record 349 million kg in 1985/86), and also from month to month (e.g. less than 1 million kg in June to 35–40 million kg in most Octobers, Novembers, and Decembers). The most important variations occur in the January to May period (Fig. VII.1). For example, although the production in the June–December portion of the 1969/70 season of 162 million kg was a record not exceeded until the 1976–77 season, the subsequent five-month period (i.e. January to May) in the 1969/70 season produced only 81 million kg which is one of the lowest January–May production totals on record. This substantial drop in production was mainly associated with drought conditions which occurred in January and February of 1970 (see Maunder 1971b).

Dairy production in the autumn month of March is also important – and very variable, as shown in Table VII.1. Of greater importance, however, are the differences between the actual production and the 'optimal' or 'ideal' production. Various methods can be employed to estimate the 'optimal' production, but a simple graphical analysis of production over several seasons can give a reasonable indication of the probable production –

FIGURE VII.1 Milkfat processed by dairy factories in New Zealand (million kg): 1964/65–1984/85

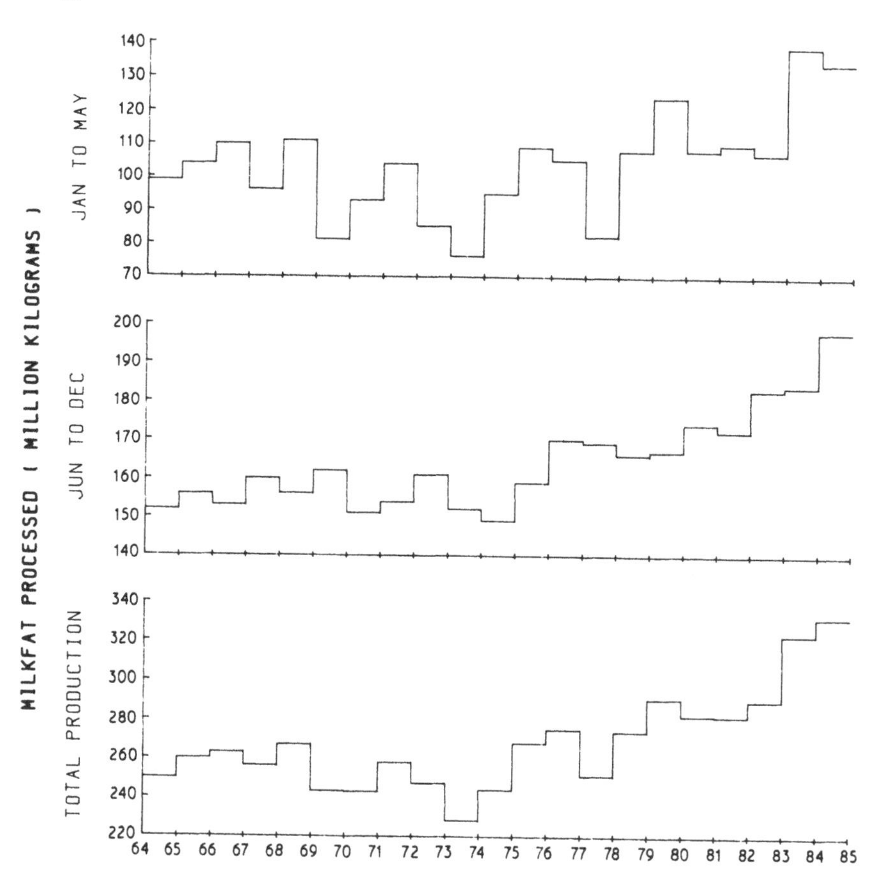

Source: Compiled from data supplied by the New Zealand and Dairy Board.

assuming both 'optimal' or 'ideal' weather *and* non-weather conditions. However, the prediction of the 'optimum' production assuming various weather and non-weather factors poses many problems and clearly is not particularly satisfactory. One solution is the use of appropriate commodity-weighted weather and climate information in a regression type analysis, in which *differences from season to season* (as shown in Table VII.1), rather than differences from some average or optimum, are used. The historical development of such an analysis for New Zealand dairy production is now given.

Table VII.1 Milkfat processed by dairy factories in New Zealand in March, and percentage changes from March–March

Year	*Production (million kg)*	*Value* ($ million)*	*Production (% of previous year)*
1968	19.5	88	79
1969	25.6	115	131
1970	13.8	62	54
1971	20.9	94	151
1972	21.7	98	104
1973	16.7	75	74
1974	16.8	76	100
1975	20.6	93	124
1976	24.1	108	114
1977	22.8	103	96
1978	17.4	78	76
1979	23.7	107	136
1980	28.0	126	118
1981	23.0	104	82
1982	24.9	112	108
1983	22.6	102	91
1984	30.8	139	136
1985	31.3	141	101

Source: Updated from Maunder (1984a). Compiled from information supplied by the New Zealand Dairy Board.
Note:
* Based on a value of $4.50/kg which is the value in May 1984 if milkfat is converted into cheese and exported. (1985 value about $5/kg.)

(c) Weighted days of soil water deficit/dairy production variations

(i) *Introduction*

Weighted soil water deficit indices are a convenient way of expressing a vast amount of weather and climate data which can be meaningful to studies relating weather and climate to agricultural production (Maunder 1974b). To assess the association between the weighted number of days of water deficit and monthly milkfat production, sample regression analyses were initially made using data for the seasons from 1964/65 to 1970/71. As previously explained the 'optimal' production in each of the January to April months was first assessed graphically. The relationship between the actual production and the days of soil water deficit in various months was then assessed for the twenty-eight January to April months from 1965 to 1971, using the following equation:

$$y = a_o + a_1x_1 + a_2x_2 + a_3x_3 + a_4x_4$$

where: y = milkfat processed by dairy factories in New Zealand expressed as a percentage of the 'optimum' for that month.

x_1, x_2, x_3, x_4 = weighted number of days of soil water deficit for the months (n − 3), (n − 2), (n − 1), and (n), respectively weighted according to the distribution of dairy cows.

Analyses were made for the six dairy regions of New Zealand as well as New Zealand as a whole. A regression analysis summary is given in Table VII.2, which shows a relatively close association between the weighted number of days of soil water deficit in the preceding two months (months (n – 1) and (n – 2)) and the milkfat production. The addition of soil water deficit data from the third month before production also permitted a better explanation of the variation in some areas. For example, for New Zealand as a whole the relevant R^2 value for the three months before production was 0.79, compared with 0.77 for the two months before production. However, the number of days of soil water deficit in the *actual* month of production accounted for only a small additional variation in milkfat production. This is an important fact, since it means that a knowledge of the soil water deficit in the month of production is *not* a prerequisite for predicting that month's milkfat production.

In the initial stages of this investigation, the actual and 'optimal' milkfat production data were used in association with the weighted soil water deficit

Table VII.2 Regression analysis summary of the association between the weighted number of days of soil water deficit and milkfat production expressed as a percentage of the 'expected' or 'optimum' production: 1964/65–1970/71

	Area				
Correlation values (R^2)	*North Auckland*	*South Auckland*	*Bay of Plenty*	*Taranaki*	*New Zealand*
Y/(n – 3)(n – 2)	0.53	0.66	0.48	0.42	0.53
Y/(n – 2)(n – 1)	0.74	0.83	0.69	0.54	0.72
Y/(n – 3)(n – 2)(n – 1)	0.80	0.91	0.71	0.59	0.79
Y/(n – 1)(n)	0.49	0.64	0.42	0.39	0.56
Y/(n – 2)(n – 1)(n)	0.78	0.85	0.71	0.62	0.77
Y/(n – 3)(n – 2)(n – 1)(n)	0.83	0.92	0.73	0.68	0.81

Source: After Maunder (1977b).
Notes:
Y = milkfat production in month (n) expressed as a percentage of the expected or optimum production.
(n – 3) = month (n – 3).
(n – 2) = month (n – 2).
(n – 1) = month (n – 1).

data to formulate predicting regression equations. However, as previously noted, it was considered that a much more useful indicator of milkfat production would be one which expressed the production in each month as a *percentage* of the production in the same months of the previous season. Typical 'difference' data for both days of soil water deficit and dairy production are shown in Table VII.3 for the main dairying area of New Zealand, for selected months, in the period 1966/67 to 1975/76. Similar calculations were made for all months for all seasons since 1950/51, for five other dairying areas, and for New Zealand as a whole.

(ii) *Predicting equations*

In order to assess the relationship between milkfat production and the seasonal differences in the weighted number of days of soil water deficit, various regression analyses were made. One of the original analyses for the seasons 1950/51 to 1971/72 is shown in Table VII.4. The squared multiple correlation coefficients (R^2) for six regions in the North Island, the South Island, and New Zealand as a whole for the same predicting equation are also shown in Table VII.5. The R^2 values indicate that in several months the *differences* in the weighted number of days with soil water deficits is a good predictor of the *differences* in milkfat production.

In the original analysis (see Maunder 1974b) data for the twenty-two seasons 1950/51 to 1971/72 were used to predict the differences in the milk-fat production between January, February, and March in 1972 and the

Table VII.3 Data for 'dairying region'* analysis – New Zealand

	*Weighted days of soil water deficit***			*Dairy production****	
Season	*Dec.*	*Jan.*	*Feb.*	*Feb.*	*Mar.*
1966–67	−0.6	0.0	−0.2	107	105
1967–68	0.0	+1.7	+12.8	92	77
1968–69	0.0	−1.1	−12.1	113	136
1969–70	+0.5	+4.7	+18.7	76	50
1970–71	+1.9	−2.0	−15.5	108	160
1971–72	−2.3	−0.1	+7.8	107	104
1972–73	+1.3	+2.1	+8.5	96	74
1973–74	−0.5	+9.9	−6.2	75	100
1974–75	0.0	−11.9	−7.7	140	130
1975–76	−0.5	−2.6	−3.9	119	115

Source: After Maunder (1977b).
Notes:
* North Auckland, South Auckland, Bay of Plenty and Taranaki Dairy Board areas. Approximately 85% of the total New Zealand milkfat production comes from these areas.
** Differences in days from previous year for same month.
*** Percentage of previous year for same month.

Table VII.4 April analysis – New Zealand (based on period 1950/51–1971/72)

Estimating equation:

$$y = a_0 + a_1 x_1 + a_2 x_2 + a_3 x_3$$
(n) (n−3) (n−2) (n−1)

$$y = 104.16 - 1.1\,x_1 - 0.9\,x_2 - 2.8\,x_3$$

Apr.	Jan.	Feb.	Mar.
S.E.*	0.9	0.5	0.6
S.L.** %	20	10	0.1

$y/x_1\,x_2\,x_3$:	R^2	= 0.71
	F ratio	= 14.4
	S.L.** %	= 0.1
Standard error of estimate		= 13.2%

Source: After Maunder (1977b).
Notes:
* Standard error.
** Significance level.

Table VII.5 Multiple regression (R^2) summary – dairying analysis

Estimating equation:

$$y = a_0 + a_1 x_1 + a_2 x_2 + a_3 x_3$$
(n) (n−3) (n−2) (n−1)

*Region**		*Months*	
	Feb.	*Mar.*	*Apr.*
North Auckland	0.81†	0.71†	0.60†
South Auckland	0.81†	0.67†	0.74†
Bay of Plenty	0.58	0.63†	0.54
Taranaki	0.26	0.49	0.67†
Gisborne/Hawke's Bay	0.58	0.73†	0.62†
Wellington	0.46	0.70†	0.62†
South Island	0.25	0.51	0.64†
New Zealand	0.69†	0.63†	0.71†

Source: After Maunder (1977b).
Notes:
* Dairy Board areas.
† Significant at 0.1% level.

Table VII.6 Milkfat production prediction for March 1973 – New Zealand

Estimating equation:					
	$y = 101.95$	$-0.7x_1$	$-0.4x_2$	$-1.5x_3$	$-0.9x_4$
	(Mar.)	(Dec.)	(Jan.)	(Feb.)	(Mar.)
Days of soil water deficit:					
1972/73		3.2	8.0	20.1	9.8
1971/72		0.6	4.9	11.5	1.8
Difference		+2.6	+3.1	+8.6	+8.0
	i.e. y = 101.95	− 1.82	− 1.24	− 12.90	− 7.20
	= 78.8				
	= 79% of 1972 March production				
% change from March 1972 production					
Forecast	= −21%				
Actual	= −24%				

Source: After Maunder (1977b).

corresponding months of 1973. A sample regression equation for the 1972–73 comparison is given in Table VII.6 for New Zealand. As shown, the predicted changes from 1972 to 1973 were very close to the actual change in milkfat production, and the ability to predict large decreases, such as that which occurred in March 1973 as a result of drier conditions in the preceding months, is noteworthy.

(iii) *Initial operational uses of the dairy production model*

The first operational use of the predicting model was in the 1972/73 season. The model has now been successively *updated each season*, so as to include more 'seasonal history' in the model, so that, for example, predictions for the 1984/85 season were based on data for the thirty-four seasons 1950/51 to 1983/84.

A specific case is now given for four areas of New Zealand using the regression equations compiled for the twenty-six seasons 1950/51 to 1975/76 which were used to predict dairy production in 1976/77. A summary of the predicted and actual production for March 1977 is given in Table VII.7. This shows a relatively small error in the March prediction for the North Island of New Zealand. However, in view of the fact that this prediction was made nearly two months *before* the actual production was known by the Dairy Board,* the error was within the 'requirements' for decision-making by the New Zealand Dairy Board.

* Actual production is not usually known until the 20th of the month following the production.

Table VII.7 Dairy production predictions for March 1977 – four areas of New Zealand

Area	*Dec.*	*Jan.*	*Feb.*	*Predicted**	*Actual***
North Auckland					
	y = 105.1 + 0.12 D – 0.10 J – 0.17 F				
Diff. days of deficit***	–7	–38	+26		
Predicted effect	+0.8	+3.8	–4.4	= 104	(97)
South Auckland					
	y = 106.0 + 0.13 D – 0.04 J – 0.19 F				
Diff. days of deficit***	0	–7	+69		
Predicted effect	0	+0.3	–13.1	= 93	(99)
Bay of Plenty					
	y = 103.4 + 0.03 D – 0.03 J – 0.11 F				
Diff. days of deficit***	–19	+38	+148		
Predicted effect	–0.6	–1.1	–16.3	= 85	(96)
North Island					
	y = 103.9 + 0.11 D – 0.03 J – 0.19 F				
Diff. days of deficit***	–6	–3	+83		
Predicted effect	–0.7	+0.1	–16.0	= 87	(94)

Source: After Maunder (1977b).
Notes:
* Issued 28 February 1977.
** Known late April 1977.
*** Difference in tenths of days of soil water deficit from same month in previous season.

It was nevertheless obvious – at this time – that there was considerable room for improvements in the model. The use of smaller time intervals (possibly ten days) for example, would clearly enable a more realistic assessment of the number of days of soil water deficit in the period preceding production. In addition one could predict the days of soil water deficit during the month before production. This would allow more lead-time to be gained, although errors would clearly increase. In Table VII.8, the results of predicting† the March days of soil water deficit (or more specifically the difference in the days of soil water deficit from the previous March) on 14 March 1977 are shown. In this case dairy production predicted for the South Auckland area of New Zealand for April varied from 119 per cent (of April

† Based on the *known* days of soil water deficit from 1–14 March, plus the minimum, 'expected', and maximum number of days of soil water deficit for the period 14–31 March, using the known soil moisture on the 'regional dairy farm' on 14 March and the five-day weather forecast.

Table VII.8 Milkfat predictions for South Auckland, New Zealand: April 1977 – as at 14 March 1977*

Estimating equation:

$$\underset{\text{(April)}}{y} = 110.6 - \underset{\text{(Jan.)}}{0.23\,x_1} - \underset{\text{(Feb.)}}{0.12\,x_2} - \underset{\text{(March)}}{0.34\,x_3}$$

Days of soil water deficit			*March*		
	Jan.	*Feb.*	*Min.*	*Average*	*Max.*
1977	0.0	7.2	11.5	15.1	29.1
1976	0.3	1.0	16.1	16.1	16.1
Difference (1977–76)	−0.3	+6.2	−4.6	−1.0	+13.0

i.e. y = 110.6 + 0.7 − 7.4 + 15.6 (minimum March)
+ 3.4 (expected March)
− 44.2 (maximum March)

= 119 (with no further days of soil water deficit in March)
107 (with 'average' number of days of soil water deficit in March)
60 (with *no* further rain in March; i.e. maximum number of days of soil water deficit)

Source: After Maunder (1977b).
Note:
* Actual production – not known until 25 May 1977 – was 80% of April 1976 production.

1976) if there were no further days of soil water deficit after 14 March (which would have occurred if there was adequate rain from *that* date), to 60 per cent if there was *no* rain from 14 to 31 March. Although the forecasting of the number of days of soil water deficit introduces errors, this is usually compensated for by the earlier availability of the dairy production predictions to the user, in this case the New Zealand Dairy Board.

(d) Operational dairy production models

The current weather-based dairy production models used in New Zealand are based on the *differences* in the commodity-weighted number of days of soil water deficit, the commodity-weighted rainfalls, and the commodity-weighted mean temperatures. Differences in these indices are then used in the various models to predict dairy production as a percentage change from the corresponding month(s) in the previous season. Typical days of soil water deficit data for New Zealand which are used in these models are given in Table VII.9, and typical dairy production data for New Zealand are shown in Fig. VII.1.

In order to quantify these relationships various regression equations are used:

$$y_n = a + bx_{(n-3)} + cx_{(n-2)} + dx_{(n-1)} + ex_{(n)} \quad (1)$$

$$y_n = a + bx_{(n-3)} + cx_{(n-2)} + dx_{(n-1)} \quad (2)$$

$$y_n = a + bx_{(n-3)} + cx_{(n-2)} \quad (3)$$

$$y_n = a + bx_{(n-3)} \quad (4)$$

where x = the weighted differences in the respective climatic variables used in the particular model, for the months (n – 3), (n – 2), (n – 1), and (n) respectively.

y_n = dairy production in month (n) as a percentage of the production in the same month of the previous season.

In practice it has been found that equation (2) is the most useful, in that it allows a prediction of production *differences* in month (n), say March (early autumn), using only the weather data *difference* for months (n – 1), (n – 2), and (n – 3), i.e. February, January and December. Further, if the (n – 1) month weather conditions (Feburary in this case) can be forecast, then the actual prediction for, in this case, March could be done as early as

Table VII.9 Weighted* number of days of water deficit – New Zealand

Season	*Jan.*	*Feb.*	*Mar.*	*Apr.*	*Jan.–Apr.*
1964/65	2	1	0	0	3
1965/66	0	1	1	0	2
1966/67	1	1	2	2	6
1967/68	2	14	11	1	28
1968/69	2	2	7	4	15
1969/70	6	18	7	1	32
1970/71	4	6	1	3	14
1971/72	5	12	2	0	19
1972/73	8	20	10	4	42
1973/74	15	13	4	1	33
1974/75	4	7	4	1	16
1975/76	1	2	15	1	19
1976/77	1	8	13	2	24
1977/78	7	15	18	7	47
1978/79	12	6	1	0	19
1979/80	1	2	0	0	3
1980/81	3	8	7	2	20
1981/82	6	11	1	2	20
1982/83	9	14	12	1	36
1983/84	3	2	1	1	7
1984/85	2	3	2	4	11

Source: Updated from Maunder (1978b).
Note: * By distribution of dairy cows.

mid-January. Similarly, equations (3) and (4) can be used, giving even more lead-time. Indeed, for many purposes, a reasonably successful forecast using equation (4), i.e. forecasting April (mid-autumn) production, using only January (mid-summer) weather information, may well be of greater value than a much more 'accurate' forecast given at the beginning of the month of production.

A sample prediction equation and the resultant predictions for April 1978 are given in Table VII.10, and this shows that using equation (4) (that is using only January weather differences to predict April production differences), the predicted New Zealand dairy production in April 1978 was a relatively low 71 per cent of the April 1977 production (see note to Table VII.10), whereas the actual April 1978 production was a very low 46 per

Table VII.10 Predicted New Zealand dairy production in April 1978 as a percentage of April 1977, using January weather information

Predicting equation:

$$y = 106.3 - 0.72\,x_1 - 0.08\,x_2 + 0.33\,x_3$$

where y = milkfat production expressed as a % of the same month last season (i.e. April 1978 as % of April 1977)

x_1 = difference in the deviation of the weighted† mean temperatures (tenths of degree) from the 1941–70 normal, between one *January* and the preceding *January* (i.e. January 1978–January 1977).

x_2 = as for x_1, but differences in the weighted† number of days of soil water deficit (tenths of days).

x_3 = as for x_1, but differences in the weighted† rainfalls expressed as a % of the 1941–70 normal.

Prediction for April 1978 as % of April 1977:

$$y = 106.3 - 0.72\,(+14^{*}) - 0.08\,(+62^{**}) + 0.33\,(-61^{***})$$

* January 1978 = +0.8°C; January 1977 = −0.6°C
Therefore difference = +1.4° = +14 units

** January 1978 = 7.0 days; January 1977 = 0.8 days
Therefore difference = +6.2 days = +62 units

*** January 1978 = 41%; January 1977 = 102%
Therefore difference = −61 units

$$\text{i.e. } y = 106.3 - 10.1 - 5.0 - 20.1 = 71.1$$

Therefore: Predicted New Zealand dairy production in April 1978 as a % of the April 1977 production is 71%

Source: After Maunder (1978b).

Notes:

† Weighted by the distribution of dairy cows.

This prediction model used weather information only up to January 1978 for predicting production in April 1978. The actual production in April 1978 (which was not known until late May) was 7.2 m.kg or 46% of the April 1977 production.

cent. However, since the actual production of 46 per cent was not known until late May 1978, the predicted production of 71 per cent, although a little 'optimistic', did give about 15 weeks' lead-time to the decision-makers of the impending *dramatic* fall in production from April 1977 to April 1978.

The importance of providing production forecasts with sufficient lead-time to enable appropriate decisions (transportation, marketing, etc.) to be made cannot be overemphasized and a summary of the progressive forecasts made during the 1977/78 season is given in Table VII.11. The dramatic 'drop' in production in February to April of the 1977/78 season compared with February to April of the 1976/77 season, shown in the table, was well anticipated (e.g. with an 11-week lead-time the March decrease in production was predicted to be 69 per cent of March 1977, compared with an actual decrease of 76 per cent). The significance of the 8- to 15-week lead-time predictions is well illustrated if one considers the cumulative production (actual and forecast) over the season. Table VII.12 gives such information for the 1977/78 season. It is considered that such weather-based forecasts are of direct benefit to the decision-makers within the dairy industry, and if correctly applied will clearly lead to a considerably more 'efficient' system 'beyond the farm gate'.

Table VII.11 Predicted milkfat production in New Zealand during 1977/78 season, with lead-time (LT) (in weeks) before actual production was known

	Month predicted													
Month of prediction†	*Oct.*		*Nov.*		*Dec.*		*Jan.*		*Feb.*		*Mar.*		*Apr.*	
	%	*LT*	*%*	*LT*	*%*	*LT*	*%*	*LT*	*%*	*LT*	*%*	*LT*	*%*	*LT*
Jul.	101*	(15)*												
Aug.	103	(11)	103	(15)										
Sep.	101	(8)	105	(11)	105	(15)								
Oct.	106	(3)	104	(8)	105	(11)	102	(15)						
Nov.			102	(3)	103	(8)	99	(11)	104	(15)				
Dec.					104	(3)	102	(8)	107	(15)	108	(15)		
Jan.							101	(3)	76	(8)	69	(11)	71	(15)
Feb.									74	(3)	63	(8)	68	(11)
Mar.											82	(3)	57	(8)
Apr.													55	(3)
Actual production	98		102		99		95		81		76		46	

Source: After Maunder (1978b).

Notes:

† End of month shown.

* i.e. Prediction was made at end of July 1977 for the New Zealand October 1977 milkfat production to be 101% of the actual production in October 1976. The actual production of 98% was not known until late November, giving a lead-time of 15 weeks.

Table VII.12 Cumulative predicted and actual milkfat production in New Zealand during 1977/78 in million kg

Month of prediction†	*Known production 77/78*	*Predicted production 77/78*	*Known + predicted production* A	*Actual production 77/78* B	*Actual production 76/77* C	$\frac{A}{C}$ (%)	$\frac{B}{C}$ (%)
	Oct.	*Oct.–Jan.*		*October to January*			
Oct.	–	159.6	159.6	151.0	153.0	104	99
	Oct.	*Nov.–Feb.*		*October to February*			
Nov.	39.5	142.0	181.5	172.6	179.7	101	96
	Oct.–Nov.	*Dec.–Mar.*		*October to March*			
Dec.	80.1	128.3	208.4	190.0	202.5	103	94
	Oct.–Dec.	*Jan.–Apr.*		*October to April*			
Jan.*	117.7	82.2	199.9	197.2	218.1	92	90
	Oct.–Jan.	*Feb.–Apr.*		*October to April*			
Feb.	151.0	44.8	195.8	197.2	218.1	90	90
	Oct.–Feb.	*Mar.–Apr.*		*October to April*			
Mar.	172.6	27.5	200.3	197.2	218.1	92	90
	Oct.–Mar.	*Apr.*		*October to April*			
Apr.	190.0	8.6	198.6	197.2	218.1	91	90

Source: After Maunder (1978b).
Notes:
† End of month shown.
* e.g. At the end of January, the known production (October to December) was 117.7 million kg, the predicted production (January to April) was 82.2 million kg, giving a 'known plus predicted' production for the October to April period of 199.9 million kg. This is 92% of the production for the same period of the previous season, and compares with 90% which was the actual percentage not known until late May 1978.

(e) The use and value of dairy production predictions

The value of being able successfully to predict milkfat production up to several months ahead is considerable, as a 10 per cent change in the New Zealand January to April monthly production is worth from \$10 to \$20 million *per month*. A comparison of the predicted and actual values of milkfat production for various areas for March 1973 is given in Table VII.13 which shows that, using commodity-weighted days of soil water deficit differences for the month of prediction (n) and the three preceding months (n – 1, n – 2, n – 3), the predicted *loss* of dairy production for New Zealand was (in 1977 dollars) worth \$11 million, compared with an actual *loss* of \$13 million.

The 'positive' value of rain or the 'negative' value of continuing dry conditions to the New Zealand dairy industry can also be assessed from the dairy production model, since the model is based on the number of weighted

Table VII.13 The value* of predicted and actual milkfat production differences between March 1972 and March 1973

Estimating equation:

$$(n) = (n - 3) + (n - 2) + (n - 1) + (n)$$

Region	*Predicted difference*** *($m)*	*Actual difference ($m)*
North Auckland	−0.9	−3.1
South Auckland	−2.8	−4.0
Bay of Plenty	−0.5	−0.5
Taranaki	−3.8	−3.1
Hawkes Bay/Gisborne	−0.2	−0.9
Wellington/Manawatu/Wairarapa	−0.5	−0.2
South Island	−0.5	−0.2
New Zealand	−11.3	−13.0

Source: After Maunder (1977b).
Notes:
* Based on the 1971/72 production and a value of $2.50/kg, which is the 1977 value if milkfat is converted into cheese and exported (1985 value about $5/kg).
** The difference between March 1973 and March 1972.

days of soil water deficit (or more specifically the *differences* in the weighted days of soil water deficit from March to March, April to April, etc.) and each weighted day of soil water deficit has a *specific value* in terms of production and hence revenue. This is given in Table VII.14 which shows the 'value' of rain in March 1977 in terms of the April 1977 milkfat production in New Zealand. Specifically, the model on the 14 March 1977, based on the *actual* days of soil water deficit in January and February and the first fourteen days of March, showed that if no rain fell during the rest of March, then the April 1977 production would be 60 per cent of the April 1976 production (see note to Table VII.14). The model also showed that rain sufficient to 'saturate' the soil would have effectively increased production by about 4 per cent *for each day* that the soil retained moisture sufficient for grass growth. For example, as shown in Table VII.14, 'saturation rain' on the 25 March would have increased the predicted April production from 60 to 82 per cent, whereas 'saturation rain' ten days earlier on the 15 March would have nearly doubled the predicted April production from 60 to 119 per cent. The monetary value of this rain, also shown in Table VII.14, indicates that an extra ten days of grass growth on the 'national dairy farm' (or more specifically an extra ten days in which there were *no* days of soil water deficit) from 20 to 30 March would have been worth about 5.67 million kg (6.51−0.84) in milkfat

Table VII.14 Value of rain in March 1977 to New Zealand dairy industry in April 1977

Date of rain to saturate soil	*Predicted April production** %	*Effective change from that predicted on 14 March* %	m.kg	$m.**
15 March	119	+59	+9.85	+24.6
16 March	113	+53	+8.85	+22.1
17 March	109	+49	+8.18	+20.5
20 March	99	+39	+6.51	+16.3
25 March	82	+22	+3.67	+9.2
30 March	65	+5	+0.84	+2.1

Source: After Maunder (1977c).
Notes:
* Percentage of 1976 production of 16.7 m.kg.
** Based on 1977 value of $2.50/kg which is value if milkfat is converted into cheese and exported. (1985 value about $5/kg.)

As of 14 March 1977 the April 1977 production was predicted to be 60 per cent of the April 1976 production, but would have increased if any rain occurred during March. Specifically, the April 1977 production could have increased to 119 per cent of the 1976 production, if rain equivalent to 17 days (31 − 14) of soil moisture (approximately 50 mm) occurred in the area. The value of this rain decreases with time from data of prediction (14 March) as shown in the table.

production, or about $14 million (in 1977 dollars and prices) on a New Zealand wide basis.

The use of successive models (that is using January only, January plus February, January plus February plus March) to predict, say, April dairy production also provides useful and valuable information to the New Zealand dairy industry. A summary of typical progressive dairy production forecasts made for April 1978 is given in Table VII.15. As shown, various assumptions were made in each of the models (for example, assuming *no* further rain in February as from 15 February), and most predictions show that the 'dramatic' drop in production in April 1978 compared with April 1977 was well anticipated. It is considered that such weather-based forecasts are of direct benefit to decision-makers within the dairy industry, and if correctly applied could lead to a considerably more efficient system. For example, the forecasts of production given in Table VII.15 would have provided the New Zealand Dairy board with considerable advance notice of the quantity of milkfat (and by inference the amount of milkfat that would have been available to convert into butter, cheese, dried milk, etc.) available for potential markets outside of New Zealand. Similarly, the forecasts would have given valuable lead-time to the several shipping companies serving New Zealand of the quantity of milkfat likely to be available for export.

Of course, such weather-based forecasts are not always accurate; indeed,

Table VII.15 New Zealand milkfat production predictions for April 1978

Date of prediction	*Model**	*Assumption*	*Predicted production differences***	
			%	*$ million*
Jan. 9	J	No further rain in Jan.	−30	−12
Jan. 31	J	None	−35	−14
Feb. 1	J + F	Normal Feb.	−19	−7
Feb. 1	J + F	As Feb. 1977	−27	−11
Feb. 1	J + F	No rain in Feb./Temp. +1.5°C	−54	−21
Feb. 1	J + F	No days of soil water deficit/Temp. −1.5°C	−14	−55
Feb. 15	J + F	No further rain in Feb.	−52	−20
Feb. 15	J + F	No further days of soil water deficit in Feb.	−31	−12
Feb. 28	J + F	None	−45	−18
Mar. 1	J + F + M	Normal Mar.	—32	−13
Mar. 13	J + F + M	No further rain in Mar.	−69	−27
Mar. 31	J + F + M	None	−62	−24

Source: After Maunder (1986b).
Notes:
* J = January (i.e. model used January weather data only); J + F = January + February (i.e. model used January and February weather data).
** April 1978 compared with April 1977.
Actual difference was −54%, 'worth' $21 million at 1978 prices.

in the 1982/83 season, although the early season was correctly forecast to be in excess of the corresponding 1981/82 production (for example, the forecasts of production for September 1982 compared with September 1981, made on 1 July, 1 August, and 1 September, of 107, 108, and 106 per cent respectively compared very favourably with the actual production of 109 per cent), some of the forecasts made for the autumn months were considerably astray. For example, although the April 1983 production of 74 per cent of the April 1982 production was correctly forecast on 1 March, the forecast issued one month earlier on 1 February of 135 per cent was seriously in error. The reason for these errors is partly an incomplete understanding of the weather/biological processes, partly the unavailability of milk production data in a usable form for shorter periods than one month, and partly the influence of other factors (such as the 'drying off' of dairy cows) which are only partially related to weather and climate factors.

The weather-based dairy production models used operationally in the New Zealand Meteorological Service are modified and improved on the basis of operational experience, the latest (1984) modification being the incorporation of the known *actual dairy production data* (specifically, *differences* in the actual monthly dairy production that is known at the time of the prediction) in the model. This additional information is providing predictions of dairy production which are more biologically based, and thereby provide improved forecasts of national dairy production. A good example of these improvements was shown in the 1984/85 season in which the likelihood of a record New Zealand milkfat production in 1984/85 was correctly forecast several months ahead of June 1985 when the final figure of 332 million kg, or 3 per cent above the previous highest production which was itself 10 per cent above the previous highest production, was officially released.

3. Wool production forecasts: A New Zealand example

(a) Wool production variations

One of the main requirements for an efficient production, processing, and marketing system in the wool industry is the accurate forward estimation of the quantity and quality of wool which the system is required to handle. However, most wool producing countries are characterized by significant

Figure VII.2 Fluctuations in New Zealand sheep population and wool production: 1965/66–1983/84 (data give percentage of the previous season)

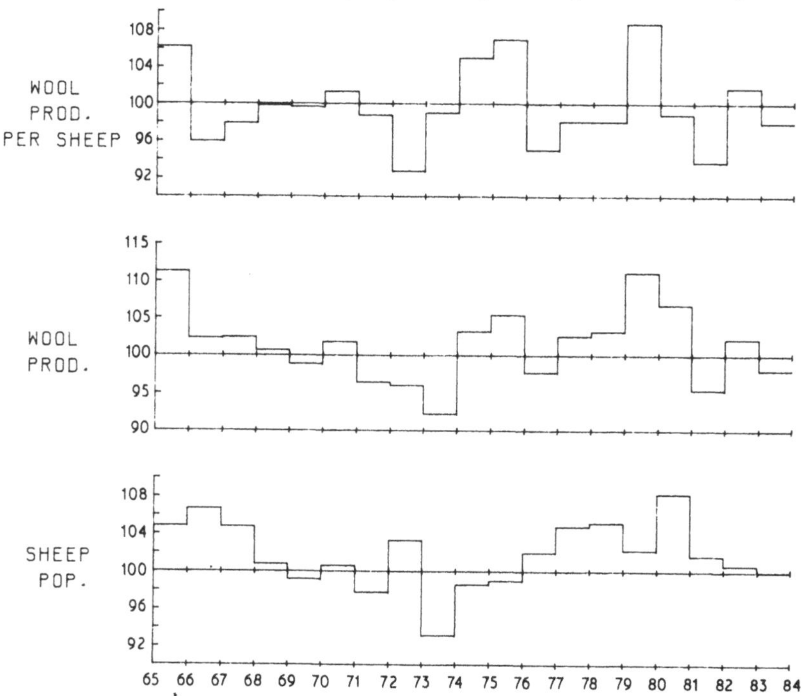

Source: Compiled from data supplied by the New Zealand Meat and Wool Boards' Economic Service.

variations in the total volume of wool which comes forward each season. It is important therefore to establish a methodology for accurately estimating in advance – preferably one season in advance – the volume of wool available, and adjusting this within the season as new information becomes available.

One of the 'customers' for this advance information on wool production in New Zealand is the Economic Service of the New Zealand Meat and Wool Boards, which is required to predict annually, with in-seasonal adjustments, the national wool clip. These estimates are necessary for making farm income, production, and export volume assessments for the New Zealand Meat Producers' Board and the New Zealand Wool Board, such estimates also being of considerable value to other organizations in New Zealand such as the Institute of Economic Research and the Reserve Bank of New Zealand. Until recently estimates of New Zealand's wool production were based largely on 'subjective expert opinion', but more recently econoclimatic models which use seasonal weather conditions – specifically, *differences* in the weather between seasons – for predicting the national wool production in the subsequent season have been developed (see Maunder 1980a). Variations in this wool production are quite considerable, as illustrated in Fig. VII.2 which shows the season-to-season fluctuations in New Zealand's sheep population, wool production, and wool production per sheep for the period 1965/66 to 1983/84.

(b) VARIABLES AFFECTING WOOL PRODUCTION*

Liveweight: Within sheep of the same breed, heavier sheep on the average produce more wool than lighter sheep since heavier sheep have a greater wool-producing surface. The effect of sheep liveweight on wool production was quantified in an early study by Coop and Hayman (1962) whose observations on 4000 Corriedale sheep showed that a 10 per cent increase in liveweight was associated with a 4.3 per cent increase in fleece weight. Thomson (1975) used this finding in an attempt to predict the national per-head wool clip in New Zealand, using as a practical indicator of liveweight the average 'cull export ewe slaughter weight'. The resulting equation satisfactorily explained changing per-head wool production over a 34-year period, but the equation was unstable through time and therefore unreliable for prediction purposes. A probable explanation of this instability is that farmers over time manage their cull ewes differently from their breeding ewes.

---

* Different kinds of factors affect different kinds of production. This summary, based on studies by Maunder (1967a) and Thomson (1975), who reviewed previous research on the subject, applies to wool production in New Zealand. Similar factors would affect wool production in other countries, and similar but necessarily different factors would have to be considered for other agricultural commodities in New Zealand as well as all other countries.

Winter–summer differences: The rate of wool growth in mid-winter is approximately one-third that in late summer/early autumn, during which time wool growth is at a maximum. Early work by Morris (1961) and Hart *et al.* (1963) showed that this seasonal difference in wool growth rate was primarily the result of the changing 'light:dark' ratios between seasons. From a *national* viewpoint, this ratio remains relatively static over time; accordingly it is not usually necessary to include the light:dark ratio as a variable affecting the *year-to-year* variations in per-head wool production.

Level of grease in wool: The level of grease in fine-woolled fleece is significantly higher than in fleeces derived from strong-woolled sheep. The proportion of fine-woolled to strong-woolled sheep in the national flock will therefore affect national wool production. However, this proportion changes only slowly over time and it can usually be ignored as a variable affecting the year-to-year variations in per-head wool production.

Changing proportion of ewes: Breeding ewes tend to grow less wool than barren ewes and wethers of the same breed. The changing proportion of ewes in the national flock could therefore affect wool production over time. As with the two previous factors, however, since the proportion of ewes in a national flock does not change significantly from year to year, it is considered that it need not be taken as a significant variable.

Farm investment: On-farm investment in one year, especially fertilizer application, obviously has effects on per-head wool production in the following year and in some cases the following years. For example, in the period from 1966/67 to 1971/72, on-farm investment in New Zealand measured by real expenditure per stock unit fell, the lagged effect of this lower investment showing up in the declining per-head wool clip through to the 1973/74 season. Although significant, its effect can best be considered as an 'additional factor', which should be taken into account after weather factors have been assessed. However, as previously noted in Chapter V, weather and climate variations do have an effect *on* farm investment and hence the two-way 'farm investment weather mix' must always be considered by decision-makers.

Facial eczema: Facial eczema is a significant sheep disease which occurs most frequently when a hot dry summer, causing premature death of pasture plants, is followed by a warm wet autumn. This disease is responsible for severe wool production losses in localized areas and clearly affects on-farm and regional production.

Double shearing: Sheep shorn twice a year tend to clip more wool annually than if they had been shorn only once a year. As a result, in years of high wool *price* expectations the incidence of second shearing increases. On a national basis, however, the *change* in the number of sheep shorn twice a year is relatively small in comparison with the total sheep shorn in New Zealand.

(c) Econoclimatic models of wool production

The discussion in the previous section of the factors affecting wool production per sheep suggests that the significant variable, at least as far as New Zealand sheep are concerned, is sheep liveweight. From a national point of view the *weather and climate* factors affecting liveweight of sheep therefore assume importance, and in the econoclimatic models of wool production used in New Zealand (see Maunder 1980a) the soil water deficit and the temperature deviations from normal proved to be useful indicators of the 'true' weather variables. For example, months in which the soil water deficit index is high imply that there is less moisture available during that month for grass growth; therefore in such months sheep could be expected to be relatively poorly fed. However, there is a considerable time lag between the lack of soil moisture for pasture growth and a change in the growth of wool. This lag is of the order of several months; indeed, it is evident that a drought, particularly in the autumn, not only has an effect on the feed available for sheep at that time, but also a latent effect on both lambing percentages and wool production in the *following* season.

The agro-economic data used in the New Zealand wool production models are derived from the New Zealand Wool Board, the sample surveys of the New Zealand Meat and Wool Boards' Economic Service, and the relevant commodity-weighted weather information as assessed by the New Zealand Meteorological Service. The national weighted number of days of soil water deficit for New Zealand for the thirty-six seasons 1949/50 to 1984/85 using the sheep population as weightings have been given earlier in Fig. IV.1. These data are for New Zealand as a whole, and can therefore be said to represent the 'national sheep farm'. The indices provide a good indication of just how variable the New Zealand 'wool climate' is; noteworthy are the very high values in the drought seasons of 1972/73 and 1977/78, which are in marked contrast to the 'green era' of the mid-1960s. For example, the 'growing season' or June–May indices, weighted by the distribution of sheep, give a *total index* of 38 days for the two consecutive seasons 1964/65 and 1965/66, this being in marked contrast to the *average index* during the six seasons 1972/73 to 1977/78 of 42 days, and the total index during the two consecutive drought seasons of 1972/73 and 1973/74 of 101 days.

The use of this weighted weather information in weather-based wool production models is now discussed. Real-time wool production forecasts based on commodity-weighted weather/wool production models have been made by the New Zealand Meteorological Service on a routine basis since 1977/78, and they continue to be modified and improved. In addition, an earlier weather-based model was developed by the New Zealand Meat and Wool Boards' Economic Service. The progressive development of these models is now outlined.

The New Zealand Meat and Wool Boards' Economic Service used soil water deficit indices in an econoclimatic model first developed in 1975 (Thomson 1975) and modified by Rich and Taylor (1977). These models included a lag term of soil moisture deficit of 4, 5, and 6 months, and R^2 values of about 0.80 were obtained (significant at the 0.1 per cent level). The models used data of the soil water deficit up to the end of December and January (that is mid-summer for the current season), and although useful, they did not give very much lead-time of the actual seasonal production. The New Zealand Meat and Wool Boards' Economic Service has progressively updated and improved this model, and in association with the models developed by the New Zealand Meteorological Service endeavours to provide decision-makers in the New Zealand wool industry with as much advance information on the probably national wool production as is possible.

The weather-based wool production models developed by the New Zealand Meteorological Service (see Maunder 1980a) are based on the *differences* in both the commodity-weighted number of days of soil water deficit and the commodity-weighted mean temperatures. These differences are used in the models to predict wool production as a *percentage change* from the previous season. Typical monthly days of soil water deficit data (shown as *differences*) used in these models are given in Table VII.16, and typical wool production data are shown in Fig. VII.2. In order to quantify

Table VII.16 Weighted* soil water deficit indices – New Zealand 'sheep farm'

Season	*Jan.*	*Feb.*	*Mar.*
1970/69	−1	+6	−4
1971/70	+1	−1	+1
1972/71	+3	+1	−2
1973/72	+4	+7	+6
1974/73	+2	−8	−4
1975/74	−6	−3	−3
1976/75	−3	−3	+11
1977/76	−5	+1	−2
1978/77	+10	+9	+5
1979/78	0	−6	−15
1980/79	−9	−5	−2
1981/80	+8	+7	+7
1982/81	+1	+1	−2
1983/82	−2	+1	+7
1984/83	−1	−9	−11
1985/84	+1	+5	+6

Source: Updated from Maunder (1980a).
Notes:
* Weighted by the distribution of sheep. The index measures differences between seasons in days (e.g. January 1978–January 1977 = +10 days).

these relationships various step-wise regression models are used involving monthly or combined monthly commodity-weighted weather data for each of the twelve months in the season before production. For example, data for the months June 1984 to May 1985 would be used for the 1985/86 production season, as well as in some cases monthly weather data for the actual production season.

Table VII.17 The first model: predicted New Zealand wool production per hectare in 1978/79 as a percentage of 1977/78, using weather information up to May 1978

Predicting equation:

$$y = 101.4 + 0.038x_1 - 0.039x_2 + 0.022x_3 - 0.048x_4$$

where y = wool production per hectare expressed as a % of the previous season (i.e. 1978/79 as % of 1977/78).

x_1 = difference in the weighted number of days (tenths of days) of soil water deficit for the period June to October between seasons (i.e. June to October 1977 compared with June to October 1976).

x_2 = difference in the deviation of the weighted cumulative mean temperatures (tenths of degree) from normal for the period June to October between seasons.

x_3 = as for x_1 but period November–January.

x_4 = as for x_1 but period February–May.

Prediction for 1978/79 as % of 1977/78:

$$y = 101.4 + 0.038\,(+5^{*}) - 0.039\,(+1^{**}) + 0.022\,(+134^{***}) - 0.048\,(+141^{****})$$

* June to October 1977 = 0.5 days
June to October 1976 = 0.0 days
Therefore difference = +0.5 days = +5 units

** June to October 1977 = −1.2°C (cumulative)
June to October 1976 = −1.3°C (cumulative)
Therefore difference = +0.1°C = +1 unit

*** November 1977 to January 1978 = 16.3 days
November 1976 to January 1977 = 2.9 days
Therefore difference = +13.4 days = +134 units

**** February 1978 to May 1978 = 38.4 days
February 1977 to May 1977 = 24.3 days
Therefore difference = +14.1 days = +141 units

Therefore $y = 101.4 + 0.2 - 0.0 + 3.0 - 6.8 = 97.8$

Therefore predicted New Zealand wool production per hectare in 1978/79 as a % of the 1977/78 production is 97.8%.

Source: After Maunder (1980a).
Notes:
(1) The weightings are by the distribution of sheep.
(2) This prediction model used weather information only up to May 1978, for predicting the total production in the 1978/79 season.

The *first weather/wool model* included data for the combined months June to October, November to January, and February to May. A summary of this model and the forecast of the *wool production per hectare* for the 1978/79 season (as a percentage of the 1977/78 season) is given in Table VII.17. This indicated that, using weather information up to May 1978, the predicted *wool production per hectare* for the 1978/79 season (that is the season *following* the weather used in the model) would have been 97.8 per cent of the 1977/78 production. The main reason for the 2.2 per cent decrease in production was – according to the model – the impact of the 1978 autumn drought.

Predictions for seasons previous to 1978/79 using the same type of model are shown in Table VII.18 together with other factual information about the New Zealand wool industry such as sheep population and total wool production, and this indicated that useful forecasts of wool production can be given *before the start of a season*. Indeed, up to fourteen months' lead-time can be given before the actual production is known. While the variations (both predicted and actual) may appear to be small, even a 2 per cent variation from season to season is equivalent to a variation in production of 6 million kg of wool, worth at least $30 million and requiring about one container ship to transport the wool from New Zealand to the overseas market.

Table VII.18 New Zealand wool production: Predicted and actual* (1973/74–1978/79)

Date of prediction	*Season*	*Predicted wool/ha*	*Actual sheep population*	*Actual production /ha*	*/Sheep*	*Total*
May 1973	1973/74	97	93	95	99	92
May 1974	1974/75	106	99	116	105	103
May 1975	1975/76	102	99	101	107	106
May 1976	1976/77	97	102	101	95	97
May 1977	1977/78	102	105	(104)	98	103
May 1978	1978/79	98	(105)	(99)	(92)	(103)

Source: Updated from Maunder (1978b). Actual data compiled from information supplied by the New Zealand Meat and Wool Boards' Economic Service, and the Department of Statistics.

Notes:

* All data expressed as a percentage of the previous season.

(1) Actual data on wool production per hectare is usually not available until at least 15 months after the end of the season. Thus the data for 1976/77 would be available in November 1978.

(2) Data on the sheep population is for the beginning of the season (i.e. 30 June).

(3) Data on the actual wool production per sheep and the total actual wool production is known at the end of each season, usually with a delay of about one month. Thus as of May 1978, when a weather-based forecast of the wool production per hectare was made for 1978/79 season, the actual production was known for the 1976/77 season, and provisionally available for 1977/78, the final production for the 1977/78 season not being available until August 1978.

(4) Values in brackets not known in May 1978 when wool production per hectare prediction made.

The *second weather/wool model* included data for the individual months June to December of the season *prior* to that being forecast, and associated combinations of months in a step-wise regression method with *wool production per sheep* in the following season. This model (details are given in the original paper; see Maunder 1980a) indicated that using weather information up to December 1977, the predicted *wool production per sheep* for the (*following*) 1978/79 season would have been 102.4 per cent of the 1977/78 production. In this case the lead-time is considerable, being six months prior to the start of the 1978/79 season and eighteen months prior to the end of the 1978/79 season.

Table VII.19 The third model: Predicted New Zealand wool production per sheep in 1978/79 as a percentage of 1977/78, using weather information up to April 1978

Predicting equation:

$$y = 999.19 + 0.725\,x_1 - 0.371\,x_2 - 0.235\,x_3 - 0.488\,x_4$$

where y = wool production per sheep expressed as a % (times 10) of the previous season (i.e. 1978/79 as % of 1977/78).

x_1 = difference in the weighted number of days (tenths of days) of soil water deficit between December of the previous summer and the previous December (i.e. December 1977 compared with December 1976).

x_2 = as for x_1 but January (i.e. January 1978 compared with January 1977).

x_3 = as for x_1 but February.

x_4 = as for x_1 but April.

Prediction of 1978/79 as % of 1977/78:

$$y = 999.19 + 0.725\,(+34^{*}) - 0.371\,(+95^{**}) - 0.253\,(+90^{***}) - 0.488\,(+2^{****})$$

* December 1977 = 4.0 days, December 1976 = 0.6 days.
Therefore difference = +3.4 days = +34 units.

** January 1978 = 10.9 days, January 1977 = 1.4 days.
Therefore difference = +9.5 days = +95 units.

*** February 1978 = 15.4 days, February 1977 = 6.4 days.
Therefore difference = +9.0 days = +90 units.

**** April 1978 = 4.8 days, April 1977 = 4.6 days.
Therefore difference = +0.2 days = +2 units.

Therefore $y = 999.19 + 24.65 - 35.25 - 22.77 - 0.98 = 964.8$

Therefore predicted New Zealand wool production per sheep in 1978/79 as a percentage of the 1977/78 production is 96.5%.

Source: After Maunder (1980a).

Notes:

(1) The weightings are by the distribution of sheep.

(2) This prediction model used weather information only up to April 1978, for predicting wool production per sheep in the 1978/79 season. The lead-time is therefore one month prior to the start of the 1978/79 season and 13 months prior to the end of the 1978/79 season.

(3) The actual wool production in 1978/79 as a percentage of the 1977/78 production was 98.1%.

The *third weather/wool model* included data for the individual months June to May of the season *prior* to that being forecast, and associated combinations of months in a step-wise method with *wool production per sheep* in the *following* season. A summary of this model using data for four of the twelve months in the forecast of the wool production per sheep for 1978/79 (as a percentage of the production in the 1977/78 season) is given in Table VII.19. This indicates that using weather information up to April 1978, the predicted wool production per sheep for the 1978/79 season would have been 96.5 per cent of the 1977/78 season. The prediction is therefore 6 per cent *lower* than that obtained using the second model (i.e. 96.5 compared with 102.4 per cent), but it does include the important late summer and autumn months which in 1978 were very dry compared with the same months in 1977. The third model therefore predicted that the 1978 late summer/autumn drought (in particular) would reduce the 1978/79 wool production per sheep by about 3.5 per cent from that produced in 1977/78. In this model, weather information is used up to April 1978; thus the lead-time is one month prior to the start of the 1978/79 season and thirteen months prior to the end of the 1978/79 season.

The *actual* wool production per sheep in 1978/79 (which was not known until August 1979) was 98.1 per cent of the 1977/78 production; therefore the weather-based forecast made in May 1978 of 96.5 per cent was reasonably close to the actual production. The same basic model as discussed above has been used to predict the wool production per sheep for the seasons following 1978/79 (that is using weather information up to April 1979 for predicting the

Table VII.20 New Zealand wool production: Predicted and actual* (1980/81–1985/85)

Date of prediction	*Season*	*Predicted total wool production*	*Predicted wool/sheep*	*Actual total wool production*	*Actual wool/sheep*
May 1980	1980/81	105	97	107	99
May 1981	1981/82	96	98	95	93
May 1982	1982/83	104	99	102	102
May 1983	1983/84	100	98	98	98
May 1984	1984/85	102	104	99	103

Source: *Actual* data compiled from information supplied by the New Zealand Meat and Wool Boards' Economic Service, and the Department of Statistics.

Notes:

* All data expressed as a percentage of the previous season.

Data on the actual wool production per sheep and the total actual wool production is known at the end of each season, usually with a delay of about one month. Thus as of May 1984, when a weather-based forecast of the wool production was made for the 1984/85 season, the actual production was known for the 1982/83 season, and provisionally available for 1983/84, the final production for the 1983/84 season not being available until August 1984. The final production for the 1984/85 season was not known until August 1985, 15 months *after* the date of prediction.

1979/80 production, etc.). Table VII.20 gives details of the weather-based predictions and wool production per sheep for the 1980/81 to 1984/85 seasons in New Zealand, as well as the *actual* production data which are not available until about August of each production season.

(d) FUTURE WORK

Forward estimation of the supply of agricultural production is a problem with which many individuals and organizations in the agricultural industry have to contend. More specifically, organizations associated with the production, transport, processing and marketing of wool have a continuing requirement for advance information of the likely availability, quality and flow of wool in the international market place. Historically this information has generally been provided after a subjective assessment of the various factors which influence the quantity and quality of wool becoming available. The accuracy of these forecasts is largely dependent on the skill and experience of the individual making them, but these subjective forecasts can at times give misleading information, with adverse consequences to a variety of decision-makers – both on and off the farm, and also within and outside of the country of production.

In this section an attempt has been made to incorporate some quantitative data into statistical models which have been designed to provide those making forward estimates affecting the wool industry with an additional weather-based opinion on the likely level of wool production in New Zealand. The models presented 'explain' a significant proportion of the annual variation in wool production, but they are by no means perfect. However, used in association with the other more traditional methods they should result in more soundly based forecasts. It must also be noted that few if any weather-based forecasts of the *quality* of wool are used operationally. Clearly the ideal wool/weather prediction model should be concerned with both the quantity and the quality of wool since both parameters influence the final price.

The quantity and quality of wool produced in New Zealand – as well as in other key wool producing countries such as Australia, South Africa, Uruguay, China, and the Soviet Union – is the result of a combination of many factors. Yet, in this industry, which is vital to many of these economies, it is important that resources should be applied to develop more efficient forecasting and monitoring techniques, if costly errors of judgement are to be prevented. What is required is greater efficiency in the whole industry (specifically in this case the wool industry, but obviously this is also applicable to most other aspects of agricultural production). This applies especially in the area 'beyond the farm gate', if the comparative agricultural advantages which a number of nations have enjoyed' in the past are to continue. The key question then, as far as New Zealand is concerned, is: Can this country, dependent on the export of a specific agricultural product, afford to see such an export earner function at anything less than peak efficiency through the whole land to market chain? It is considered that further work in these areas will pay high dividends.

D. WEATHER- AND CLIMATE-BASED FORECASTS OF BUSINESS ACTIVITY INDICATORS

1. Business activities: Sensitivity analysis

The sensitivity to weather and climate of the non-agricultural activities of most countries is usually more difficult to assess than those of the agricultural sectors. However, several sectors – notably energy, transport, and business – are potentially as weather- and climate-sensitive as are the more obvious agricultural sectors. In particular, it is evident that both weather and climate fluctuations are producing major economic, social, and political consequences. Moreover, it is evident that many aspects of the monitoring and prediction of productivity and consumption *are* ultimately associated with the availability, use, and interpretation of information on the past, present, and future weather and climate.

One specific aspect of the climate-economy mix is the growing importance of the availability of 'leading economic indicators' in near real-time, and both Presidents and Prime Ministers and their advisers have found that they provide essential background information for their day-to-day decision-making. This is particularly so in countries like the United States, Japan, and France, but in most countries of the world there is a growing appreciation of the need for accurate and up-to-date economic data on national productivity and related matters. For example, in New Zealand, in the Department of Statistics *Monthly Abstract of Statistics*, monthly, quarterly, and twelve-monthly production and consumption data are given for various areas of New Zealand, as well as for the nation as a whole. In a like manner, in countries like the United States, agencies such as the Department of Commerce publish similar types of data on consumption and production on a weekly, monthly, quarterly, or annual basis. In addition, several business organizations collate and publish specialized economic information relating to specific sectors of their national economies.

As previously noted, the 'traditional' response of the climatologist to interpreting such data has usually been luke-warm; however, several national meteorological services now not only observe, collect, and process data in real-time but also analyse it in real-time on a commodity basis. This means that key economy-related weather and climate indices cannot only be computed and made available regionally and nationally, but also, if required, on an international basis. Clearly, this is a new vision for the traditional climatologist to focus on.

One specific example of the value and use of such economy-related indices in New Zealand is given by Maunder (1979), and a recent analysis applying the technique to the United States has been prepared by Maunder (1981b). This latter study assessed the influence of weather on housing starts, itself a leading indicator of United States economic activity. The results gave a strong indication that in critical spring months, such as April, the effect of weather fluctuations on housing starts can be quantified for certain parts of the United States. For example, the study showed that in both the northeast and north-central regions of the United States, housing starts in April increased in a significant manner over the housing starts in the previous April *if* the month of April was warmer than the previous April. However, in the north-central region an increase in the April precipitation indicated a decrease in April housing starts.

It is evident that the availability of 'climate productivity indices' for key commodities is a reality. Furthermore, the availability of such information through videotex systems offers a tremendous challenge not only to meteorologists and climatologists in terms of the marketing potential of their services, but also to the decision-makers who have the potential to use such information. One specific aspect of these new products is their use in 'weather-adjusting' national economic indicators, so that the true 'economic climate' of a nation may be ascertained. In the next section a specific example of adjusting a national economic index for weather conditions, namely the *Business Week Index* of United States weekly economic activity, is examined, to emphasize the point (already noted in Chapter VI) that some aspects of economic activity such as electricity consumption reflect – at least in part – environmental conditions rather than pure economic 'strength'.

2. National economic indicators: A United States example

Many business activities rely heavily on the compilation of *national* weekly, monthly and seasonal business indicators. In this context one advantage of looking at a country like the United States is that its sheer size and *the importance of the private sector* in that country mean that there is a real need for national indicators. Indeed, several such national economic indicators are published in the United States for periods as short as a week. The availability of such short-period indices not only allows relevant weather-impact studies to be made using United States data, but also permits the transfer of such experience to other countries. Indeed such countries may well find that collations of national economic indicators for as short a period as a week are not only possible but also essential if 'close monitoring' and/or 'fine-tuning' of an economy is to take place, either by governments or by sectors of the society. In both situations, there is a clear need to provide guidance to economists and decision-makers on what impact the weather and climate has had, or will have, on indicators of national economic activity.

In the specific case of the United States, *Business Week* computes and publishes each week their *Business Week Index*, and a similar index of United States economic activity is published by the *US News and World Report*. Both indices are computed by weighting 'seasonally adjusted' economic activities such as raw steel production, rail freight traffic, and electric power production. The *Business Week Index* includes twelve components, and the important electric power production component comprises 17.3 per cent of the total activity weighting. The index uses a base year of 1967 = 100, and a comparison of the index for the summers of 1979 and 1980 shows average values of 150 in 1979 and 138 in 1980. This decrease of 12 points from 1979 to 1980 reflects the downturn in the United States economy that occurred at that time, but of greater importance in many respects are the *variations from week to week* within each summer, and the *trend* within each summer. Indeed, although the downward trend was more marked in the 1980 summer than in the 1979 summer, it is considered that the index would have decreased *even more rapidly* if electricity production had not been at record levels. These high levels of electricity production were primarily the result of the extremely hot conditions which prevailed in the 1980 summer, a factor which kept the index 'artificially' high. This is because compilers of the index automatically assume that an increase in electric power consumption *from any cause* – including the use of air conditioners in a heat wave – must be associated with an increase in national economic activity.

For example, during the summer of 1980, the *Business Week Index* varied from 139 to 136 (1967 = 100) with quite significant week-to-week variations. Fig. VII.3 shows some of these variations together with the fluctuations in the United States weekly electric power production expressed as a percentage of the comparable week(s) in 1979. Because electric power production is such a significant factor in the index a question to be asked is 'Would the "downturn" in United States economic activity have been more pronounced if the heat wave (which caused a substantial increase in electric power consumption during this period) had *not* occurred?' More importantly, since *the severity of the heat wave was well known in real-time*, and indeed *forecast* several weeks in advance, could the real-time and forecast weather conditions have been used by the relevant authorities to monitor more correctly and even to predict United States economic activity? It is considered that this is a clear example of *weather information* (actual and forecast) being able to be used to monitor and forecast an *economic activity* well before the official trend in economic activity is known. For example, actual temperature data – which are available in real-time – could be used to adjust the actual electric power production data *before* they are incorporated in any index of 'true' economic activity. In addition, a correctly used forecast of temperatures (assuming the forecast was a 'useful' indication of likely conditions) could be employed to anticipate the index of 'true' economic activity well before the actual event.

FIGURE VII.3 The *Business Week Index* and US electric power production (1980 as percentage of 1979), June–September. 1980

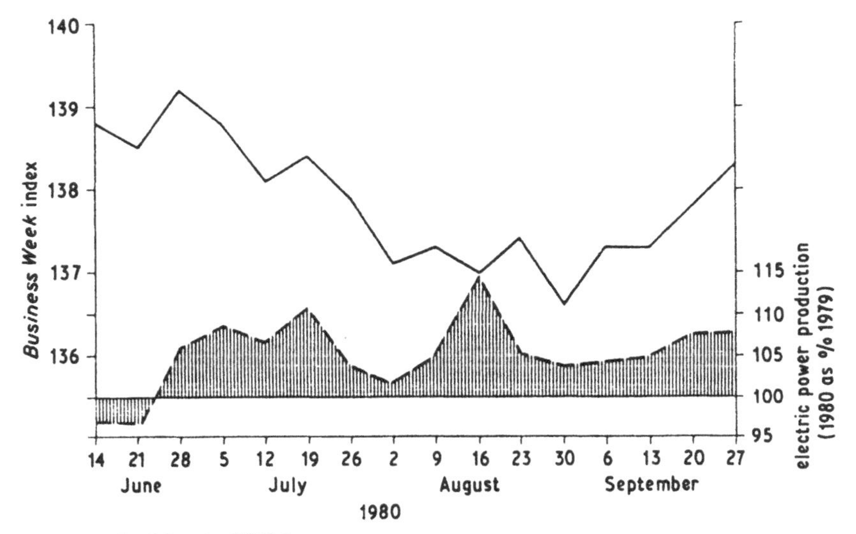

Source: After Maunder (1982a).

3. Adjusting the Business Week Index

(a) INTRODUCTION

As explained in the previous section it is considered that the *Business Week Index* of United States economic activity should be adjusted to take into account weather factors which increase electric power production in a 'non-productive' way. This adjustment was first examined in Chapter VI where it was shown that for each 1 per cent increase in the requirements for cooling (over and above what would be expected on the basis of average temperatures), the *Business Week Index* should be adjusted or reduced by a factor of 0.009 per cent. It was also shown that for each population-weighted cooling degree day (above the average for the specific week), the *Business Week Index* should be adjusted or reduced by a factor of 0.015 per cent. The specific application of these adjustment factors is now examined.

(b) ADJUSTMENTS FOR PERCENTAGE CHANGES IN COOLING DEGREE DAYS

As a first example of the use of these weather reduction factors, consider the impact of the *percentage* increase in the national population weighted cooling degree days for the United States for the week ending 13 September 1980. These totalled 53, or 36 per cent more than the *average* for this week of 39, and the published *Business Week Index* was 137.3. It is considered that this index should therefore be adjusted or decreased by a factor of 0.32 per cent (i.e. the 0.009 per cent 'adjustment factor' × 36). The weather-adjusted *Business Week Index* for that week would therefore have been 136.9 (i.e. (100 − 0.32)/100 × 137.3), or 0.4 index points *lower* than that published. Similar adjustments were made for the other weeks of 1980, and the published and adjusted indices are shown in Table VII.21 and Fig. VII.4.

Of considerable importance in any analysis or interpretation of national weekly economic indicators are the *week-to-week variations*, and the overall effect of the weather adjustments may be described as the 'weather impact'. For example, in the consecutive weeks ending 9 August and 16 August 1980, the difference in the 'published index' was −0.3 (137.0 in the week ending 16 August compared with 137.3 in the week ending 9 August), whereas the difference in the 'weather adjusted index' was +0.1 (136.9 compared with 136.8). Thus the impact of the percentage weather adjustments on the published *Business Week Index* is to change it from a *decline* of 0.3 to a *small increase* of 0.1.

The major reasons for these differences were the very hot conditions in the week ending 9 August 1980, in which the population-weighted cooling degree days were 44 per cent above the average for that week, and the

Table VII.21 *Business Week Index:* Percentage weather adjustments

	Population-weighted cooling degree days				*Business Week Index*		
Week ending 1980*	*1980***	*Normal*	*% Normal*	*Weather adjustment (%)*	*Published*	*Weather adjustment*	*Weather-adjusted index*
June 28	83	63	132	−0.29	139.2	−0.4	138.8
July 5	77	69	112	−0.11	138.8	−0.2	138.6
July 12	90	73	123	−0.21	138.1	−0.3	137.8
July 19	110	76	145	−0.41	138.4	−0.6	137.8
July 26	84	77	109	−0.08	137.9	−0.1	137.8
Aug. 2	97	76	128	−0.25	137.1	−0.3	136.8
Aug. 9	105	73	144	−0.40	137.3	−0.5	136.8
Aug. 16	75	72	104	−0.04	137.0	−0.1	136.9
Aug. 23	73	64	114	−0.13	137.4	−0.2	137.2
Aug. 30	81	55	147	−0.42	136.6	−0.6	136.0
Sept. 6	75	47	160	−0.54	137.3	−0.7	136.6
Sept. 13	53	39	136	−0.32	137.3	−0.4	136.9
Sept. 20	39	32	122	−0.20	137.8	−0.3	137.5
Sept. 27	33	25	132	−0.29	138.3	−0.4	137.9

Source: After Maunder (1982a).
Notes:
* Week ending Saturday.
** Week ending Sunday.

FIGURE VII.4 The *Business Week Index*, as published and weather adjusted (top), and (bottom) actual and average US population-weighted cooling degree days, June–September, 1980

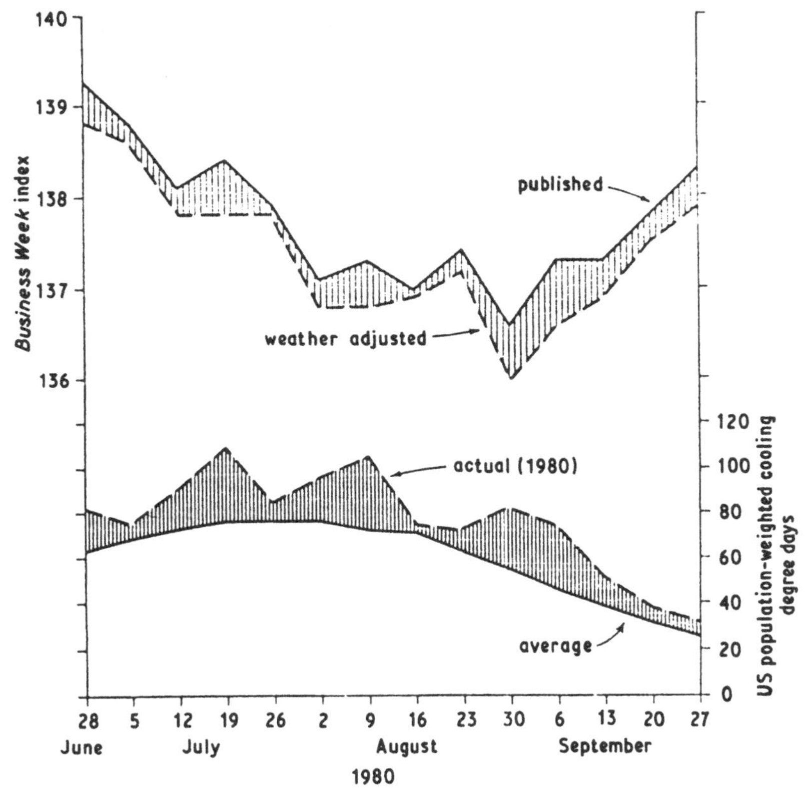

Source: After Maunder (1982a).

near-average temperature for the week ending 16 August 1980 in which the population-weighted cooling degree days were only 4 per cent above the average. Accompanying the very hot conditions was a near-record electric power consumption in the United States for the week ending 9 August of 51,834 million kWh, whereas in the following week electric power consumption was closer to the average at 49,507 million kWh. In this instance it is important to emphasize that – on the basis of the arguments presented – the *decrease* in electric power production in these two successive weeks was 'incorrectly' interpreted by the published *Business Week Index*, in that the index showed a *decrease* from 137.3 to 137.0. As previously shown, however, the published index of 137.3 for the very hot week ending on 9 August was already 'artificially' high, and *if* weather adjusted would have been *reduced* by a factor of 0.5 to 136.8, whereas the week ending 16 August needed a reduction of only 0.1 to 136.9.

The conclusion, therefore, is that *if* the published *Business Week Index* of 137.3 and 137.0 for the two successive weeks had been weather adjusted, it would have produced indices of 136.8 and 136.9, indicating a slight weather-adjusted *increase* and *not* the 'published' decrease in overall business or economic activity. It is considered that the *Business Week Index* may therefore have given a 'false' impression of the direction of the United States economy in the week ending 16 August 1980, in that a published 'decline' was indicated, whereas if weather adjustments had been made, the economy would have shown a 'slight recovery'.

(c) ADJUSTMENTS FOR ACTUAL CHANGES IN COOLING DEGREE DAYS

The impact of the *actual* change (rather than the percentage change) in the national population weighted cooling degree days may also be assessed. Again considering the week ending 13 September 1980, the actual national population weighted cooling degree days of 53 were 14 above the average of 39 for that week, and the published *Business Week Index* was 137.3. If this index is adjusted or decreased by a factor of 0.21 per cent (i.e. the 0.015 per cent 'adjustment factor' × 14), then as a first approximation the (actual) weather-adjusted *Business Week Index* for that week would have been 137.0 (i.e. (100.00−0.21)/100 × 137.3), or 0.3 points lower than that published. This may be compared with a reduction of 0.4 points if the percentage change, rather than the actual change, was considered.

(d) COMPARISON OF ACTUAL WEATHER AND PERCENTAGE WEATHER ADJUSTMENTS

A comparison of the adjustments that could or should have been made using both the actual weather differences and the percentage weather differences is given in Table VII.22. This suggests that the use of the *actual weather differences* from average gives a more 'realistic' correction to the published *Business Week Index* at the beginning and end of the air-conditioning season, whereas the *percentage differences* from the average give a more 'realistic' correction during the peak periods. Whatever method is used, however, it is evident that weather adjustments of weekly indicators of economic activity are significant, particularly in their ability to 'fine tune' more correctly the trend in national business on a weekly basis. For example, as shown in Table VII.23, although the published comment on United States economic activity between the weeks ending 23 August and 30 August 1980 was 'dipped very sharply', a more 'true' picture reflecting the *change in the weather between those two weeks* may well have been 'gained slightly'. It is considered that 'weather-based fine tuning' is very important in correctly interpreting economic trends.

Table VII.22 *Business Week Index:* Adjustments for weather: Summer 1980

Week ending 1980	*Published*	*Week/Week difference*	*Percentage weather adjustment*	*Week/Week difference*	*Actual weather adjustment*	*Week/Week difference*
June 28	139.2		138.8		138.8	
		−0.4		−0.2		−0.2
July 5	138.8		138.6		138.6	
		−0.7		−0.8		−0.9
July 12	138.1		137.8		137.7	
		+0.3		0.0		0.0
July 19	138.4		137.8		137.7	
		−0.5		0.0		0.0
July 26	137.9		137.8		137.7	
		−0.8		−1.0		−1.0
Aug. 2	137.1		136.8		136.7	
		+0.2		0.0		−0.1
Aug. 9	137.3		136.8		136.6	
		−0.3		+0.1		+0.3
Aug. 16	137.0		136.9		136.9	
		+0.4		+0.3		+0.3
Aug. 23	137.4		137.2		137.2	
		−0.8		−1.2		−1.1
Aug. 30	136.6		136.0		136.1	
		+0.7		+0.6		+0.6
Sept. 6	137.3		136.6		136.7	
		0.0		+0.3		+0.3
Sept. 13	137.3		136.9		137.0	
		+0.5		+0.6		+0.7
Sept. 20	137.8		137.5		137.7	
		+0.5		+0.4		+0.4
Sept. 27	138.3		137.9		138.1	

Source: After Maunder (1982a).

4. The way ahead

A further refinement of such adjusting is that it is also potentially possible to use real-time commodity-weighted weather and climate information to provide a *forecast* of the tendency of national economic indicators. That is, since weather information *is* available in real-time, whereas most weekly national economic indicators have a time publication delay of two or three weeks (or months, depending on the economic parameter), weather-based forecasts of economic activity can be made available well in advance of the availability of 'official' production/consumption information.

Table VII.23 *Business Week Index:* Published and percentage weather adjusted

Week ending 1980	*Published index*	*Week/ Week diff.*	*Published comment*	*Weather adjusted index*	*Week/ Week diff.*	*Weather adjusted comment*
June 28	139.2			138.8		
		−0.4	fell slightly*		−0.2	fell slightly*
July 5	138.8			138.6		
		−0.7	took a moderate drop		−0.8	dipped sharply
July 12	138.1			137.8		
		+0.3	strengthened		0.0	no change
July 19	138.4			137.8		
		−0.5	declined slightly		0.0	no change
July 26	137.9			137.8		
		−0.8	dipped again		−1.0	fell sharply
Aug. 2	137.1			136.8		
		+0.2	rose slightly		0.0	no change
Aug. 9	137.3			136.8		
		−0.3	declined slightly		+0.1	gained slightly
Aug. 16	137.0			136.9		
		+0.4	edged upwards		+0.3	gained slightly
Aug. 23	137.4			137.2		
		−0.8	dipped sharply		−1.2	dipped very sharply
Aug. 30	136.6			136.0		
		+0.7	gained moderately*		+0.6	gained moderately*
Sept. 6	137.3			136.6		
		0.0	unchanged		+0.3	strengthened
Sept. 13	137.3			136.9		
		+0.5	moved up modestly		+0.6	gained moderately
Sept. 20	137.8			137.5		
		+0.5	gained moderately		+0.4	gained moderately
Sept. 27	138.3			137.9		

Source: After Maunder (1982a).
Note: * Holiday week.

Indeed, it cannot be overemphasized that, although last week's or last month's weather *is* history, the *measurement* of its economic impact has usually hardly begun at that time using conventional economic techniques. But a forecast of economic data – which at the time the forecast is made *do not actually exist*, or more correctly have not yet been compiled, collected, or analysed – can be made, using the *unique* availability of real-time weather and climate data. Thus, where and when weather and climate is a *limiting factor* its economic impact can be assessed immediately.

Attempts to 'weather adjust' national economic indicators are difficult and controversial; nevertheless such attempts not only provide economists, decision-makers, and publishers of economic indicators with realistic adjustment factors, but they also provide appropriate 'guidance' for the economic forecaster whose track record in many countries has been often criticized. For example *Business Week* (14 May 1981), in noting that the official announcement that the United States real gross national product rose at an annual rate of 6.5 per cent in the first quarter of 1981, said that it was an acute embarrassment to many economists, since the difference between the actual result and the consensus forecast of more than forty 'blue chip' economists was 'an enormous' 7.6 percentage points.

In the light of the above comment, perhaps it is time for the applied climatologist and the weather forecaster to provide some assistance to the economic forecaster. But to do this will require the concerted effort of a number of people, for much of the necessary thinking involves disciplinary areas peripheral to the normal activities of meteorologists, climatologists, and economists. The problems are real, however, and it is believed that appropriate analysis of the climate–economic mix can contribute in a very positive and useful way in assisting decision-makers to understand better some of the world's important problems associated with the supply and utilization of food, fibre, and energy. The key factor, however, is the 'leading role' which must be provided by the economic climatologist. The more specialized viewpoints and expertise of people outside of meteorology and climatology must also be encouraged in order that 'real' problems are solved.

From the foregoing it should be obvious that climate-impact studies cannot and indeed *must not* be confined either to the analysis of atmospheric information or to the analysis of economic information. Rather, such studies must consider the human *problems* of food, health, energy and well-being, and of all disciplines it is perhaps the geographically or environmentally trained meteorologist or climatologist who can best cast such studies in a real-world framework. Indeed, as Mather *et al.* (1981) have noted: 'to the degree that this is done climatology will not only meet the challenge of the eighties, and so contribute to the solution of a wide range of human problems, but will also come of age.'

VIII Forthcoming challenges and opportunities

A. Weather and Climate: The Challenge of Operational Decision-Making

While it is comparatively easy to make assessments of the general relationship between weather and climate factors and some aspects of production or consumption, or in a few cases prices, the more precise relationships necessary for operational decision-making are much more difficult to formulate. Moreover, even with a perfect weather- and climate-economic model, major problems still exist in the acceptance by decision-makers of this new aid.

The acceptance and use of commodity-weighted weather and climate information, and forecasts of production resulting from this information, clearly offers a challenge to both the meteorologist and the climatologist, as well as the user of weather and climate information, and a necessary first step is to place greater emphasis on the meteorological and climatological aspects of planning and development.

The question remains – where do we go from here? The issues are clear.

First, the impacts of weather and climate on productivity and consumption must be assessed and presented in terms of production figures, costs, or other similar measures which can be used directly by decision-makers – including economists, agriculturalists, planners, and politicians. Second, national meteorological services or their equivalent should actively encourage personnel who have a background that will allow them to become 'development' or 'application/marketing' meteorologists and climatologists. Indeed, one could and indeed must comment, following Bernard (1976), that the purely physical and mathematical approach of conventional meteorologists and climatologists results in their being too impervious to the scientific and technical applications of meteorology and climatology for socio-economic progress. However, it must also be stated that the comments by Bernard a decade ago *have* been acted upon by a number of national meteorological services including Canada, Japan, France, Sweden, United Kingdom, United States, and New Zealand; indeed, it is now clearly recognized that the days of the 'purely physical and mathematical approach of conventional meteorologists' are rapidly becoming a thing of the past.

Clearly, the opportunities provided through *real* operational decision-making in the weather and climate business have (or will) become key issues in the meteorological and climatological scene. In this regard the 'lead' role of the World Meteorological Organization should be emphasized, and in particular the positive response of the WMO Commission for Climatology to these issues. For example, at the Ninth Session of the Commission for Climatology held in Geneva in December 1985, it was agreed that to improve the usefulness of application activities, national Meteorological Services should develop capacities for the extensive implementation of professional knowledge for using and interpreting the complexity of the data/information package. It was also noted that for these user-oriented activities, meteorologists have to understand the problem from the users' point of view; further, with a view to achieving mutual understanding, professional dialogues should be established and maintained continuously. Similarly, the Commission agreed that for many operational purposes new approaches are needed to tailor short-term weather forecasts to meet the requirements of specific users, and that both governments and specific users need to be better informed as to the relevance and value of both weather and climate services.

B. WEATHER AND CLIMATE: THE INFORMATION OPPORTUNITY

As previously noted, the sensitivity of the commodity markets to weather and climate information is a clear indication that, in the real world, weather and climate sensitivity is a reality. There is also realism in the very difficult areas of disaster relief, and agricultural and energy policies. However, the real sensitivity – in economic, social, and political terms – of nations, sectors of nations, and commodities to weather and climate variations and changes has to be better understood. Indeed, the need for such understanding offers the most important challenge to the meteorological and climatological community (see Mather *et al.* 1981), and a necessary step is to educate both the *producers* of weather and climate information, and the potential *users* of these products, in the *specific* applications of weather and climate information to problems.

The connections between the difficulties of weather (and climate?) forecasting and the equally complex problems which politicians and social scientists face point clearly to an even more difficult problem when one tries to link the meteorological and climatological system *with* the economic, political, and social system. It can of course be done, indeed it must be done, and Gordon McKay had some very pertinent comments to make in an editorial in *Climatic Change* (vol. 2, no. 1, 1979):

> Most applications of climatology have advanced because of recognised value rather than through theoretical conjecture. Where value can be demonstrated clearly, the product will be demanded. Our challenge is to produce practical information that can be readily understood and integrated in a smooth and timely fashion into the planning process. The chances of success in this regard are improved when the planning process is understood – they are much improved when the user is convinced and involved.

These comments underline the difficulty facing applied climatologists and applied meteorologists in convincing the potential decision-maker of a weather- or climate-sensitive operation, that there *is* much more to meteorology and climatology than tomorrow's forecast or the average monthly or annual rainfall in Auckland, Amsterdam or Algiers. McKay noted further in the editorial that closer involvement with the user is essential to ensure viability, relevance, and real benefits from new information. Indeed, their interest is, says McKay, 'in more useful information, not in answers to complex problems that they do not understand or complex answers they have to suspect. Usefulness is the prime criterion'.

C. CLIMATE CHANGE AND POLITICAL REALITIES

In discussing 'climate change and political realities', a host of questions are raised – including what is meant by 'climate'. This has already been discussed in Chapter II, but it is appropriate here to re-emphasize that the present thinking in a number of quarters is that *all* 'atmospheric' data for *all* time scales (that is current, to near real-time, to 'traditional' climate time scales) should be considered 'climate data'. This implies that climate is *not* restricted by any time scale such as the conventional but surely now outdated thirty-year 'normal' period.

A specific aspect of climate change and political realities is the agricultural consequences of a temperature change. Indeed, the impacts of a large cooling or a large warming (through one or several causes including, but not restricted to, an increasing amount of 'greenhouse' gases such as carbon dioxide in the atmosphere) on key agricultural crops in various parts of the world have been assessed by a number of investigators. On such study by the United States National Defense University (1980) showed that a specific warming could increase the Canadian spring wheat yield by about 8 per cent, whereas a similar warming could decrease the Australian winter wheat yield by about 4 per cent (Fig. VIII.1). If only a large warming is considered (from *whatever* case), countries 'gaining' most, according to the report, would be Canada for an increased spring wheat yield, and the USSR for increases in both spring wheat and winter wheat. In contrast a large warming would adversely affect the production of Australian winter wheat, Indian winter wheat, and Argentinian winter wheat and corn. On the other hand, a large cooling would adversely affect the production of Canadian spring wheat and USSR winter and spring wheat, but significantly increase yields of Australian winter wheat and Argentinian winter wheat and corn.

It is obvious that should such a warming or cooling take place there would be significant and far-reaching economic, social, and political consequences including those associated with famine relief, possible large-scale migration, and international trade and aid programmes. Some of these consequences have been studied by the International Institute of Applied Systems Analysis

FIGURE VIII.1 Potential impacts of a large cooling and a large warming on the expected values of the normalized relative annual yields of key crops

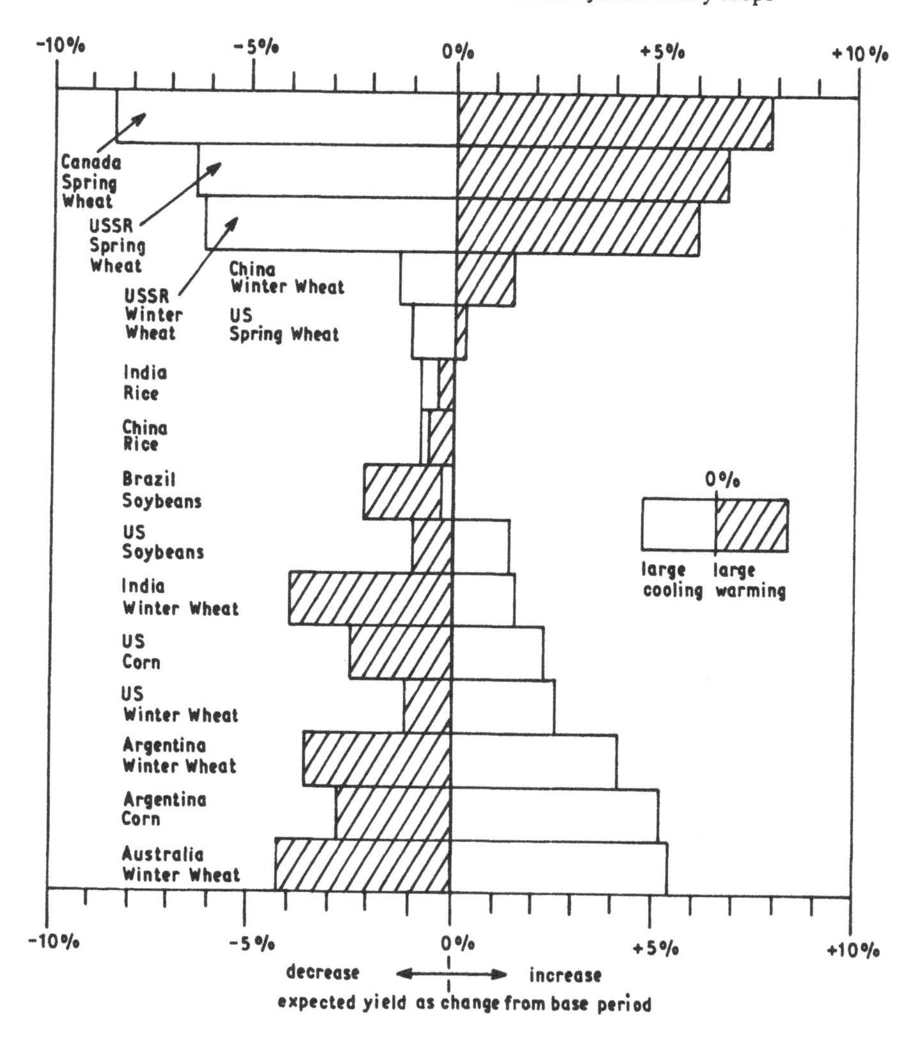

Source: After Maunder (1985a), from National Defense University (1980).

(see Parry 1985b) in association with the United Nations Environment Programme (UNEP) and the World Climate Impact Programme of WMO, and clearly the results of these studies are of importance to all countries.

Another important question relates to a relatively slow *trend* in a climate variable over time. However, whatever the political, economic, and social impact of any long-term trend, it is evident that much more significant year-to-year variations occur (see, for example, Hare 1985a). Indeed, it is of considerable economic significance to note that in most cases trends in a climatic series over time – which *are* able to be adjusted to – are usually very small by comparison with short-term variations. Moreover, even if there is *no* significant cooling, warming, or change in the precipitation, it is clear that the *impact* of short-term variations in the available atmospheric resources must continue to increase in importance because of the growing demand for food and the increasing cost of energy.

In any discussion of the 'economic climate', it is essential therefore to understand the increasing *impact* climate variations must have on economic, social, and political activities. Further, as has already been noted in early chapters, since many political regimes – both developing and developed – have a 'thinking lead-time' of only a few years, any real economic, agricultural, or political significance of a longer term trend in the climate is very difficult to infiltrate into the decision-making process. It is important therefore to appreciate the political realities of a situation whatever the 'climatological niceties'.

D. POLITICAL AND STRATEGIC ECONOCLIMATOLOGY

Politicians – as distinct from economists and climatologists – have to look at a *very wide* range of responses, and their decisions are primarily based on political goals, viewing such things as climate and economic concerns in *combination* with other current issues. Sewell and MacDonald-McGee (1983), in discussing this situation in Canada, suggest that, inevitably, governments will become increasingly involved in the management of climate resources, but that it is unclear whether present legislation, policies, and administrative structures and procedures are adequate to meet this challenge. In a similar vein, Meyer-Abich (1980), in a paper discussing the transformation of climatological facts into political facts, commented:

> Approached by a climatologist who is reporting his latest results in terms of temperature, pressure, and humidity, the politician will ask: 'All this may be so, but what difference does it make to me? I am concerned with achieving goals in a given situation, and how does the climate come in here?' Obviously the climatologist cannot answer this question within the framework of his scientific discipline. He may, however, hand over the question to the geographer and to the ecologist, since these researchers are concerned with the habitat of mankind and other species, and can interpret climatological parameters through their disciplines as impacts on the conditions of life.

Meyer-Abich's comment that the climatologist 'cannot answer the question', is perhaps a little unfair, for there are climatologists who can answer these questions. Nevertheless, the point is valid, and should give food for thought to many conservative-thinking climatologists.

E. WEATHER AND CLIMATE MODIFICATION

1. An overview

Weather and climate affect human activities in pervasive ways. For example, the profitability in certain economic activities depends very much on the weather and the climate, obvious examples being agriculture and air transportation. Similarly, some human behaviour patterns are influenced by weather variations. It is natural, therefore, that people should try to find ways of not only adjusting to the weather and the climate, but also even altering the atmospheric environment in which they live.

Two major types of action can be taken in adjusting to the weather and the climate: first, the adoption of techniques which insulate activities from weather and climate variations, such as air conditioning, home heating, storm windows, or drought-resistant crops; second, temporary alterations in patterns of activity may be made, such as the modification of a harvesting schedule, the postponement of a shopping trip, or temporary movements away from the area, as in the case of evacuation from the path of a hurricane. Alternatively, or in addition, people may deliberately or inadvertently alter the amount or temporal distribution of particular weather elements in relation to a 'small' area, and there have been suggestions (see, for example, Gray *et al.* 1976) that weather modification of larger areas such as river basins, parts of oceans, or even countries is feasible.

For the most part people and nations have concentrated their attention on the first alternative; that is, moving to those locations where the weather and the climate is most acceptable for personal comfort, and where economic activities can be pursued at reasonable cost, as well as trying to develop ways of reducing the impact of variations in the weather and the climate. However, there now seems to be agreement that in certain areas, under certain circumstances and at certain times, weather modification in the form of an increased precipitation of the order of 10 per cent can be achieved, and that the dispersal of cold fog is feasible. There is less conclusive evidence about success in the suppression of lightning and of hail, but results of some experiments seem to suggest that ways may soon be found to accomplish these types of modification as well. In this regard, it should be emphasized that on the microscale, weather modification is a reality. However, there is still considerable doubt as to just how 'successful' the weather modifiers have been, considering the *overall* effects in an area.

2. Weather modification and tomorrow's weather: Essential linkages

It is evident that modification of the atmosphere – whether intentional or otherwise – can be done at least in some areas and on some time scales. Little thought, however, has gone into some fundamental questions regarding weather modification. The basic scientific question may well be: can you modify the weather? The more fundamental question is: should you do so, where should you do so, and what safeguards are there either in the form of laws or compensation to provide for the 'errors' which will occur? Some of these aspects were first explored in a symposium held at Boulder, Colorado in July 1965 which culminated in the publication of a monograph *Human Dimensions of Weather Modification* (Sewell 1966). As previously noted, the National Center for Atmospheric Research subsequently established a Task Group on the Human Dimensions of the Atmosphere to explore these matters in greater detail.

It may of course be argued that weather modification and weather forecasting are very far apart. However, as previously noted in Chapter IV, it must be remembered that modification of tomorrow's weather may be very dependent on the weather that is forecast for tomorrow, particularly if decisions as to whether to modify are going to be based on the predicted weather. It is also important to re-emphasize that *to the client* accurate weather forecasting and successful weather modification may both have similar effects, for a client such as a retail store manager is not usually concerned about *why* it rains, but only that it *does* or does not rain. That is, the consumer is concerned with *what* it does, not *why*.

Weather and climate modification is also considered when there is a need to improve the economy of a region by increasing water resources for agricultural use, for cities, or for hydro-electric power generation. However, in assessing the benefits to some segments of the population, losses to other groups must also be considered, together with possible legal and compensatory factors.

Precipitation enhancement should therefore be viewed from the overall aspect of total water resources management, since it may well be difficult or impossible to ameliorate drought conditions *when they occur*. This is because in most droughts clouds suitable for seeding are normally scarce. Thus replenishing aquifers with water, which can be pumped to the surface when it is needed, or filling reservoirs, are obviously more appropriate measures to take because the timing of the precipitation is not so crucial. Accordingly changes in agricultural practices, with appropriate conversion to storage and irrigation, may well be just as important and in many cases much more important than any precipitation enhancement.

Several major experiments have been conducted in various types of cloud systems, including orographic, winter convective, and summer convective clouds. Some of these have provided either statistical or physical indications that seeding may have affected precipitation. But many meteorologists consider that there are few cloud-seeding experiments which have combined physical evidence in support of a seeding hypothesis *with* persuasive statistical evidence of increases in precipitation over an area. However, the practical application of weather modification activities to a number of agricultural and energy sectors has been considered in some detail, and two case studies applicable to the electric power industry were discussed in Chapter V.

3. The carbon dioxide question

During the last decade there has been increasing concern over the impact of carbon dioxide and other greenhouse gases on the global atmospheric temperature. The subject is *not* without controversy, and it is perhaps relevant – at this point in time – to give a brief insight into some of these viewpoints.

In a research review, 'The CO_2 climate controversy: an issue of global concern', Idso (1984) made the following comments:

> The CO_2 greenhouse hypothesis was first put forward by Tyndall (1861), and since that time it has provided a provocative stimulus for many students of the atmosphere. Although a wide range of investigators subsequently produced a wide range of estimates relative to the magnitude of the phenomenon, a consensus was finally forged by a group of climate modellers meeting under the auspices of the US National Research Council (NRC) in 1979. Based on computer calculations of the general circulation of the atmosphere, this group concluded (NRC, 1979), that a 300 to 600 ppm doubling of the atmospheric CO_2 concentration would lead to a 3 ± 1.5°C increase in mean global air temperature.
>
> Almost immediately, however, this consensus was attacked by Newell and Dopplick (1979) and Idso (1980), who claimed that the surface air temperature sensitivity of the models was fully an order of magnitude too great, based on several independent analyses of real-world data. The NRC responded two years later with a second report (NRC, 1982), which essentially reiterated the stance it had taken in 1979. . . . And now we have the latest response of the American scientific establishment in the form of two hefty volumes from the NRC (1983) and the US Environmental Protection Agency (EPA, 1983), entitled *Changing Climate* and *Can We Delay a Greenhouse Warming*?

Idso then discussed in detail many aspects of the CO_2 issue, and concluded his paper as follows:

> In summary, there is no *a priori* reason to believe that the computer models of the atmosphere currently employed to investigate the effects of increasing atmospheric CO_2 are reasonable representations of reality. Indeed, by the admissions of their own creators, they fall far short of that ideal. And on top of that, the few predictions which they have made relative to the greenhouse effect of CO_2 appear to be contrary to what is observed in nature.
>
> On the other hand, it is well documented fact that atmospheric CO_2 enrichment will tremendously boost agricultural productivity and reduce crop water requirements, while at the same time increasing streamflow and regenerating groundwater reservoirs (Bouwer, 1984). Thus, it seems only logical to want to encourage the release of CO_2 to the atmosphere, rather than curtail it.

In contrast, Tucker (1985), in discussing 'The global CO_2 problem' with particular reference to Australia, takes a more middle-of-the-road viewpoint. Tucker, for example, comments on the work of Monteith (1981) on the sensitivity of crops to climatic variation, and quotes from Monteith as follows:

> Until the predictions from climatic models become more reliable, I see little point in developing 'scenarios' for agricultural production based on numerous insecure premises. Crop models need to be improved too, particularly in the light of carefully conducted experiments showing that crop growth may respond to levels of CO_2 in dim as well as bright light – contrary to classical ideas about limiting factors.

Tucker nevertheless points out that the range of plausible future CO_2 – climate scenarios is approaching the stage where a programme of more detailed response assessment could be commenced. He further notes that care must be taken however to sift diverse estimates of future climate, and to consider as the basis for consequential studies only those that are founded on sound experiment and reasoning, supported wherever possible by some form of corroborative evidence. Tucker then concludes:

> It seems inevitable that life on earth will be affected by a continuing increase in atmospheric CO_2 concentrations but it is too soon yet to give more than a general indication of the size of the effects and of their net beneficial or deleterious nature. Focussed research programmes involving relevant aspects of the carbon budget, the sensitivity of the climate system, and the response of crops to changes in ambient conditions are essential if adequate warning in terms of timeliness and accuracy is to be given.

Although there is no 'official' viewpoint on the 'carbon dioxide question' it is important that reference be made to the joint UNEP/WMO/ICSU conference on 'An assessment of the role of carbon dioxide and of other greenhouse gases in climate variations and associated impacts' held in Villach (Austria), 9–15 October 1985. At this conference scientists from twenty-nine developed and developing countries assessed the role of increased carbon dioxide and other radiatively active constituents of the atmosphere (collectively known as greenhouse gases and aerosols) on climate changes and associated impacts. They concluded that as a result of the increasing concentrations of greenhouse gases it is now believed that in the first half of the next century a rise of global mean temperature could occur *which is greater than any in man's history*.

Following the conference a statement* was released to the press which included the following three conclusions:

* See: United Nations Environment Programme (UNEP) 1985; *Joint UNEP/WMO/ICSU Statement on the 1985 Villach Conference*, on 'An assessment of the role of carbon dioxide and of other greenhouse gases in climate variations and associated impacts'.

> Many important economic and social decisions are being made today on major irrigation, hydro-power and other water projects, on drought and agricultural land use, on structural designs and coastal engineering projects, and on energy planning, all based on assumptions about climate a number of decades into the future. Most such decisions assume that past climatic data, without modification, are a reliable guide to the future. This is no longer a good assumption since the increases of greenhouse gases are expected to cause a significant warming of the global climate. It is a matter of urgency to refine estimates of future climate conditions to improve these decisions.
>
> Climate change and sea level rises due to greenhouse gases are closely linked with other major environmental issues, such as acid deposition and threats to the Earth's ozone shield, mostly due to changes in the composition of the atmosphere by man's activities. Reduction of coal and oil use, and energy conservation undertaken to reduce acid deposition will also reduce concentrations of greenhouse gases, reduction in emissions of chloro-fluorocarbons (CFCs) will help protect the ozone layer and will also slow the rate of climate change.
>
> While some warming of climate now appears inevitable due to past actions, the rate and degree of future warming could be profoundly affected by governmental policies on energy conservation, use of fossil fuels, and the emission of some greenhouse gases.

Clearly the 'carbon dioxide question' will be a significant research and impact problem in the years ahead, and the many investigators in the area will need to provide guidance to decision-makers in a much more positive manner if, as it is predicted, a significant warming does begin to occur. However, it is considered that time is on the meteorologists' side, and this is one case where perhaps more research is needed, *before* the politicians and the planners become too closely involved in what undoubtedly *will* become an extremely important issue.

4. The acid rain problem

Much has been written about 'acid rain'. Indeed, the editor of *Weatherwise*, introducing a special section on acid rain in the October 1984 issue of the journal, noted that:

> So much has been written about acid rain – what it is, where it comes from, what damage it is doing or not doing – that it is difficult to distinguish fact from conjecture. In this issue of *Weatherwise*, the scientists and the experts sort through the myths and the conjecture to bring our readers the latest information on the causes, the processes, and the effects of acid rain.

In a survey paper in this special issue of *Weatherwise*, Miller (1984) says that acid rain has become one of the principal environmental issues of our time. But he notes that this is not just a local problem affecting a handful of people in isolated factory towns, but rather a phenomenon of national – even international – scope that involves very large regions of North America as well as Europe. He continues:

> By some accounts these affected regions have grown in size from year to year, and the acidity of the rain in some of them may have been increasing as well. The economic and environmental implications of these increases have propelled public debate on the issue to a fever pitch. Politicians have been inundated with appeals for legislative action to curb the problem.

Miller (1984) notes that there are three important 'impact' aspects of acid rain, namely: the general degradation of sensitive lakes and rivers, the damage to forests, and the surface 'erosion' of buildings. Clearly, there is much discussion on the *real* causes of these changes in the environment, and further research is needed. For example, Miller (1984) notes that there has been a decline in several species of trees over the last twenty years in both Europe and North America, and continues:

> Trees may be injured either by direct acidic impaction on leaves and needles, or by changes in soil chemistry through leaching by acidic wet and dry deposition. However, whether this decline is due to acid by itself, or to a synergism between acid rain and other factors, among them ozone damage, climatic changes, or natural cycles of forest maturation, remains to be clarified. Certainly, if acid rain is implicated in any way as a contributory cause of forest damage, this is a very serious matter.

The 'serious' matter of damage to forests has indeed been to the forefront of several intergovernmental 'exchanges'; indeed it is not unusual for the 'acid rain problem' to be the main news story in daily newspapers and on network television news in both North America and Europe. Several countries have positive programmes in this area of environmental concern and perhaps it is not surprising that Switzerland (which is clearly vulnerable to acid rain destruction of her alpine forests) has issued public information brochures on the problem, such as the one entitled 'Ses dernières pousses nous lancent un SOS', published by the Federal Department of the Interior in December 1984.

5. Implications of a 'nuclear winter'

The impact of any future nuclear war on the climate of the earth has been the subject of many reports and research papers since the early 1980s. Elsom (1984) in a general summary of the situation stated that prior to 1982 climatologists believed that even a large-scale nuclear exchange involving tens of thousands of nuclear detonations totalling 5000 or 10,000 megatons (MT) equivalent of TNT would have little effect on the global climate. He further noted that at that time it was considered that ground bursts would volatize and inject enormous quantities of soil particles into the atmosphere, somewhat analogous to a massive volcanic eruption such as Tambora in 1815 or Krakatoa in 1883, thereby causing a (modest) global cooling of approximately 1°C (National Academy of Sciences 1975b). However, Elsom (1984) emphasized that as of 1984 such a view was no longer held to be valid, and he stated:

> Crutzen and Birks (1982), Ehrlich *et al.* (1983), and Turco *et al.* (1983) have suggested that a large-scale nuclear exchange may cause profound changes to the climate not only of the northern hemisphere, in which such a future nuclear war would most likely be fought, but to the climate of the southern hemisphere too. The change in the climate which may occur in the weeks, months and years following a nuclear exchange may so profoundly disrupt the biological base that sustains human life that the consequences to mankind may be as serious, or even more serious, than the immediate effects of blast, fire and ionising radiation during the nuclear exchange itself.

A more recent study (Cess 1985) further summarized the position as of the mid-1980s. Cess discusses the significant studies on the impact of a nuclear war on the climate – including the study by Turco *et al.* (1983) noted earlier, as well as the studies of McCracken (1983), Aleksandrov and Stenchikov (1983), and Covey *et al.* (1984). Cess discusses each of these papers and notes that it has been suggested that following a nuclear exchange there might be a significant reduction in surface temperature over land areas, due to the impact upon the radiation budget of the surface atmosphere system of smoke produced by fires and of dust injected into the stratosphere by ground bursts. Cess then addresses several aspects of this possible radiative perturbation, such as the unusual nature of the climate response to the perturbation, a description of differences which are inherent within existing model studies, an evaluation of radiative transfer assumptions which have been employed in existing model studies, and illustrative latitudinal and diurnal variability of the smoke-dust impact upon solar radiation. His concluding remarks are as follows:

> It is important to reemphasize that the type of climate response associated with existing nuclear-war climate studies is unique with respect to our experience with climate models, since past investigations, such as those pertaining to increasing levels of atmospheric CO_2, involve very modest climate perturbations for which the conventional surface-troposphere structure is not altered. This raises the question as to the validity of existing climate models when the models are forced into a mode of strong atmospheric static stability. Because of this, and because of different climate forcing employed in existing model studies . . . it would seem prudent to define a 'standard experiment' by which models could conveniently be intercompared.

Finally, in this brief review of the implications of a 'nuclear winter', mention should be made of the report prepared by a 'National Academy' in the Southern Hemisphere, namely the Royal Society of New Zealand. This report describes the present (1985) threat of nuclear war, possible ways of relieving this threat, and the role of scientists in these endeavours. One chapter of this publication (Royal Society of New Zealand 1985), on the climatic effects on the Southern Hemisphere, discusses the implications for New Zealand of nuclear explosion in various parts of the world (including Australia, and New Zealand), and it states:

> the long term effects on New Zealand of a northern hemisphere nuclear war with some smaller extension south of the equator cannot be given with any certainty but are likely to be much greater than had been thought in recent years. The greatest environmental hazard in the northern hemisphere from a nuclear war is thought to come from the reduction in sunlight produced by smoke clouds. There is a possibility that the southern hemisphere may be exposed to this 'nuclear winter' effect though at a reduced level.

F. THE WEATHER ADMINISTRATOR OF THE FUTURE*

1. The 'Ministry of Atmospheric Resources'

Good morning! It is a sunny 15 March 1994. Welcome to another episode in the award-winning series 'A Day in the Life of the Decision Maker', an intimate, in-depth look at what is going on in our government corporations. It is brought to you by New Zealand Independent Television's Channel 15. It is live, unscripted and unrehearsed.

This time we're going to drop in on Dr Brown, Administrator of the Ministry of Atmospheric Resources (MAR). It is one of our newest and most important agencies. MAR was set up in 1990.

It now has a budget of over $160 million and a technical staff of more than 1000. We know it best as 'the agency that brought the atmosphere in from the cold', and which gave it status along with other resources such as water, agricultural land, forests and fisheries. It has brought an enlightened approach to resources management in New Zealand. It is seen by similar agencies overseas as bold and imaginative. Little wonder that MAR now regards itself as 'the agency that looks upon the nation from above'.

The Ministry of Atmospheric Resources consists of four main branches. These are responsible for Atmospheric Intelligence, Atmospheric Marketing, Atmospheric Management, and International Affairs. Each is in the

---

* This discussion is based on part of the text (see Sewell and Maunder 1985) of a futuristic discussion between 'Dr Brown' and 'Dr Jones', presented by Dr W. R. D. Sewell and Dr W. J. Maunder at the Fifth Conference of the Meteorological Society of New Zealand, held in Wellington on 11 October 1984. The editor of *Weather and Climate* (in which the paper was published), commented in a footnote that 'although the views expressed are not necessarily based on current mainstream thinking on the expected technical progress in meteorology, we publish the article as a provocative statement to encourage our readers to think about the direction atmospheric science may take in the next ten years'.

hands of a Director who oversees its overall management and who reports to a Corporate Affairs Group headed by the Administrator of MAR.

The Ministry is a very practical body. You can depend upon it for an accurate forecast of the weather, not merely for the next week but also for the next three months. It helps you plan your ski trips and your Pacific Island vacation. It gives the Featherston Street [New Zealand's Wall Street] financiers up-to-date, weather-based pastoral and horticultural indices for key areas – both within and outside of New Zealand. It determines the effects of waste disposal into the atmosphere and issues permits to factories and to car owners for effluent discharge. It also collects the fees for such permits. Besides this, MAR is a referee. Its Clouds Jurisdiction Division sorts out disputes between individuals who have attempted to modify the weather and those who claim compensation from those who have done so.

The International Affairs Division provides New Zealand's input into the New International Meteorological Organization (NIMO), which was established in 1990 following the break-up of the UN-based World Meteorological Organization. It also deals with atmospheric resource disputes with other countries, and between the State Governments of North and South Zealand.* In addition, it is New Zealand's 'nuclear winter' monitoring agency.

We take you now to the office of Dr Brown, located on the seventh floor of the Ad Astra Building. This magnificent edifice on the hills overlooking Kapiti Island† was opened only in 1989, as part of the government's Decentralization of Head Offices Programme. Buildings, you see, now move to the staff rather than vice versa.

The Administrator is talking with Dr Jones, his Chief Executive Officer.

Dr Brown. Good morning, Dr Jones. What have we got on the agenda today?

Dr Jones. It looks like a very busy day. Lots of interesting items: some of them quite urgent.

Dr Brown. All right then – what's the first** item?

2. Seasonal forecasts and climate impacts

Dr Jones. Due to be released today is the forecast for winter 1994 and spring 1994. The forecast for winter indicates severe cooling in eastern areas of New Zealand, and a significant reduction in the rainfall in the major hydro-electric catchment areas. These forecasts will be disseminated through the usual channels, including the new environmental television and videotex service.

* New Zealand which comprises two main islands (the North Island and the South Island) has currently (1986) one central government.

† Kapiti Island is a delightful bird sanctuary 60 km north of the capital city of Wellington.

** Three items were discussed in the original paper; only two are discussed here.

You will recall that seasonal forecasts for rainfall, temperature, and soil moisture have been made operationally since autumn 1988. Since then their track record has been very good – so much so that there is an embargo on the release of today's seasonal forecast until 10 p.m. This will allow the Special Intelligence Unit of the Prime Minister's Department to make any appropriate recommendations to the Prime Minister as to which part of the forecast, if any, should be restricted.

DR BROWN. Does it look as though we have got a class A seasonal forecast for issue tonight, or is it likely to be unrestricted?

DR JONES. Well, the nation will certainly face problems if the forecast is correct – and the Special Long Range Forecasting Unit is very confident. What it indicates is that there is a 90 per cent probability of a second spring season in succession with very cold temperatures. It is expected that these cold conditions (following on the cold and dry autumn which we are now experiencing), will reduce the 1994/95 and 1995/96 wool production by as much as 25 per cent with a significant impact on the fertilizer and shipping industries. In addition, a substantial increase in unemployment is anticipated (mainly as a result of the extremely harsh growing conditions which are now entering their second season). As a result the Minister of Employment will have to be advised. Besides this, the Environment Planning Council will be called into emergency session.

The National Energy Authority in its meeting tomorrow will indicate to the fifty local energy authorities that their requests to the North and South Zealand State Governments for energy must be increased by at least 30 per cent. Because of the financial difficulties that this will place on domestic users, the Ministers of Finance and Domestic Affairs will be advised that provision for supplementary benefits to the local authorities must be provided for in the 1994 budget.

DR BROWN. It looks pretty serious then, especially if we also take into account the atmospheric dust problem.

DR JONES. It sure does. In fact, this dust problem is partly the reason for the cold conditions being forecast. It looks as though we could be reaching a crisis point over this. I have prepared a brief for the Minister reminding him that, while atmospheric carbon dioxide has continued to increase during the last decade, it is now growing extremely rapidly. This has resulted from the serious and widespread forest fires in Africa and South America in the early 1990s. However, and this is the significant factor, the eruptions of Mount Etna, El Chichon and Krakatoa, all within three weeks of each other in 1993, have meant that the expected significant warming through carbon dioxide accumulation has been replaced (at least on a temporary basis) by an extremely cold spell in the Northern Hemisphere.

It seems that this cold spell is now spreading into parts of the Pacific.

This has been shown in the cold summer, and the already cool autumn. The forecast cold winter and spring emphasize it even more.

Dr Brown. Presumably, we will have to call a media conference about this?

Dr Jones. Well, in addition to the items I have already mentioned in regard to wool production, shipping, employment, and energy, I have been in touch with the Minister of Health. It seems we will need to alert the Long Range Forecasting Unit and the Climate Monitoring Unit to provide a confidential review to the Minister of the likelihood of the cold conditions continuing next summer and even into next autumn. The health problems associated with continued cooling could be quite serious. The Minister of Pastoral Production has also requested an urgent discussion of the impact statements issued by the New International Meteorological Organization (NIMO) Impact Monitoring Unit. The statements indicate that the global cooling may produce worldwide shortages of wool, meat and dairy products. You will recall that the NIMO Impact Monitoring Unit was established in the old Meteorological Office building in Wellington in January 1990, so it's an exciting expansion of what the 'old' Meteorological Service was trying to do in the 1970s and 1980s.

Dr Brown. Yes, it certainly is. I remember well how New Zealand pioneered some of the work on the impact of weather and climate on socio-economic activities. Our research obviously made some impression on the Governing Council of NIMO. The problem in those days was that we did most of the work and others took the credit. However, with the headquarters of NIMO now in Perth, Australia, the north now looks to the south; quite a contrast with the old Geneva days!

3. Decision time with Tropical Cyclone 'Omega'

Dr Brown. Today's agenda is certainly anything but routine. It's decisions, decisions, decisions. Not like the old days of the 'Muldoon/Lange'* era when all we had to do was 'refer', or 'comment'. What is the next problem breathing down our necks?

Dr Jones. We have just had a call from Air Commodore Raoul of the New Zealand Air Force in reference to Tropical Cyclone 'Omega', which is heading our way. He suggests that we should activate the Cyclone Steering Committee and have a Video Display Interaction call this morning, as soon as we can get at least three of the committee together. He sounded as though it was a RAC (Red Alert Crisis) situation.

---

* Sir Robert Muldoon was Prime Minister of New Zealand from 1975 to 1984. Mr David Lange succeeded Sir Robert as Prime Minister in July 1984 following a 'snap' election.

Dr Brown. All right, you had better explain what it's all about.

Dr Jones. You will recall the emergency plans for the 1993/94 tropical cyclone season (as agreed on 1 September 1993) indicated that the following courses of action should be followed. First, that any tropical cyclone of potential hurricane force intensity entering the New Zealand protection zone must be considered as a potential target for modification. Second, any such modification is subject to a stringent set of procedures – political, legal, and strategic.

Dr Brown. What is the present position and track of the tropical cyclone?

Dr Jones. This morning Tropical Cyclone 'Omega' was located by the joint Australian New Zealand weather satellite ANZAC II at 8.08 a.m., 52 nautical miles south-southeast of Norfolk Island. The expected path – as forecast by the Southern Hemisphere Tropical Cyclone Centre in Darwin, Australia – is southeast at 16 knots. The path takes Omega over New Plymouth at 11.30 p.m. tomorrow [Fig. VIII.2].

Dr Brown. What can we do? With a storm approaching at that rate, have we got time to activate the tropical cyclone steering procedures? Remember, any mistake in our judgement could put our jobs at risk.

Dr Jones. The question is whether we should attempt to alter the course of the storm. You will know that the Atmospheric Management Branch, through the deployment of New Zealand based aircraft, has the capability of altering the path of a tropical cyclone by as much as 10 degrees, but with an error of 5 degrees. The decision to do so, however, requires the joint

Figure VIII.2 Forecast path of Tropical Cyclone 'Omega', and the projected paths if the cyclone is steered

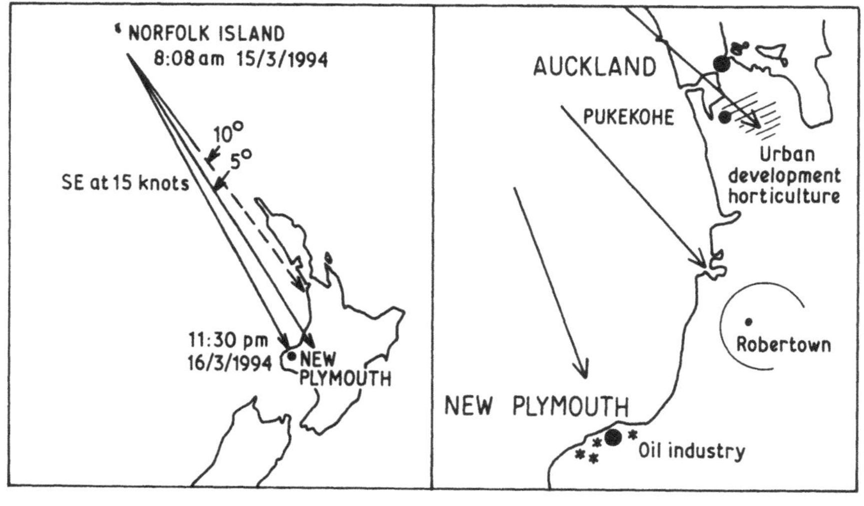

Source: After Sewell and Maunder (1985).

agreement of the Ministers of the Environment and Atmospheric Resources, and the Attorney General, in consultation with the Prime Minister.

Dr Brown. Well, what is our course of action?

Dr Jones. The Special Weather Prediction Unit is being convened at 11.30 a.m. They will be faced with two choices.

The first is to leave tropical Cyclone 'Omega' to follow its predicted course which will bring considerable destruction to New Plymouth, now New Zealand's third largest urban area, as well as to the surrounding oil fields. At least 30,000 homes and apartments are expected to be damaged, with possibly 50 lives lost and as many as 500 injured. In addition, the multi-billion dollar oil industry will suffer such severe losses that New Zealand will have to import oil from Indonesia. We have not had to do this since 1988. It also seems quite possible that emergency contracts will also have to be entered into with the Southern Hemisphere Antarctic Energy Consortium to obtain additional supplies.

The second choice is to deflect the cyclone by 5 degrees. This will make it pass over the King Country close to Robertown (formerly Te Kuiti). In this case the total population affected will be less than 10,000 and the major damage will be to rural holiday homes and retirement villages. The larger part of the resident population could be evacuated, but there will be severe social disruptions particularly among the very elderly.

We therefore have these two choices – either do nothing or steer the tropical cyclone – but the problem of uncertainty remains, relating to the 5-degree error if steering is attempted.

Dr Brown. How serious is this?

Dr Jones. Based on experience over the last four seasons – the average steering error is 3 degrees, with a maximum error of 6 degrees. The key problem therefore is that, if the deflection was 5 degrees more than intended, the resulting track would take Tropical Cyclone 'Omega' just south of Auckland. In this case, it is anticipated that at least 3000 lives would be at risk, and the horticultural industry of South Auckland would be virtually eliminated.

Dr Brown. All right Dr Jones, get onto Air Commodore Raoul right away, tell him that there will be a video conference at 11 a.m. with throughput to the Prime Minister's Office. Also, give Dr Black a ring and get her to activate the Cyclone Steering Committee immediately.

Dr Jones. I'm already on the way.

G. The Challenge Ahead

As we enter the late 1980s both weather and climatic fluctuations are producing major economic, social, and political consequences. Indeed, our vulnerability to such fluctuations has undoubtedly grown as the world population has increased and the use of available resources has become more intense. Many examples of the importance that both weather information and climate information have on national economies have been given in previous chapters. Clearly, it is now well recognized that particularly in regard to food, energy, and commodity flows, the monitoring and prediction of local, regional, national, and international productivity and consumption are ultimately associated with the availability, use, and interpretation of the past, present and future weather and climate.

The foregoing suggests possible goals for weather- and climate-related activities during the next decades. Many of the concerns involve work in areas peripheral to the normal activities of many meteorologists and climatologists. The problems are nevertheless real, and it is believed that geographers (physical and human), meteorologists (dynamic and physical), and climatologists, plus experts in other disciplines, can contribute in a very positive and useful way to solving some of the world's important problems associated with the utilization and supply of food, fibre, and energy. For example, there is a clear need for the development of more rigorous techniques for assessing the *specific* impacts of weather and climate, and in particular determining the degree of impact that climate variability has upon various national economies. This will clearly involve the development of new areas of interdisciplinary research requiring the skills of a wide range of people including meteorologists and climatologists, as well as contributions from the fields of geography, agriculture, forestry, economics, planning, marketing, political science, and sociology. In addition, the more specialized viewpoints and expertise of people on the 'fringe' of the disciplines mentioned must be encouraged in order that 'real-world' problems are solved.

The multidisciplinary nature of the climate system also calls for a mechanism to make available concise, condensed, easily readable information to research scientists and technical/managerial/governmental personnel. These comments, made by Unninayar (1983) in a review of the need for better climate monitoring systems, highlight a key aspect of the future of meteorology and climatology. These and other futuristic aspects of the subject are listed in Table VIII.1.

The list reflects in part a personal bias; however, whatever the bias, the ultimate question must surely be 'what does it all mean?' To answer such a question, it is clear that assessments must be made using all available meteorological and non-meteorological information. Of course, skilled assessors with vision, knowledge, and an appreciation of real problems are needed to do this, for as pointed out several times in this book, although it is comparatively easy to make assessments of the *general* relationship between weather factors and production or consumption, or in some cases prices, the more *precise* relationships necessary for operational decision-making are much more difficult to formulate.

Table VIII.1 Climatology – future trends

Observations	Automatic stations Transmission in real-time Analyses in near real-time Immediate availability to users
Computer-to-computer exchange	Climatic historic data Near real-time data
Global/regional monitoring	Enhanced global telecommunications and World Weather Watch system
Real-time exchange	Data Derived products Mapped parameters Commodity-weighted indices
Videotex services	Past information Present data Future indicators
Applications	Area National Regional Global
Research	Continuous monitoring (linked to past) Diagnostic analyses Large-scale modelling Predictions
Impacts	National/regional Advice to governments Political implications Social implications Strategic implications

Source: After Maunder (1984b).

The concerns and implications of the wide variety of items mentioned in this book highlight three major factors. First, climate must be recognized as a resource, and not solely as a factor which imposes limitations on settlement patterns and economic activities. Second, there is a need for a much improved understanding of the many interrelationships between climate and society and especially the manner in which changes in one may result in shifts in the other. Third, it must be fully appreciated that many of the issues of concern to the meteorological and climatological community have potential if not actual political implications.

If we are to live within the limit of our 'climatic income' or our 'élite atmospheric resource', however, appropriate meteorological and climatological planning must be involved. For this to be accomplished, the politician and the planner must become more weather and climate orientated, for only then will optimum use be made of the climate resources of the 1990s. Such planning has already been recognized by the World Meteorological Organization in its World Climate Programme, and central to this planning is the need for a much more comprehensive monitoring and analysis of the world's climate, both to detect and predict changes. The consequences of such changes must also be better understood. Indeed, the need for more relevant and more timely information about the weather and the climate offers the most important challenge to the meteorological community.

A key factor in all or most aspects of the whole 'uncertainty business' as discussed in this book, and more specifically noted above, is the question of *how* will the real importance of weather and climate *in* the market place be accomplished. Clearly, meteorologists and climatologists have an important role to play in educating and influencing governments, the community, as well as the key weather- and climate-sensitive sectors within each community, in the importance of the 'élite' resource which we call the atmosphere.* But perhaps the final influencing factor will *not* be the meteorologists or the climatologists, or the decision-makers who may become more aware of the significance of weather and climate in their activities; rather, it is suggested that the atmosphere itself may well have the final say if people do not learn to live within their 'climatic income'.

From the foregoing it is obvious that meteorological and climatological studies of the future cannot and indeed must not be confined entirely to the analysis of atmospheric data. Rather, such studies must consider the social aspects of food, health, energy, and well-being, and of all disciplines it is the geographically, environmentally, or economically trained meteorologist or climatologist who can perhaps best cast such studies in frameworks that consider the earth's surface and the atmospheric boundary layer, *as well as* the economic, social, and political implications of the climate understanding so developed. To the degree that this is done climatology and meteorology will not only meet the challenge of the eighties and the nineties, and so contribute to the solution of a wide range of human problems, but will also emerge as one of the most powerful disciplines of the twenty-first century.

---

* The need for appropriate 'climatic guidance' to politicians and decision-makers during times of major climatic aberrations, such as the recent droughts in the Sahel, Ethiopia, Brazil, and Australia, as well as the 1982–3 El Niño event (see Hare 1985b), is now becoming much more accepted by all parties. However, the need for climatic guidance during 'near normal' times must also be given due recognition.

Appendix: Guidelines for further investigation

A. OVERVIEW

The eight chapters of this book have in the main been a *personal* view of '*The Uncertainty Business: Risks and Opportunities in Weather and Climate*' and clearly it has not been possible to review all the growing literature on the subject. Appropriate mention has of course been made in the book of selected publications and it is hoped that the text is neither too personal nor too parochial. Nevertheless, it is important that the reader be guided to other works on this subject, and the following 'abstract' of a recent SCOPE publication should assist those who wish to develop further the ideas presented in this book. In addition, a topic guide to a *selection** of the literature on the subject of this book is given in Section C of the Appendix. The literature cited includes (for completeness) the 280 references specifically cited in the text, together with the 416 references not specifically cited. All 696 references are listed in full in the Bibliography.

B. 'SCOPE' CLIMATE IMPACT ASSESSMENT

A significant monograph, *Climate Impact Assessment: Studies of the Interaction of Climate and Society*, commissioned by the Scientific Committee on Problems of the Environment (SCOPE) of the International Council of Scientific Unions (ICSU) was published in 1985. This publication was edited by Professor R. W. Kates, assisted by J. H. Ausubel and M. Berberian. As it is almost impossible to summarize the findings of the more than 600 pages of this monograph, the following slightly edited extracts are given (with the kind permission of the editors, SCOPE and the publishers, John Wiley) to give the reader, who may not otherwise see this important monograph, an insight into 'Climate Impact Assessment' from an international viewpoint.

---

* The selection of references includes some of the pioneering papers in the various fields, together with a representative selection of the contemporary literature.

1. Preface (to the SCOPE publication)

The decade of the 1970s was marked by a growing climate consciousness, both popular and scientific. The new interest was sparked by a series of extreme climate events and related disruptions, and by scientific speculation as to increased climate variability and possible climate change. Two sets of events during this period attracted both scientific and public interest. The first, in 1972, was the apparent simultaneous occurrence of unfavourable weather in many parts of the globe and its speculative relationship to a wide variety of socioeconomic events, including the quadrupling of various commodity prices around the world, food shortages in the Sahel of West Africa and in South Asia, a drastic fall in the anchovy fishery of the Pacific, and even changes in government in Ethiopia and Niger. The second was an emerging scientific consensus that human-induced alterations in the chemical constituents of the atmosphere could lead to large regional, and even global, changes of the atmosphere in the form of more acidic rain, greater ultraviolet radiation, and altered temperatures.

The early eighties again found persistent drought in northeast Brazil and in many countries of Africa; a warning of sea-surface Pacific temperatures leading to the most remarkable 'El Niño' event recorded to date; and the warmest years in a century of northern hemisphere temperatures. A scenario for a new and most serious set of climatic consequences following a major international exchange of nuclear weapons, the 'nuclear winter', was postulated. The science of human-induced alterations of the atmosphere became more complex with the slowing of the rate of fossil fuel use and with improved understanding of the way the many 'greenhouse' gases contribute to global warming, while at the same time reducing the net destruction of the ozone shield against ultraviolet radiation.

In sum, the diversity of novel climatic experience continues unabated, the recognition of potential sources of human-induced alternation has increased, and the pace and degree of change are questioned and debated. Nonetheless, it is widely agreed that one such change, a long-term global warming derived from the enrichment of the atmospheric content of the 'greenhouse' gases, is underway. Within the time period of the projected global average warming, measured in tens of hundreds of years, sustained variation of climate will occur in many places, and lesser periods of favourable or unfavourable climate will occur in most places – a function of normal variability. Where these changes are large – the extremes greater than what is customary – where people and places are vulnerable, or where human activity meshes poorly with natural opportunity, significant climate impacts to people, ecosystems and societies are likely to occur. How to respond to such impacts – adjusting to changing climate, coping with extremes, matching human needs to climate endowment – are issues of considerable importance. The scientific study of climate and society should guide societal response. Concepts of climate impact assessment are new, the methods are still under development. This volume is an authoritative review of these methods and concepts, a contribution of the Scientific Committee on Problems of the Environment to the World Climate Programme.

The World Climate Programme (WCP) was initiated in February 1979 at a meeting of 350 scientific experts held under the aegis of the World Meteorological Organization. The Programme is directed at four goals: (1) improving our understanding of the physical climate system; (2) improving the accuracy and availability of climate data; (3) expanding the application of current climate knowledge to human betterment; and (4) advancing our understanding of the relation between climate and human activities.

Organized to address the fourth goal was the World Climate Impact Programme (WCIP), of which this study is an initial effort. . . . The Scientific Committee on Problems of the Environment (SCOPE) of the International Council of Scientific Unions [accordingly] undertook to prepare the authoritative review of the methodology of climate impact assessment called for in the World Climate Impact Programme. The objectives of the review were: (1) to examine existing methodology; (2) to foster the development of new methodological approaches; and (3) to inform a broad range of disciplines as to the available concepts, tools and methods beyond their own specialty.

The individual papers transcend professional boundaries and examine climate impact assessment in a non-disciplinary fashion as a set of linked analytic components, as techniques of case study and modeling, and as reviews of past experience. Each author has sought to review the state of his or her art, not for peers, but for scientific colleagues who are interested in climate impact assessment but schooled in a different discipline or lacking experience in a particular technique. The achievements, weaknesses, and capabilities of the various methodologies are set forth with candor, tempered by empathy. It is our hope that workers new to climate impact assessment will be realistic in their expectation of the various methods, sympathetic with the common scientific problems faced, and challenged both by their practical necessity and intellectual adventure.

2. Part I: Overviews

Running as a thread through the entire volume, linking together the sectoral studies, the analytic methods and the case examples, are conceptual models of the interaction of climate and society, definitions of climate variability and change, and assumptions as to the state of knowledge concerning climate processes. These concepts and definitions, presented by Kates [see Kates 1985] in Chapter 1, and Hare [see Hare 1985a] in Chapter 2, provide to all authors a common vocabulary for describing climatic events, consequences and human responses, a common framework for linking climate and societal impacts, and a common interest in both industrialized and developing countries. Within this framework, climate variability and change provide three types of events of interest: extreme weather events, persistent periods, and 'little ages'. These events impact on exposed social, areal, or activity units of human or ecological organization, leading to ordered biophysical, social, or ecological consequences. In turn these impacts are modified by cultural adaptation and adjustment responses that may amplify or dampen the consequences of climate events. In the simplest of frameworks, the links between events, units, and consequences of climate impact models are linear. In the more realistic and complex interaction model framework, causality is jointly determined by climate and society. As with all such frameworks, relationships are linked in ordered flows that belie the reality and simultaneity of the real world.

The overview on research by Riebsame [see Riebsame 1985] in Chapter 3 serves a different function, providing a common conceptual and historical review of climate–society research organized under four key concepts: climate as setting, as determinant, as hazard, and as resource. Riebsame's view that research, both past and future, flows directly from these different, but not exclusive, concepts of climate-society interaction serves not only to organize the diverse literature of this interdisciplinary field, but to analyze its structure as well.

The final overview, Chapter 4 by Maunder and Ausubel [see Maunder and Ausubel 1985], links directly to the rest of the volume by posing the question of how one begins to undertake specific climate impact studies. They suggest that one major way to begin is by assessing the overall climate sensitivity of activities, places, or groups of interest. Past experience and current methods for determining overall sensitivity are presented. It emerges from many studies that agriculture and water resources are activities and sectors that are clearly sensitive to climate. Methods appropriate to the study of these and other sensitive sectors follow in Part II.

3. Part II: Biophysical impacts

In the simplest of impact models, climate affects places, people and their activities, leading to a set of ordered consequences. The first set of conseqences is designated as biophysical impacts, because their causal mechanisms (where known) or transfer functions (where inferred) are in the realm of physical and non-human biological relationships: crops grow, rainfall runs off, cattle graze, fish feed, and buildings cool.

As indicated in Chapter 4 on climate sensitivity [see Maunder and Ausubel 1985], all five of the sectoral chapters deal with highly climate-sensitive activities. But they do not exhaust the set of such activities; indeed they only illustrate them. Particularly missing are chapters on unmanaged, natural ecosystems, sensitive industrial activities such as construction and transportation, and service activities such as insurance and recreation. However, the five chapters well illustrate the range of methods available for determining first-order impacts, the comparative precision and validity of these methods, and some common directions for future investigation.

In Chapter 5, Nix [see Nix 1985] suggests five methods for analyzing the impacts of climate on agriculture: trial and error, analogy, correlation, simulation modeling, and systems analysis. This typology serves well to describe not only methods employed in the agricultural sector, but also to compare the methods available in other sectors as well. Agriculture (Chapter 5), water resources (Chapter 8), and energy resources (Chapter 9) employ the full range of methodologies. The analysis of pastoralism (Chapter 7), as distinct from modern ranching, is with a few exceptions still at the level of analogy and simple correlation. The link between climate and fisheries (Chapter 6), with the possible exception of 'El Niño' types of phenomena, is still trial and error, uncertain analogy and correlation.

Greater predictability of transfer functions is found in physical than biological functions and in small-scale, controllable human activities. Thus it is easier to predict temperature–energy or rainfall–runoff relationships than the climate yields of agriculture, pastoralism, or fisheries. Similarly it is easier to predict the energy demand for a building than for a city, or the yield of a field than that of a grazing area or an ocean current. The chapter authors have not limited themselves to first-order impacts. Although concerned with biophysical impacts, all the authors work from a framework in which climate impact relationships are constrained or changed by human action. Thus as Nix (Chapter 5) shows, agricultural models must be confined to a particular crop, place and technology or specifically include changes in technology as a major variable in yield functions. Kawasaki [see Kawasaki 1985] (Chapter 6) explores the still indeterminate issue of whether fluctuations in fishery catch reflect natural (including climate) cause or simply overfishing. A similar issue is posed in Chapter 7 by Le Houérou [see Le Houérou 1985] relative to the success of human adjustment to drought. Nováky, Pachner, Szesztay, and Miller [see Nováky, Pachner, Szesztay, and Miller 1985] (Chapter 8) argue for the need to consider water-related impacts of climate clearly within the context of socially related management activities. Similarly Jäger [see Jäger 1985] (Chapter 9) notes that climate–energy demand relationships are changed rapidly as conservation methods come into widespread use.

Thus a common research question for methodological development is posed by the need, in even the simplest of impact models, to allow for the interaction of climate and society. A second issue is posed most clearly in the analysis of Le Houérou, but is also evident in the chapters on water, agriculture, and energy. Le Houérou traces the systematic amplification and dampening of impacts along the causal chain of the impacts model, suggesting that variability in climate is amplified in primary productivity of grazing lands, but in turn is dampened in livestock yield and in human impact.

Similarly Chapter 8 shows that fluctuations in rainfall are amplified by fluctuations in streamflow but are dampened by water resource management measures. The systematic comparison of climate-yield-impact ratios for different sectors might well lead to improved understanding of the processes of interaction.

4. Part III: Social and economic impacts and adjustments

As the impact analyst moves from first-order biophysical impacts to higher-order consequences, the possibilities, outcomes, and human choices attached to each link increase, and the causal chain becomes less distinct. Part III reflects that complexity. Three of the chapters, written from disciplinary perspectives, deal with second-order consequences on human health (Chapter 10), economy (Chapter 12), and society (Chapter 13). Three of the chapters deal with case study methods to assess so-called 'natural experiments', focusing on historical study (Chapter 11), climatically and economically marginal places (Chapter 14), and extreme weather and climate events (Chapter 15). Finally, two chapters deal with adjustment responses and mechanisms for their perception and choice (Chapters 16 and 17).

When studying self-provisioning societies it is convenient to think of human nutrition and related health effects as second-order consequences – climate-related yields of food leading to various health and demographic impacts. But there are few fully self-provisioning societies, therefore biological increases or decreases in the availability of foodstuffs must be traced through a network of existing social and economic relationships. Escudero's [see Escudero 1985] proposal for a case comparison of both shores of the Windward Passage is a case in point.

Economic and social relationships are the substance of the social sciences, particularly economics and sociology. Within the framework of market economies, but with some applicability to all economies, there are a set of well-defined quantitative approaches designed to answer two fundamental questions of how economies interact with climate. Given a change in either the mean or distribution of climate events, how will the allocation of resources change and which persons or places will lose or gain from such resource changes? As Lovell and Smith [see Lovell and Smith 1985] point out in Chapter 12, robust methods exist to answer these questions, but rarely have been applied. Nor have they been adapted to the special qualities of climatic variability and change: stochastic nature, large-area impact, and long time-horizon over which consequences take place.

Ironically, the more diffuse, less-defined social impacts appear to be better illustrated, with interesting and recent case examples from studies of weather modification and extreme events. With a broader view than economic analysis, social impact analysis, as Farhar-Pilgrim [see Farhar-Pilgrim

1985] notes in Chapter 13, is a class of policy analysis that arose from a concern with the hidden costs of societal undertakings. Thus there is a strong emphasis on identifying the many different stakeholders affected by climatic variability or change and assessing the differential impact upon them.

Historical analysis, de Vries [see de Vries 1985] informs us in Chapter 11, employs the full array of social science methods and is limited only by the availability of data from the past. Climate adds to the rationale of historical explanation of human events, and historical events allow us to expand the stock of relevant climates and societies to examine. (See also the extensive discussion in Chapter 21.) In the latter case, historical analysis is often employed for the two types of natural experiments that are emphasized in this volume – a focus on vulnerable margins or groups and a focus on extreme climatic events, usually of interannual or decadal length.

Natural hazard research, described by Heathcote [see Heathcote 1985] in Chapter 15, provides a rich body of relevant methodology to study the extreme events of the past as well as those of current experience. To illustrate, Heathcote reviews such methods within the ordered sequence of impacts of violent storm or persistent drought which are most easily identified and measured; the long-term impacts, however, are much in doubt. The margins of climatically sensitive activities that Parry [see Parry 1985a] documents in Chapter 14 seem to be more sensitive barometers of the impacts of longer-term changes in climate.

Woven throughout Part III are specific issues of adjustment and adaptation. Although some studies of biophysical impact attempt to ignore or to constrain societal interaction, studies of social and economic impacts are always interactive, analyzing the differential societal impacts of climate change and variability in the light of the differential ability to cope with or take advantage of such change. Indeed, several authors point out that the differential between societies and their resources is much greater than the differential between climatic regions or epochs in their impact on human beings.

At the level of individual and small-group decision-makers, studies of perception, alluded to in Chapters 11 and 15 and given full treatment by Whyte [see Whyte 1985] in Chapter 16, serve as a major link to studies of adjustment of the type described by Jodha and Mascarenhas [see Jodha and Mascarenhas 1985] in Chapter 17. Reports of the nature of adjustment to climate change and variability are scattered throughout the text. Lists of adjustments are given in most of the sectoral chapters. Given the bias towards industrialized nation experience, however, Jodha and Mascarenhas examine specifically developing country adjustment, drawing on their rich experience in South Asia and Africa.

When this set of disciplinary methods and case study opportunities for identifying human social and economic impacts is compared, a strong negative bias emerges. Most methods and case studies focus on climate as hazard; only scattered efforts have been made to study climate as a resource. Within the focus on climate as hazard, the balance of effort has been to identify the residual damages and losses caused by the impact of climate events on vulnerable groups or regions; less effort has been expanded on identifying and measuring the social cost of adaptation and adjustment. Assessing the social cost of climate adjustment and the opportunities of climate as a resource are important items for a research agenda.

5. Part IV: Integrated assessment

Studies that combine several links in the chain of sensitivity studies, biophysical impact studies, social and economic impact studies, and adjustment responses are integrated assessments. Examples of integrated assessments and problems of linkage between types of studies are found throughout the volume. Part IV explores in depth one major technique for providing linkage in integrated assessment – the use of modeling and simulation. In addition it reviews the experience with both historical and recent integrated assessment.

Integrated assessments involve a scale of activity and a set of complex linkages that encourage the use of modeling and simulation techniques. Such techniques provide an orderly and systematic way to store and analyze large arrays of data, to link data set together and to translate different disciplinary approaches into common mathematical language. A special attraction lies in the parallelism with general circulation models (GCMs), the favoured tool for exploring dynamics of climate at a global scale. These models, representing the apogee of causal explanation, scale of detail, and massive data handling and computation requiring the most sophisticated of computers, establish a criterion for modeling towards which many biological and social scientists working on integrated assessment seem irresistibly drawn. Thus the opening chapter of Part IV begins with an exploration of global modeling and simulations (Chapter 18).

In Chapter 18, Robinson [see Robinson 1985] examines some twenty global models for their potential in climate impact assessment use. She is cautious in her conclusions. Global social, economic, and environmental models have not been designed for climate impact analysis, are at best pioneering efforts, and are difficult to use. Nonetheless they can provide insight, data, and answers to restricted questions. The pioneering quality of global models is demonstrated by the coevolution biosphere model of the Computing Center of the USSR Academy of Sciences. Moisseiev, Svirezhev, Krapivin, and Tarko [see Moisseiev *et al.* 1985] describe (in Chapter 19) their ambitious, but incomplete, integrative model of climate ecosystem and society – a model attuned to a time horizon of centuries.

A more modest and limited form of modeling is presented by Lave and Epple [see Lave and Epple 1985] (Chapter 20) under the rubric of scenario analysis. They assert three virtues of scenario analysis: stretching the imagination to encompass a wide range of actions and implications: formal modeling of the causes and consequences of climate change and potential adjustments; and interdisciplinary integration to transcend the parochialism of professional method and tradition. Scenarios, and indeed all modeling to date, appear to be exploratory tools, not to be used for reliable prediction but rather to explore the bounds of both the unusual and the possible.

Large-scale modeling is an appealing tool, but still not a broadly realized one. What, then, can be said of other efforts at integrated climate assessment? Wigley, Huckstep, Ogilvie, Farmer, Mortimer, and Ingram [see Wigley *et al.* 1985] offer their evaluation of historical climate impact assessment (Chapter 21), considering some 24 examples of historical case studies. To do so they review extensively the methodological underpinnings of historical study, and Chapter 21 should be read jointly with the chapter on historical analysis by de Vries (Chapter 11). Wigley *et al.* note the attractiveness of historical case studies, seemingly free from the complications and confusions of oft polarized current historical explanation. But ironically they find the field suffering from polarities of a different sort, with exaggerated claims and rebuttals for the role of climate in history. Yet within the seeming excess of rhetoric, perhaps half of the studies examined handle both data and assumptions thoughtfully and carefully.

Glantz, Robinson, and Krenz [see Glantz, Robinson, and Krenz 1985] (Chapter 22) examine only five examples of major assessments of climate impacts of recent or future experience, but do so in considerable depth. They focus on comparative issues of study design and length, research staff, and linkages between the individual study components, and public presentation of findings. It is clear from these experiences that major climate impact assessments are substantial undertakings requiring extensive research and time, flexibility in design, and repetition as new data and methods become available. Assessments with strong scientific leadership can advance the state of the art; those organized on a constructive basis can at best attempt only to answer the questions addressed.

The review of integrated assessment concludes with the Chinese proverb: to know the road ahead, ask those coming back. The road ahead is unfolding. In the four years that this volume has been in preparation a second generation of integrated studies has been undertaken, and more are planned. Some of these studies simply repeat the past, with little evidence of having sought the advice of those coming back. But most of these half dozen studies evidence a high degree of methodological sophistication, scientific clarity, flexibility in their design, and excellence in their scientific leadership. Another final chapter is being written.

C. Topic Guide to Literature

Topic	Literature
Acid rain	Miller (1984) Rhodes and Middleton (1983) Weatherwise (1984)
Agriculture/climatology	McQuigg (1979) Maunder (1966b)
Agricultural frontiers	Parry (1981a)
Agricultural incomes	Maunder (1965) Maunder (1968a)
Agricultural impacts	Anderson (1979) Nix (1985)
Agricultural margins	Parry (1985)
Agricultural meteorology	Baier (1974) McQuigg (1972c)
Agricultural productivity	McKay and Allsopp (1977)
Agricultural technology	Chang (1981)
Agroclimatic modelling	Baier (1983)
Agroclimatic probabilities	Baier (1972)
Agroclimatological information	Sakamoto, Strommen, and Yao (1979)
Air conditioning	Robinson (1974)
Airlines	Beckwith (1966) Bollay (1962)
Applied climatology	Griffiths (1976)
Archaeology	Wright (1976)
Atmospheric constituents	Tyndall (1861)
Atmospheric 'economic climate'	Maunder (1979)

Avalanche hazards	Prowse, Owens, and McGregor (1981)
Barley and climate	Williams and Oakes (1978)
Beekeeping	Hill (1984)
Benefits of meteorological services	Tolstikov (1968)
Bias of science	Martin (1979)
Biological activity	Thompson (1981)
Biosphere	Caldwell (1974)
Biosphere impacts	Tarko (1977)
Biosphere models	Moisseiev *et al.* (1985)
Building/construction	Maunder (1972b)
Business	Hallanger (1963) Linden (1959) US Conference Board (1977) White (1964)
Business/famines	Meillassoux (1974)
Calamities	Hewitt (1983)
Canadian economy	Hare (1981) Hare (1983)
Carbon dioxide	Budyko and Efimova (1981) Clark (1982) Department of Energy (1980) Hare (1980) Idso (1980) Idso (1984) Kellog and Schware (1981) MacCracken and Luther (1985) National Research Council (1979) National Research Council (1982) Pearman (1980) Revelle and Waggoner (1983) Tucker (1985) World Climate Programme (1981)

Changing world climates	Tyson (1977)
Choice or chance	Maunder and Sewell (1968)
Civilization	Huntington (1945)
Climate	Hare (1979a) Hare (1982)
Climate/Canada	Hare and Thomas (1974)
Climate/history	Fisher (1980) Lamb (1979)
Climate/society	Riebsame (1985) SCOPE (1978)
Climate/year 2000	Gasser (1981)
Climate and history	de Vries (1980) Lamb (1982) Oliver *et al.* (1975) Ogilvie (1981b) Rotberg and Raab (1981) Wigley, Ingram, and Farmer (1981)
Climate and risk	Massachusetts Institute of Technology (1980)
Climate change	Budyko (1982) Epstein (1982) Food and Agriculture Organization (1979) Gribbin (1978) Mason (1976a) National Academy of Sciences (1975c) National Research Council (1983) Sawyer (1980) Smith (1961) Yared (1984)
Climate change?	Landsberg (1975) Winstanley (1985)

Climate change (CO_2)	Manabe and Wetherald (1980)
Climate diagnostics	Climate Analysis Center (CAC) (1982)
Climate impacts	Alexander (1974)
Climate mandate (the)	Roberts and Lansford (1979)
Climate modelling	Global Atmospheric Research Programme (1975) World Meteorological Organization (1975)
Climate modification	Gray *et al.* (1976)
Climate monitoring	Unninayar (1983)
Climate resources	Curry (1962) Freedman (1980) Maunder (1973b) Maunder (1974c) Miller (1956) Taylor (1974)
Climate variability	Hare (1985a) Flohn (1981c)
Climatic awareness	Hare and Sewell (1985)
Climatic constraints	Maunder (1977d)
Climatic determinism	Lorenz (1968)
Climatic fluctuations	Wallén (1984)
Climatic futures	Kraemer (1978)
Climatic 'normals'	Todorov (1984)
Climatic problems	Matthews (1976)
Climatic variations	Karl and Riebsame (1984)

Climatological challenges	Mather *et al.* (1981)
Climatology/applications	Mather (1974)
Climatology/history	Hann (1882)
Climatology viewpoints	Maunder (1984b)
CO_2 (agriculture)	Rosenberg (1982) Santer (1984) Waggoner (1983)
Commodity markets	Labys (1978)
Community responses	Kirkby (1974)
Construction activities	Benjamin and Davis (1971) Johnson and McQuigg (1972) Martin (1970)
Construction industry	Russo (1966) US Department of Commerce (1966)
Construction sites	Wahlibin (1984)
Consumer sales	Petty (1963)
Consumers	Maunder (1969)
Corn/weather relationships	Mostek and Walsh (1981) Terjung, Liverman, and Hayes (1984)
Cost/loss ratios	Murphy (1976)
Costs of bad weather	Taylor (1971)
Costs of cold weather	Tremayne (1967)
Costs of climatic variations	Oram (1982)
Crop–climate models	Biswas (1980) Wigley and Tu Oipu (1983)

Crop/risk management	Amerasinghe (1984)
Crop–weather models	Baier (1977) Oury (1965) Stanhill (1977)
Crop development	Robertson (1973)
Crop forecasting	Shapley (1976)
Crop growth	Monteith (1981)
Crop insurance	Dandekar (1976)
Crop production	Monteith (1977) Uchijima (1978)
Crop research	Nix (1980)
Crop yield estimation	Sakamoto (1978)
Cultural aspects	Arnold and Galbraith (1978)
Daily weather indices	Richardson (1984)
Dairy industry	Boyer (1975)
Dairy production	Maunder (1966a)
Dairy production forecasting	Maunder (1974b) Maunder (1977b)
Decision processes	Slovic, Kunreuther, and White (1974)
Demographic aspects	Caldwell (1977)
Department store	Steele (1951)
Desert communities	Szarec (1979)
Desertification	Biswas and Biswas (1980) Hare (1977) Hare (1985b)

Development strategies	Berry and Kates (1980) Seifert and Kamrany (1974)
Disasters	Foster (1976)
Drought	Anonymous (1966)
Drought/Australia	Campbell (1968)
Drought/decision-making	Goulter (1982) Oguntoyinbo and Richards (1978)
Drought/Great Plains	Warrick (1980) Warrick and Bowden (1979)
Drought/history	Foley (1957)
Drought/North America	Rosenberg (1978)
Drought/people	Garcia (1981)
Drought/Sahel	Kates (1981a) Tooze (1984)
Drought/Sudan	Hulme (1984)
Drought/water	Palutikof (1983b)
Drought adjustments	Riebsame (1981)
Drought definition	Glantz and Katz (1977)
Drought economics	Australian Bureau of Agricultural Economics (1969) Duloy and Woodland (1967) Harshbarger and Duncan (1977) McIntyre (1973) Maunder (1971b)
Drought forecasting	Glantz (1982b)
Drought impacts	Kates (1981b) Newman (1978) Perkey, Young, and Kreitzberg (1983) Warrick and Bowden (1981)

Drought rains	Perry (1962)
Drought recovery	Cumming (1966)
Drought strategies	Maunder (1982b)
Ecoclimates	Le Houérou and Popov (1981)
Econoclimate	Perry (1971)
Econoclimatic models	Maunder (1968b) Maunder (1972d) Maunder (1974a)
Econometrics	Freedman (1981)
Economic activity	d'Arge (1979) Gruza (1979)
Economic analysis	Maunder (1972c)
Economic aspects	Burroughs (1978) Flohn (1981b)
Economic benefits	Berggren (1971) Gibbs (1968) World Meteorological Organization (1968)
Economic climates	Ausubel (1980) Maunder (1981a)
Economic development	Lambert (1975) Oury (1969) Maunder (1973d)
Economic forecasting	Theil (1966)
Economic history	Hoskins (1964)
Economic impacts	Commonwealth Scientific and Industrial Research Organization (1979) Eddy *et al.* (1980)

Electricity consumption	Clark (1985) Le Comte and Warren (1981)
Electricity demand	Crocker (1976) McQuigg, Johnson, and Tudor (1972)
Electricity loads	Fleishman (1954)
Electricity production	Nye (1965)
Electricity supply	Davies (1960)
Energy	Joyce (1982) Won (1980)
Energy conservation	Murphy, Schulman, and Mahoney (1977)
Energy consumption	Cohen (1981)
Energy demand	Lawford (1981) McKay and Allsopp (1980)
Energy emergencies	Quirk (1981a)
Energy grid systems	McQuigg and Johnson (1973)
Environmental impacts	Decker and Sakamoto (1981)
Energy policy	Hudson and Jorgenson (1974)
Energy sector	Warren and Le Duc (1981)
Energy sources	Jäger (1985)
Energy supply	Quirk (1981b)
Energy systems	Jäger (1983) Quirk and Moriarty (1980)
Environment	Holdgate, Kassas, and White (1983)

Environmental analysis	Kalkstein (1979)
Environmental economics	Kneese (1977) Maler (1974)
Evaluation of weather forecasts	Murphy and Brown (1984a)
Evolution	Pearson (1978)
Expert judgements	Stewart and Glantz (1985)
Exports	Flemming (1982) Walsh (1981)
Extended range forecasts	Hipp (1972)
Extreme events	Heathcote (1985)
Famine policies	Jodha (1975)
Famines	Apeldoorn (1981) Mascarenhas (1973) Tudge (1979)
Farm costs	Taylor (1977b)
Feats/famines	Le Roy Ladurie (1971)
Fisheries	Caviedes (1981) Cushing (1982) Kawasaki (1985) Murray, Le Duc, and Ingham (1983)
Food/climate	National Academy of Sciences (1976)
Food and society	Ruttenberg (1981)
Food crops	Food and Agriculture Organization (1983) Food and Agriculture Organization (1985)

Food energy	Pimentel (1981)
Food production	Hayes, O'Rouke, Terjung, and Todhunter (1982) McQuigg (1975a)
Food reserves	Laurmann (1976)
Food supplies	Food and Agriculture Organization (1977)
Food systems	Duckham (1974)
Football attendances	Cairns (1984)
Forecast terminology	Curtis and Murphy (1985) Murphy and Brown (1983)
Forecaster/computer mix	Bristor (1976)
Forecasting/society	Maunder (1970b)
Forestry	Baumgartner (1979)
Fossil fuels	Nelson (1976)
Frost economics	Margolis (1980)
Future climates	White (1979)
Future scenarios	McHale (1981)
Future trends	Sewell and Maunder (1985)
Futures	Hughes (1982)
Gambling/insuring	Edwards (1978)
Game scenarios	Robinson and Ausubel (1983)
Genesis strategy	Schneider (1976)
Geographers	Terjung (1976)

Global 2000	US Council on Environmental Quality (1980)
Global Atmospheric Research Programme	Ashford (1982)
Global climate system	World Meteorological Organization (1985b)
Global disasters	Susman, O'Keefe, and Wisner (1983)
Global grain yields	Sakamoto, Le Duc, Strommen, and Steyaert (1980)
Global models	Robinson (1985)
Global survival	Schneider and Mesirow (1976)
Global warming	Flohn (1981a)
Grain production	CIA (Central Intelligence Agency) (1976) Kogan (1982) Thompson (1975)
Grain reserves	Cochrane and Danin (1976) Johnson and Sumner (1976)
Grain robbery	Trager (1975)
Grains/oil seeds	Rojko and Schwartz (1976)
Grassland farming	Sears (1961)
Greenhouse gases	Newell and Dopplick (1979) United Nations Environment Programme (1985)
Greenhouse warming	US Environmental Protection Agency (1983)
Growing season	Newman (1980)

Growth limitations	Meadows, Meadows, Randers, and Behrens (1972)
Hail suppression	Changnon, Farhar, and Swanson (1978)
Haymaking	Dyer and Baier (1981)
Hazards	Burton, Kates, and White (1978)
Health	Chambers (1982) Landsberg (1984)
Health/disease	Weihe (1979)
Heating demands	Mitchell *et al.* (1973)
Heatwave costs	Center for Environmental Assessment Services (1981)
Historical aspects	Abel (1980) Anderson (1981) Ogilvie (1981a) Pfister (1978) Bryson and Padoch (1980) Wigley *et al.* (1985)
Historical data	Moodie and Catchpole (1976)
History/climate	de Vries (1980) de Vries (1985)
Horticulture	Hurnard (1979)
Household budgets	Crocker, Eubanks, Horst, and Nakayama (1975)
Housing starts	Maunder (1981b) Musgrave (1968)
Human activities	Ausubel and Biswas (1980) Kellogg (1977) Mason (1981)

Human affairs	Schneider and Temkin (1978)
Human aspects	Aspen Institute (1977) Changnon (1979) Kates (1983) Maunder (1980b)
Human dimensions	Sewell (1966) Sewell *et al.* (1968)
Human ecology	Wisner (1977)
Human environment	Oliver (1973)
Human impacts	Pittock *et al.* (1981) Ingram, Farmer, and Wigley (1981)
Human populations	Bowden *et al.* (1981)
Human strategy	Fedorov (1979)
Hunger	McQuigg (1974) Mayer (1976)
Hurricane economics	Sugg (1967)
Hydrologic applications	Changnon (1981)
Ice-age earth	CLIMAP Project Members (1976)
Impact analysis	Carter, Konijn, and Watts (1984) Wigley (1983)
Impact assessments	Ausubel (1983) Glantz, Robinson, and Krenz (1985) Hard and Broderick (1976) Kates, Ausubel, and Berberian (1985) Parry and Carter (1983)
Impact studies	Glantz, Robinson, and Krenz (1982)

Impacts/cold areas	Parry (1983)
Impacts/management	Maunder (1986a)
Impacts/perception	Kates *et al.* (1984)
Impacts of climatic change	Parry (1981b)
Industrial growth	Wilson (1966)
Industrial production	Palutikof (1983a)
Industry	Maunder (1973c)
Information uses	McQuigg (1972a)
Input-output studies	Leontief (1977a)
Instability and chaos	Pippard (1982)
Institutional aspects	Bernard (1976) Bastian (1982) Epstein (1976) Hickman (1982) Jordan (1975) Ward (1977)
Institutional response	Agency for International Development (1976) (1967)
Intelligence activities	CIA (Central Intelligence Agency) (1974a) CIA (Central Intelligence Agency) (1974b)
Judgement decisions	Kahneman, Slovic, and Tversky (1982)
Lambing percentages forecasting	Rich and Taylor (1977)
Land use	Oguntoyinbo and Odingo (1979)
Land prices	Borland and Snyder (1975) Johnson and Haigh (1970)

Legal aspects	Barton (1975)
Life/property	Foster (1980)
Livestock shelters	Hahn and McQuigg (1970)
Living with climate change	Science Council of Canada (1976) Thompson (1977)
Load dispatching	Dryar (1949)
Load forecasting	Thomas and Drummond (1953)
Long-range forecasting	Marchuk (1979)
Magic/hypnosis	Neale (1985)
Man and climate	Landsberg (1970)
Management	Haigh (1977) Sewell (1968)
Managing risks	National Research Council (1981)
Marginal areas	Parry and Carter (1983)
Marketing	Atlas (1975) Bates (1976) Boyer (1966) Elling (1969) McKay (1979)
Marginal agriculture	Parry (1975)
Mass media	Morentz (1980)
Meat production	Maunder (1967b)
Meat and wool industry	Thomson and Taylor (1975) Ward (1978)
Meat industry forecasting	Taylor (1977a)
Media	Harrison (1982)

Merchandising	Linden (1962)
Microeconomic analysis	Lovell and Smith (1985)
Migration	Haurin (1980)
National Climate Programmes	Phillips (1985)
National climates	Phillips and McKay (1980)
National economy	Heathcote (1967) Mason (1966)
Natural disasters	Dworkin (n.d.) Glantz (1976b) Islam and Kunreuther (1973)
Natural hazards	Cochrane (1975) Kates (1979) White and Haas (1975)
Natural resource	Landsberg (1946)
News media	Riebsame (1983)
Nowcasting	Liljas (1984)
Nuclear exchange	Elsom (1984)
Nuclear war	Aleksandrov and Stenchikov (1983) Cess (1985) Covey, Schneider, and Thompson (1984) Crutzen and Birks (1982) Ehrlich *et al.* (1983) McCracken (1983) Royal Society of New Zealand (1985) National Academy of Sciences (1975a) Turco *et al.* (1983)
Nutrition	Escudero (1985)

Operational decision-making	Maunder (1977a)
Outdoor recreation	Paul (1972)
Outdoor work	McQuigg and Decker (1962)
Ozone	Dotto and Schiff (1978) World Meteorological Organization (1982)
Pastoralism	Le Houérou (1985)
Pastoral production	Maunder (1963) Radcliffe (1974)
Perceived risk	Slovic, Fischhoff, and Lichtenstein (1982)
Perception studies	Bickert and Browne (1966) Heathcote (1969) Whyte (1985)
Plant growth	Evans (1963)
Plant yield	Watson (1963)
Political/economic issues	Maunder (1978a)
Political/social aspects	Post (1980)
Political aspects	Glantz (1976) Primack and von Hippel (1972)
Political climates	Bandyopadhyaya (1983)
Political climatology	Meyer-Abich (1980)
Politics and science	White (1982)
Policy, design	Heal (1984)
Policy options	National Defense University (1983)
Population shifts	Diaz and Holle (1983)

Power consumption	Stephens (1951)
Predictability	Tapp (1984)
Predicting climatic variations	Mason (1976b)
Private v. public sector forecasts	Golden (1984)
Probabilistic forecasts	Murphy (1977)
Production subsidies	Taylor (1978)
Public goods	Brookshire, d'Arge, Schulze, and Thayer (1981)
Public perception	Whyte and Harrison (1981)
Public policy	Office of Technology Assessment (1983)
Public sector costs	Sassone (1975)
Quality of life	Hoch (1974)
Railways	Wintle (1960)
Raisin industry	Kolb and Rapp (1962) Lave (1963)
Rangeland management	Wallén and Gwynne (1978)
Real-time data	Burnash (1984)
Recreation	Clawson (1966)
Recreational weather	Leech (1985) Taylor (1979)
Remote sensing	Yates *et al.* (1984)
Research/services dilemma	Tucker (1976)
Retail sales	Shor (1964)

Retail trade	Maunder (1973a)
Risk decisions	Kahneman and Tversky (1979)
Risk management	US National Research Council (1981)
Risk perception	Whyte (1983)
Risk ratings	Slovic, Fischhoff, and Lichtenstein (1979)
Risks/society	Marsh and McLennan Companies (1980)
Risks/uncertainty	Maunder (1985b) Waddell (1983)
Risks/water	Cohon (1982)
Road construction	Maunder, Johnson, and McQuigg (1971a) Maunder, Johnson, and McQuigg (1971b)
Rural poverty	Chambers, Longhurst, and Pacey (1981)
Sales	Zeisel (1950)
Satellite/vegetation index	Tucker, Fung, Keeling, and Gammon (1986)
Scenario analysis	Lave and Epple (1985)
Scientific investment	Studt (1971)
Seasonal systems	Troughton (1977)
Self-provisioning	Spitz (1980)
Sensitivity analysis	Katz (1979)
Sensitivity identification	Maunder and Ausubel (1985)

Settlement	Parry (1978)
Sheep	Coop and Hayman (1962)
Ship routing	Evans (1968)
Simulation models	McQuigg (1967)
Snow impacts	Weisbecker (1974)
Social analysis	Farhar-Pilgrim (1985)
Social aspects	Beltzner (1976) d'Arge *et al.* (1975)
Social assessments	Sewell *et al.* (1973)
Social impact assessment	Vlachos (1977)
Social impacts	Haas (1975) Center for Environmental Assessment Services (1982) Connor (1977)
Social measures	Climate Impact Assessment Program (CIAP) (1975)
Social science	Chen, Boulding, and Schneider (1983)
Social science forecasting	Encel (1977)
Social science research	Sewell and MacDonald-McGee (1983)
Social systems	Logothetti (1974)
Social values	Glantz (1977b)
Societal consequences	US Department of Energy (1980)
Societal response	Warrick and Riebsame (1981)

Society/climate	Kates (1985)
Society/vulnerability	Timmerman (1981)
Society adjustments	Jodha and Mascarenhas (1985)
Society impacts	Center for Environmental Assessment Services (1980)
Society and meteorology	World Meteorological Organization (1985a)
Society's economy	Young and Wilson (1978)
Socio-economic aspects	Maunder (1985a) US Weather Bureau (1964)
Socio-economic impacts	Garcia (1978) Liebhardt (1981) McKay (1980)
Socio-economic values	Bergen and Murphy (1978)
Socio-technical systems	Starr (1972)
Soil moisture	US Department of Agriculture *et al.* (1959)
Southern oscillation	Nicholls (1985) Ward (1985) Wright (1978)
Soybeans	Ravelo and Decker (1981) Thompson (1970)
Sports	Hentshel (1964)
Stability of climate	Burgos (1979)
Statistical aspects	Brooks and Carruthers (1953) Feyerherm and Bark (1965) Feyerherm and Bark (1967) Murphy and Katz (1985)

Strategic aspects	National Defense University (1980)
Stratospheric flight	National Academy of Sciences (1975b)
Storm decisions	Cochrane and Howe (1976)
Summer variability	Parry (1976)
Synoptic data banks	Adams and Seager (1977)
Technology/crops	Thompson (1969)
Television weather	Hunt (1982)
Temperature modification	Nicodemus and McQuigg (1969)
Time series analysis	Parthasarthy and Mooley (1978)
Topoclimatology	Thornthwaite (1953)
Tornadoes/twisters	Elsom and Meaden (1985)
Tourism	Perry (1972)
Traffic accidents	Sherretz and Farhar (1978)
Transportation industry	Maunder (1978b)
Tree-ring records	Stockton and Meko (1983)
Tropical agriculture	Fukui (1979) Mattei (1979)
Universities/government	Seaborg (1972)
Urban climates	Landsberg (1981)
Use of a climate forecast	Glantz (1982a)
Use of weather forecasts	Cressman (1971)
User requirements	Murphy and Brown (1984)

Value of climatic predictions	Landsberg (1985)
Value of frost forecasts	Katz, Murphy, and Winkler (1982)
Value of weather information	Gabe (1985) McQuigg (1964) Maunder (1972e) Stewart, Katz, and Murphy (1984)
Value of Meteorological Services	Thompson (1967)
Value of the weather	Australian Bureau of Meteorology (1965) Maunder (1970a)
Value of weather forecasts	Brown, Katz, and Murphy (1985) Doll (1971) Glantz (1977a) Thompson (1969) Wilks and Murphy (1985)
Value of weather information	Gabe (1985) McQuigg (1964) Maunder (1972e) Stewart, Katz, and Murphy (1984)
Vegetation (satellite) indices	Tarpley, Schneider, and Money (1984) Taylor, Dini, and Kidson (1985)
Videotex (viewdata) systems	Parker (1978)
Volcanic activity	Bryson and Goodman (1980)
Water	Foster and Sewell (1980)
Water balance	Coulter (1973) Mather (1978)
Water conservation	Bouwer (1984)
Water resources	Novky, Pachner, Szesztay, and Miller (1985)

Weather shortages	Meier (1977)
Weather economics	Taylor (1970)
Weather games	Maunder (1983)
Weather indexes	Doll (1967)
Weather information	McQuigg (1970) McQuigg and Thompson (1966) Maunder (1971a) Maunder (1984a) Rapp and Huschke (1964) Robertson (1974)
Weather food programmes	Maunder (1975)
Weather forecasting trends	Sawyer (1967)
Weather lore	Kidd (1984)
Weather modification	Hosler (1974) Smith (1971)
Welfare policy aspects	Schelling (1983)
Weather terminology	Canadian Meteorological Service (1949)
Weathercasting	Brentzel (1985)
Weighted indices	McQuigg Consultants (1981) Maunder (1972a)
Weighting methodology	Gum, Roefs, and Kimball (1976)
Welfare analysis	McFadden (1984)
Wheat/weather relationships	Stewart (1983) Steyaert, Le Duc, and McQuigg (1978) Thompson (1962)
Wheat prices	Beveridge (1922)

Bibliography

Abel, W. (1980) *Agricultural Fluctuations in Europe from the Thirteenth to the Twentieth Century*, London, Methuen.

Adams, R. J. and Seager, Judith M. (1977) 'Agrometeorological use of the synoptic data bank in plant disease warning services', *Meteorological Magazine*, 106: 112–16.

Agency for International Development (AID) (1976, 1967) *A Report to Congress: Proposal for a Long-Term Comprehensive Development Program for the Sahel*. Part II, Technical Background Papers, Washington, DC, AID.

Aleksandrov, V. V. and Stenchikov, G. L. (1983) 'On the modelling consequences of the climatic consequences of nuclear war', in *The Proceedings on Applied Mathematics*, Moscow, The Computing Centre of the USSR Academy of Sciences.

Alexander, T. (1974) 'Ominous changes in the world's weather', *Fortune*, February: 90–5, 142, 146, 150–2.

Amerasinghe, A. R. B. (1984) *Crop Risk Management in the South Pacific*, Wellington, Bowring, Burgess, Marsh & McLennan Ltd.

Anderson, J. L. (1981) 'Climatic change in European economic history', *Research in Economic History*, 6: 1–34.

Anderson, J. R. (1979) 'Impacts of climatic variability on Australian agriculture', in *The Impact of Climate on Australian Society and Economy*, Mordialloc, Australia, Commonwealth Scientific and Industrial Research Organization (CSIRO).

Anonymous (1966) 'Drought', *Current Affairs Bulletin* (Dept of Adult Education, University of Sydney), 38 (4): 51–64.

Apeldoorn, G. J. V. (1981) *Perspectives on Drought and Famine in Nigeria*, London, Allen & Unwin.

Arnold, G. W. and Galbraith, K. A. (1978) 'Cultural and economic aspects. Case study one: Climatic changes and agriculture in Western Australia', in Pittock, A. B., Frakes, L. A., Jenssen, D., Peterson, J. A., and Zillman, J. W. (eds) *Climatic Change and Variability: A Southern Perspective*, New York, Cambridge University Press, 297–300.

Ashford, O. M. (1982) 'The launching of GARP', *Weather*, 37: 265–72.

Aspen Institute (1977) *Living with Climatic Change*, McLean, Virginia, MITRE Corporation.

Atlas, D. (1975) 'President's page – Selling atmospheric science', *Bulletin of the American Meteorological Society*, 56: 688–9.

Australian Bureau of Agricultural Economics (1969) 'An economic survey of drought affected properties – New South Wales and Queensland 1964/65 to 1965/66', *Wool Economics Research Report*, 15.

Australian Bureau of Meteorology (1965) *What is Weather Worth?*, papers presented to the Productivity Conference, Melbourne, Australia, 31 August–4 September.

Ausubel, J. H. (1980) 'Economics in the air', in Ausubel, J. H. and Biswas, A. K. (eds) *Climatic Constraints and Human Activities*, IIASA Proceedings Series, vol. 10, Oxford, Pergamon Press, 13–59.

Ausubel, J. H. (1983) 'Can we assess the impacts of climatic changes?', *Climatic Change*, 5: 7–14.

Ausubel, J. H. and Biswas, A. K. (eds) (1980) *Climatic Constraints and Human Activities*. IIASA Proceedings Series, vol. 10, Oxford, Pergamon Press.

Baier, W. (1972) 'An agroclimatic probability of the economics of tallow-seeded and continuous spring wheat in Southern Saskatchewan', *Agricultural Meteorology*, 9: 305–21.

Baier, W. (1974) 'The challenge to agricultural meteorology', *WMO Bulletin*, 24: 221–4.

Baier, W. (1977) *Crop-weather models and their use in yield assessments,* WMO Technical Note no. 151, Geneva, World Meteorological Organization.

Baier, W. (1983) 'Agroclimatic modeling: An overview', in Cusack, D. (ed.) *Agroclimatic Information for Development: Reviving the Green Revolution*, Boulder, Colorado, Westview Press.

Bandyopadhyaya, J. (1983) *Climate and World Order: An Inquiry into the Natural Cause of Underdevelopment*, Delhi, South Asian Publishers.

Barber, R. F. and Chavez, F. P. (1983) 'Biological consequences of El Niño', *Science*, 222: 1203–10.

Barton, G. P. (1975) 'Law and environment', *New Zealand Science Review*, 33: 1976.

Bastian, C. (1982) 'The formulation of federal policy', in Ward, R. (ed.) *Stratospheric Ozone and Man*, Boca Raton, Florida, CRC Press.

Bates, C. C. (1976) 'Industrial meteorology and the American Meteorological Society – a historical overview', *Bulletin of the American Meteorological Society*, 57: 1320–7.

Baumgartner, A. (1979) 'Climatic variability and forestry', in *Proceedings of the World Climate Conference*, Geneva, World Meteorological Organization, 581–607.

Beckwith, W. B. (1966) 'Impacts of weather on the airline industry: The value of fog dispersal programs', in Sewell, W. R. D. (ed.) *Human Dimensions of Weather Modification*, Research Paper no. 105, Department of Geography, University of Chicago.

Beltzner, K. (ed.) (1976) *Living with Climatic Change. Proceedings of Toronto Conference*, Ottawa, Science Council of Canada.

Benjamin, N. B. H. and Davis, C. C. (1971) *Impact of Weather on Construction Planning*. Paper presented at the National Meeting on Environmental Engineering, American Society of Civil Engineers, St Louis, 21 October 1971.

Bergen, W. R. and Murphy, A. H. (1978) 'Potential economic and social value of short-range forecasts of Boulder windstorms', *Bulletin of the American Meteorological Society*, 59: 29–44.

Berggren, R. (1971) *Preliminary Report on Economic Benefits of Climatological Services*. Submitted to the Commission for Climatology, Geneva, World Meteorological Organization.

Bernard, E. A. (1976) *Costs and Structure of Meteorological Services with Special Reference to the Problem of Developing Countries*, WMO Technical Note no. 146, Geneva, World Meteorological Organization.

Berry, L. and Kates, R. W. (eds) (1980) *Making the Most of the Least: Alternative Ways to Development*. New York, Holmes & Meier.

Beveridge, W. H. (1922) 'Wheat prices and rainfall in western Europe', *Journal of the Royal Statistical Society*, 85: 411–78.

Bickert, C. Von E. and Browne, T. D. (1966) 'Perception of the effect of weather on manufacturing: A study of five firms', in Sewell, W. R. D. (ed.) *Human Dimensions of Weather Modification*, Research Paper no. 105, Department of Geography, University of Chicago, 307–22.

Biswas, A. K. (1980) 'Crop-climate models: A review of the state of the art', in Ausubel, J. and Biswas, A. K. (eds) *Climatic Constraints and Human Activities*, Oxford, Pergamon Press, 75–92.

Biswas, M. R. and Biswas, A. K. (1980) *Desertification*, Oxford, Pergamon Press.

Bollay, E. (1962) *Economic Impact of Weather Information on Aviation Operations*, Washington, DC, Federal Aviation Agency.

Borland, S. W. and Snyder, J. J. (1975) 'Effects of weather variables on the prices of Great Plains cropland', *Journal of Applied Meteorology*, 14: 686–93.

Bouwer, H. (1984) 'Water conservation in agricultural and natural systems', in *Conference Proceedings – Water for the 21st Century: Will It be There?*, Dallas, Texas, Southern Methodist University.

Boville, B. W. and Doos, B. R. (1981) 'Why a world climate programme?', *Nature and Resources*, 17: 2–7.

Bowden, M. J. *et al.* (1981) 'The effect of climate fluctuations on human populations: Two hypotheses', in Wigley, T. M. L., Ingram, M. J. and Farmer, M. G. (eds) *Climate and History*, Cambridge, Cambridge University Press, 479–513.

Boyer, A. (1966) 'Expanding industrial meteorology', *Bulletin of the American Meteorological Society*, 47: 528.

Boyer, M. G. (1975) 'The value of weather forecasts in economic planning for the dairy industry', in *Symposium on Meteorology and Food Production*, Wellington, New Zealand Meteorological Service, 213–15.

Brentzel, Francie (1985) Article in the *Evening Post* (Wellington), 24 August.

Bristor, C. L. (1976) 'Relations of operational forecasters and applied meteorologists to "the computer types"', *Bulletin of the American Meteorological Society*, 57: 700–2.

Brooks, C. E. P. and Carruthers, N. (1953) *Handbook of Statistical Methods in Meteorology*, London, HMSO.

Brookshire, D. S., d'Arge, R. C., Schulze, W. D., and Thayer, M. A. (1981) 'Experiments in valuing public goods', in Smith, V. K. (ed.) *Advances in Applied Microeconomics*, vol. 1, Greenwich, Connecticut, JAI Press.

Brown, B. G., Katz, R. W., and Murphy, A. H. (1985) 'A case study of the economic value of seasonal precipitation forecasts in the fallowing/planting problem', in Preprints, *Ninth Conference on Probability and Statistics in Atmosphere Sciences* (Virginia Beach, Virginia), American Meteorological Society.

Bryson, R. A. and Goodman, B. M. (1980) 'Volcanic activity and climatic changes', *Science*, 207: 1041–4.

Bryson, R. A. and Padoch, C. (1980) 'On the climates of history', *Journal of Interdisciplinary History*, 10: 583–97.

Budyko, M. I. (1982) *The Earth's Climate: Past and Future*, New York, Academic Press.

Budyko, M. I. and Efimova, N. A. (1981) 'The CO_2 effects of climate', *Meteorologiya i Hydrologiya*, no. 2: 1–10.

Burgos, J. J. (1979) 'Renewable resources and agriculture in Latin America in relation to the stability of climate', in *Proceedings of the World Climate Conference*, Geneva, World Meteorological Organization, 525–51.

Burnash, R. J. C. (1984) 'The meaning and challenge of real-time data and analysis systems to future public service programs', *Bulletin of the American Meteorological Society*, 65: 338–42.

Burroughs, W. (1978) 'Cold winter and the economy', *New Scientist*, 77 (19 January): 146–8.

Burton, I., Kates, R. W., and White, G. F. (1978) *The Environment as Hazard*, New York, Oxford University Press.

Business Week (1975) 15 December.

Business Week (1976) 2 August.

Business Week (1977) 31 January.

Business Week (1980) 13 October.

Business Week (1981) 14 May.

Cairns, J. A. (1984) 'The effect of weather on football attendances', *Weather*, 39: 87–90.

Caldwell, J. C. (1977) 'Demographic aspects of drought: An examination of the African drought of 1920–74', in Dalby, D., Harrison Church, R. J., and Bezzaz, F. (eds) *Drought in Africa*, vol. 2: 93–102, London, International African Institute.

Caldwell, M. M. (1974) *Impacts of Climatic Change on the Biosphere*. CIAP Monograph 5, Washington, DC, US Dept of Transportation, Climatic Impact Assessment Program.

Campbell, D. (1968) *Drought: Causes, Effects, Solutions*, Melbourne, F. W. Cheshire.

Canadian Meteorological Service (1949) 'University of Toronto poll of students on weather terminology', *Bulletin of the American Meteorological Society*, 30: 61–2.

Carter, T. R., Konijn, N. T., and Watts, R. G. (1984) *The Role of Agroclimatic Models in Impact Analysis*, Working Paper, WP – 84–98 Laxenburg, Austria, International Institute for Applied Systems Analysis.

Caviedes, C. N. (1981) 'The impact of El Niño on the development of the Chilean fisheries', in Glantz, M. H. and Thompson, J. D. (eds) *Resource Management and Environment Uncertainty*, New York, Wiley, 351–68.

Center for Environmental Assessment Services (1980) *Guide to Environmental Impacts on Society*, Washington, DC, US Department of Commerce.

Center for Environmental Assessment Services (1981) *Retrospective Evaluation of the 1980 Heatwave and Drought*, Washington, DC, US Department of Commerce.

Center for Environmental Assessment Services (1982) *US Economic and Social Impacts of the Record 1976/77 Winter Freeze and Drought*, Washington, DC, US Department of Commerce.

Cess, R. D. (1985) 'Nuclear war: Illustrated effects of atmospheric smoke and dust upon solar radiation', *Climatic Change*, 7: 237–51.

Chambers, R. (1982) 'Health, agriculture and rural poverty: Why seasons matter', *Journal of Development Studies*, 18: 217–38.

Chambers, R., Longhurst, R., and Pacey, A. (1981) *Seasonal Dimensions to Rural Poverty*, London, Frances Pinter.

Chang, Jen-Hu (1981) 'A climatological consideration of the transference of agricultural technology', *Agricultural Meteorology*, 25: 1–13.

Changnon, S. A., Jr (1979) 'How a severe winter impacts on individuals', *Bulletin of the American Meteorological Society*, 60: 110–14.

Changnon, S. A., Jr (1981) 'Hydrologic applications of weather and climate information', *Journal of the American Water Works Association*, October: 514–18.

Changnon, S. A., Jr, Farhar, B. C., and Swanson, E. R. (1978) 'Hail suppression and society', *Science*, 200: 387–94.
Chen, R. C., Boulding, E., and Schneider, S. H. (1983) *Social Science and Climate Change*, Dordrecht, Holland, Reidel.
CIA (Central Intelligence Agency) (1974a) *A Study of Climatological Research as it Pertains to Intelligence Problems* (August), Washington, DC, CIA.
CIA (Central Intelligence Agency) (1974b) *Potential Implications of Trends in World Population, Food Production, and Climate* (August), Washington, DC, CIA.
CIA (Central Intelligence Agency) (1976) *USSR: The Impact of Recent Climate Change on Grain Production* (October, ER 76-10577 U), Washington, DC, CIA.
Clark, C. (1985) 'The effect of temperature and insulation on the consumption of electricity', *Weather*, 40: 85–8.
Clark, W. C. (ed.) (1982) *Carbon Dioxide Review, 1982*, New York, Oxford University Press.
Clawson, M. (1966) 'The influence of weather on outdoor recreation', in Sewell, W. R. D. (ed.) *Human Dimensions of Weather Modification*, Research Paper no. 105, Department of Geography, University of Chicago.
CLIMAP Project Members (1976) 'The surface of the ice-age earth', *Science*, 191 (19 March): 1131–7.
Climate Analysis Center (CAC) (1982–) *Monthly Climate Diagnostic Bulletin*, Washington, DC, Climate Analysis Center, NOAA.
Climate Impact Assessment Program (CIAP) (1975) *Economic and Social Measures of Biological and Climatic Change*, Monograph 6, Washington, DC, US Department of Transportation.
Cochrane, H. C. (1975) *Natural Hazards and their Distributive Effects: A Research Assessment*, Boulder, Colorado, Institute of Behavioral Science, University of Colorado.
Cochrane, H. C. and Howe, C. W. (1976) 'A decision model for adjusting to natural hazard events with application to urban snow storms', *Review of Economics and Statistics*, February: 50–8.
Cochrane, W. W. and Danin, Y. (1976) *Reserve Stock Grain Models for the World, 1975–1985*, in Eaton, D. J. *et al.* (eds) *Analysis of Grain Reserves, A Proceedings*, USDA Economic Research Service Report no. 634, Washington, DC, US Dept of Agriculture.
Cohen, S. J. (1981) 'Climatic influences on residential energy consumption', Dissertation, Department of Geography, University of Illinois, Urbana-Champaign, Illinois.
Cohon, J. I. (1982) 'Risk and uncertainty in water resources management', *Water Resources Research*, 18(1): 1.
Commonwealth Scientific and Industrial Research Organization (CSIRO)

(1979) *The Impact of Climate on Australian Society and Economy*, Mordialloc, Australia, CSIRO.

Connor, D. M. (1977) 'Social impact assessment: The state of the art', *Social Impact Assessment*, 13/18 (June): 4–7.

Coop, I. E. and Hayman, B. I. (1962) 'Liveweight productivity relationships in sheep', *New Zealand Journal of Agricultural Research*, 5: 165–72.

Coulter, J. D. (1973) 'A water balance assessment of the New Zealand rainfall', *Journal of Hydrology* (NZ), 12: 83–91.

Covey, C., Schneider, S. H., and Thompson, S. L. (1984) 'Global atmospheric effects of massive smoke injections from a nuclear war: Results from a general circulation model', *Nature*, 308: 21–5.

Cressman, G. P. (1971) 'The uses of public weather forecasts', *Bulletin of the American Meteorological Society*, 52: 544–6.

Crocker, T. D. (1976) 'Electricity demand in all-electric commercial buildings: The effect of climate', in Ferrar, T. A. (ed.) *The Urban Costs of Climate Modification*, New York, Wiley.

Crocker, T., Eubanks, L., Horst, R., and Nakayama, B. (1975) 'Covariation of climate and household budget expenditures', in CIAP *Economic and Social Measures of Biological and Climatic Change*, Washington, DC, US Department of Transportation.

Crutzen, P. and Birks, J. W. (1982) 'The atmosphere after a nuclear war: Twilight at noon', *Ambio*, 11: 114–25.

Cumming, J. N. (1966) 'The effects of the 1965/66 drought on sheep numbers and the expected rate of recovery', *Quarterly Review of Agricultural Economics*, 19: 169–76.

Curry, L. (1962) 'The climate resources of intensive grassland farming: the Waikato, New Zealand', *Geographical Review*, 52: 174–94.

Curtis, J. C. and Murphy, A. H. (1985) 'Public interpretation and understanding of forecast terminology: some results of a newspaper survey in Seattle, Washington', *Bulletin of the American Meteorological Society*, 66: 810–19.

Cushing, D. H. (1982) *Climate and Fisheries*, London, Academic Press.

Dandekar, V. M. (1976) 'Crop insurance in India', *Economic and Political Weekly*, 11(6): A 61–A 80.

d'Arge, R. C. (1979) 'Climate and economic activity', in *Proceedings of the World Climate Conference*, Geneva, World Meteorological Organization, 652–81.

d'Arge, R. C. *et al.* (eds) (1975) *Economic and Social Measures of Biological and Climatic Change*, CIAP Monograph no. 6, Washington, DC, US Department of Transportation.

Davies, M. (1960) 'Grid system operation and the weather', *Weather*, 15: 18–24.

Decker, W. L. and Sakamoto, C. M. (1981) 'Simulation of environmental

impacts on productivity', in Mendt, W. J. (ed.) *Strategies of Plant Reproduction*, BARC Symposium no. 6, Totowa, New Jersey, Allanheld, Osmun, 307–21.
Demsetz, H. (1962) *Economic Gains from Storm Warnings: Two Florida Case Studies*, Memo RM 3168 (NASA), Santa Monica, California, The Rand Corporation.
de Vries, J. (1980) 'Measuring the impact of climate on history: The search for appropriate methodologies', *Journal of Interdisciplinary History*, 10, 4 (Spring): 599–630.
de Vries, J. (1985) 'Analysis of historical climate–society interaction', in Kates, R. W., Ausubel, J. H., and Berberian, M. (eds) *Climate Impact Assessment: Studies of the Interaction of Climate and Society*, SCOPE, 27, New York, Wiley, 273–92.
Diaz, H. F. and Holle, R. L. (1983) 'The relative effects of US population shifts (1930–80) on potential heating, cooling and water demand', *Journal of Climate and Applied Meteorology*, 23: 445–8.
Doll, J. P. (1967) 'An analytical technique for estimating weather indexes from meteorological measurements', *Journal of Farm Economics*, 49: 79–88.
Doll, J. P. (1971) 'Obtaining preliminary bayesian estimates of the value of a weather forecast', *American Journal of Agricultural Economics*, 53: 651–5.
Dotto, L. and Schiff, H. (1978) *The Ozone War*, New York, Doubleday.
Dryar, H. A. (1949) 'Load dispatching and Philadelphia weather', *Bulletin of the American Meteorological Society*, 30: 159–67.
Duckham, A. N. (1974) 'Climate, weather and human food systems – a world view', *Weather*, 29: 242–51.
Duloy, J. H. and Woodland, A. D. (1967) 'Drought and the multiplier', *Australian Journal of Agricultural Economics*, 11: 82–6.
Dworkin, J. (nd) *Global Trends in Natural Disasters 1947–1973*, Natural Hazards Research Working Paper no. 26, Boulder, Colorado, Institute of Behavioral Science, University of Colorado.
Dyer, J. A. and Baier, W. (1981) 'The use of weather forecasts to improve hay-making reliability', *Agricultural Meteorology*, 25: 27–34.

Eberly, D. L. (1966) 'Weather modification and the operations of an electric power utility: The Pacific Gas and Electric Company's test program', in Sewell, W. R. D. (ed.) *Human Dimensions of Weather Modification*, Research Paper no. 105, Department of Geography, University of Chicago, 209–26.
The Economist (1976) 7 August.
The Economist (1977a) 15 January.
The Economist (1977b) 9 April.
The Economist (1977c) 11 June.

The Economist (1977d) 13 August.
The Economist (1985) 7 December.
Eddy, A. *et al.* (1980) *The Economic Impact of Climate*, vol. 1 (subsequent vols 1–22: 1980, 1981, 1982, 1983, 1984, 1985), Norman, Oklahoma, Oklahoma Climatological Survey.
Edwards, C. (1978) 'Gambling, insuring and the production function', *Agricultural Economics Research*, 30: 25–8.
Ehrlich, P. R., Harte, J., Harwell, M. A., Raven, P. H., Sagan, C., Woodwell, G. M., Berry, J., Ayensu, E. S., Ehrlich, A. H., Eisner, T., Gould, S. J., Grover, H. D., Herrera, R., May, R. M., Mayr, E., McKay, C. P., Mooney, H. A., Myers, N., Pimentel, D., and Teal, J. M. (1983) 'Long-term biological consequences of nuclear war', *Science*, 222: 1293–300.
Elling, K. A. (1969) *Introduction to Modern Marketing: An Applied Approach*, New York, The Macmillan Company.
Elsom, D. M. (1984) 'Climatic change induced by a large-scale nuclear exchange', *Weather*, 39: 268–71.
Elsom, D. M. and Meaden, G. T. (1985) 'Devils and twisters screw things up', *New Scientist*, 27 June: 44–7.
Encel, S. (1977) 'Forecasting and the social sciences', *Search*, 8: 185–9.
Epstein, E. S. (1976) 'NOAA policy on industrial meteorology', *Bulletin of the American Meteorological Society*, 57: 1334–40.
Epstein, E. S. (1982) 'Detecting climatic change', *Journal of Applied Meteorology*, 21: 1172–82.
Escudero, J. C. (1985) 'Health, nutrition, and human development', in Kates, R. W., Ausubel, J. H., and Berberian, M. (eds) *Climate Impact Assessment: Studies of the Interaction of Climate and Society*, SCOPE 27, New York, Wiley, 251–72.
Evans, L. T. (ed.) (1963) *Environmental Control of Plant Growth*, New York, Academic Press.
Evans, S. H. (1968) 'Weather routing of ships', *Weather*, 23: 2–8.
Evening Post (Wellington) (1981) 22 August.
Evening Post (Wellington) (1985) 24 August.

Farhar-Pilgrim, B. (1985) 'Social analysis', in Kates, R. W., Ausubel, J. H., and Berberian, M. (eds) *Climate Impact Assessment: Studies of the Interaction of Climate and Society*, SCOPE 27, New York, Wiley, 323–50.
Fedorov, E. K. (1979) 'Climatic change and human strategy', in *Proceedings of the World Climate Conference*, Publication no. 537, Geneva, World Meteorological Organization, 15–26.
Feyerherm, A. M. and Bark, L. D. (1965) 'Statistical methods for persistent precipitation patterns', *Journal of Applied Meteorology*, 4: 320–8.
Feyerherm, A. M. and Bark, L. D. (1967) 'Goodness of fit of a Markov

chain model for sequences of wet and dry days', *Journal of Applied Meteorology*, 6: 770–3.

Financial Times (1983) 3 November.

Financial Times (1985) 14 December.

Fisher, D. H. (1980) 'Climate and history: Priorities for research', *Journal of Interdisciplinary History*, 10: 820–30.

Fleishman, E. (1954) 'How outdoor temperature affects electrical system loads', *Electrical West*, June: 92–6.

Flemming, A. K. (1982) 'An industry in crisis?', *4 Quarter*, 3: 15–17.

Flohn, H. (1981a) *Life of a Warmer Earth, Possible Climatic Consequences of a Man-Made Global Warming*, Laxenburg, Austria, International Institute for Applied Systems Analysis.

Flohn, H. (1981b) 'Short-term climatic fluctuations and their economic role', in Wigley, T. M. L., Ingram, M. J., and Farmer, G. (eds) *Climate and History*, Cambridge, Cambridge University Press, 310–18.

Flohn, H. (1981c) 'Climate variability and coherence in time and space', in Bach, W., Pankrath, J., and Schneider, S. H. (eds) *Food–Climate Interactions*, Dordrecht, Holland, Reidel, 423–41.

Foley, J. C. (1957) *Droughts in Australia – Review of Records from Earliest Years of Settlement to 1955*, Australian Bureau of Meteorology Bulletin no. 43.

Food and Agriculture Organization (1977) *Food Quarterly Outlook*.

Food and Agriculture Organization (1979) 'Scanning the future for climatic change', *CERES*, January–February: 7.

Food and Agriculture Organization (1983) *Special Report on Foodcrops and Shortages*, 15 April.

Food and Agriculture Organization (1985) *Foodcrops and Shortages*, January issue.

Foster, H. D. (1976) 'Assessing disaster magnitude: A social science approach', *Professional Geographer*, 28: 241–7.

Foster, H. D. (1980) *Disaster Planning: The Preservation of Life and Property*, Springer Series of Environmental Management, New York, Springer Verlag.

Foster, H. D. and Sewell, W. R. D. (1980) *Water: The Emerging Crisis in Canada*, Ottawa, Canadian Institute for Economic Policy.

Freedman, A. M., III (1980) 'The hedonic price technique and the value of climate as a resource', paper presented at the *Climate and Economics Workshop*, 24–25 April, Fort Lauderdale, Florida, Washington, DC, Resources for the Future.

Freedman, D. (1981) 'Some pitfalls in large econometric models: A case study', *Journal of Business*, 54: 479–500.

Fukui, H. (1979) 'Climatic variability and agriculture in tropical moist regions', in *Proceedings of the World Climate Conference*, Geneva, World Meteorological Organization, 436–74.

Gabe, Masanobu (1985) 'Weather information – valuable economic tool in an era of low growth', *Tokyo Newsletter* (August), Tokyo, Corporate Communications Office, Mitsubishi Corporation, 1–4.

Garcia, R. (1978) 'Climate impacts and socioeconomic conditions', in National Academy of Sciences *International Perspectives on the Study of Climate and Society*, Washington, DC, NAS, 43–7.

Garcia, R. V. (1981) *Drought and Man: The 1972 Case Study*. Vol. 1: *Nature Pleads not Guilty*, Oxford, Pergamon Press.

Gasser, W. R. (1981) 'Climate change to the year 2000 and possible impacts on world agriculture: A review of the National Defence University Study', paper presented to the Institute for Energy Analysis/NCAR Workshop on Improving the Science of Climate-Related Impact Studies, Oak Ridge, Tennessee, 30 June–2 July, 1981.

Gibbs, W. J. (1968) 'Benefits of meteorological services in Australia', in *Economic Benefits of Meteorology*, *WMO Bulletin* 17: 181–6.

Glantz, M. H. (ed.) (1976a) *The Politics of Natural Disaster: The Case of Sahel Drought*, New York, Praeger.

Glantz, M. H. (1976b) 'Nine fallacies of natural disaster', in Glantz, M. H. (ed.) *The Politics of Natural Disaster: The Case of the Sahel Drought*, New York, Praeger.

Glantz, M. H. (1977a) 'The value of a long-range weather forecast for the West African Sahel', *Bulletin of the American Meteorological Society*, 58: 150–8.

Glantz, M. H. (1977b) 'The social value of a reliable long-range forecast', *Ekistics*, 43: 305–13.

Glantz, M. H. (1982a) 'The use of a climate-related forecast: The case of El Niño', Paper presented at the *WMO Symposium on Education and Training in Meteorology with Emphasis on Climatic Change and Variability*, San José, Costa Rica, 6–10 December 1982.

Glantz M. H. (1982b) 'Consequences and responsibilities in drought forecasting: The case of Yakima, 1977', *Water Resources Research*, 18: 3–13.

Glantz, M. H. and Katz, R. W. (1977) 'When is a drought a drought?', *Nature*, 267: 192–3.

Glantz, M. H., Robinson, J., and Krenz, M. E. (1982) 'Climate-related impact studies: A review of past experiences', in Clark, W. (ed.) *Carbon Dioxide Review*, New York, Oxford University Press.

Glantz, M. H., Robinson, J., and Krenz, M. E. (1985) 'Recent assessments', in Kates, R. W., Ausubel, J. H., and Berberian, M. (eds) *Climate Impact Assessment: Studies of the Interaction of Climate and Society*, SCOPE 27, New York, Wiley, 565–98.

Global Atmospheric Research Programme (GARP) (1975) *The Physical Basis of Climate and Climate Modelling*, WMO-ICSU Joint Organizing Committee, Report of the International Study Conference in Stockholm,

29 July–10 August 1974. GARP Publication, Series no. 16. Geneva, World Meteorological Organization.
Golden, J. H. (Moderator) (1984) 'Joint Panel Session: On the role of the private sector in disseminating hurricane forecasts and warnings', *Bulletin of the American Meteorological Society*, 65: 972–80.
Goulter, J. (1982) 'Disaster country', *Christchurch Star*, 1 September.
Gray, W. M., Frank, W. M., Corrin, M. L., and Stokes, C. A. (1976) 'Weather modification by carbon dust absorption of solar energy', *Journal of Applied Meteorology*, 15: 355–86.
Gribbin, J. (ed.) (1978) *Climatic Change*, Cambridge, Cambridge University Press.
Griffiths, J. F. (1976) *Applied Climatology: An Introduction*, 2nd edn, Oxford, Oxford University Press.
Gruza, G. V. (1979) 'Fluctuations of climate and man's economic activity', *Hydrometeorology*, vol. 3 (translation from the Russian by the National Science Foundation).
Gum, R. L., Roefs, T. G., and Kimball, D. B. (1976) 'Quantifying societal goals: Development of a weighting methodology', *Water Resources Research*, 12: 617–22.

Haas, J. E. (1975) 'Social impact of induced climate change', in d'Arge, R. C. *et al.* (eds) *Economic and Social Measures of Biological and Climatic Change*, Climate Impact Assessment Program (CIAP) Monograph no. 6, Washington, DC, US Department of Transportation.
Hahn, L. and McQuigg, J. D. (1970) 'Evaluation of climatological records for national planning of livestock shelters', *Agricultural Meteorology*, 7: 131–41.
Haigh, P. A. (1977) 'Separating the effects of weather and management on crop production', MS Report to Charles F. Kettering Foundation, Dayton, Ohio.
Hall, A. L. (1978) *Drought and Irrigation in North-east Brazil*, Cambridge, Cambridge University Press.
Hallanger, N. L. (1963) 'The business of weather: its potential and uses', *Bulletin of the American Meteorological Society*, 44: 63–7.
Hamilton, D. (1970) 'Winter's warmth in jeopardy', *New Scientist*, 29 October: 213–14.
Hann, J. (1883) *Handbook of Climatology* (English translation by R. de C. Ward of the original German publication), London, The Macmillan Co., 1903.
Hard, T. M. and Broderick, A. J. (eds) (1976) *Proceedings of the Fourth Conference on the Climatic Impact Assessment Program*, Springfield, Virginia, NTIS, US Department of Commerce.
Hare, F. K. (1977) 'Climate and desertification', in UN Conference on

Desertification Secretariat (eds) *Desertification: Its Causes and Consequences*, Oxford, Pergamon Press, 63–120.

Hare, F. K. (1979a) 'Focus on climate', *Environmental Science and Technology*, 13: 156–9.

Hare, F. K. (1979b) 'The vaulting of intellectual barriers: The Madison thrust in climatology', *Bulletin of the American Meteorological Society*, 60: 1171–4.

Hare, F. K. (1980) *The Carbon Dioxide Question: Canadian Perspectives*, paper prepared for the Climate Planning Board of Canada.

Hare, F. K. (1981) 'Future climate and the Canadian economy', *Climatic Change Seminar Proceedings*, Regina, Saskatchewan. Downsview, Ontario, Atmospheric Environment Service, 92–122.

Hare, F. K. (1982) 'Climate: The neglected factor?', *The International Journal*, 36: 371–87.

Hare, F. K. (1983) 'Future climate and the Canadian economy', in Harrington, C. R. (ed.) *Climate Change in Canada 3*, National Museum of Natural Science Syllogeus, Series 49: 15–49. Ottawa, National Museums of Canada.

Hare, F. K. (1985a) 'Climate variability and change', in Kates, R. W., Ausubel, J. H., and Berberian, M. (eds) *Climate Impact Assessment: Studies of the Interaction of Climate and Society*, SCOPE 27, New York, Wiley, 37–68.

Hare, F. K. (1985b) *Climate Variations, Drought and Desertification*, WMO Publication no. 653, Geneva, World Meteorological Organization.

Hare, F. K. and Sewell, W. R. D. (1985) 'Awareness of climate', in Burton, I. and Kates, R. W. (eds) *Geography, Resources and Environment. Essays in Honor of Gilbert F. White*, Chicago, University of Chicago Press.

Hare, F. K. and Thomas, M. K. (1974) *Climate Canada*, Toronto, Wiley.

Harris, D. W. (1964) 'The relationship between relative humidity, temperature and demand for electric power at peak periods', *New Zealand Electrical Journal*, 37(7): 169.

Harrison, M. R. (1982) 'The media and public perceptions of climatic change', *Bulletin of the American Meteorological Society*, 63: 730–8.

Harshbarger, C. E. and Duncan, M. (1977) 'The economic realities of drought', *Federal Reserve Bank of Kansas City, Monthly Review*, May: 3–13.

Hart, D. S., Bennet, J. W., Hutchison, J. C., and Wodzicka-Tomaszewka, M. (1963) 'Reversed photo-periodic seasons and wool growth', *Nature*, 198: 310–11.

Haurin, D. B. (1980) 'The regional distribution of population, migration, and climate', *Quarterly Journal of Economics*, 95: 293–308.

Hayes, J. T., O'Rourke, P. A., Terjung, W. H., and Todhunter, P. E. (1982) 'A feasible crop yield model for worldwide international food

production', *International Journal of Biometeorology*, 26: 239–57.
Heal, G. (1984) 'Interactions between economy and climate: A framework for policy design under uncertainty', in Smith, V. K. and Witte, A. D. (eds) *Advances in Applied Micro-economics*, vol. 3. Greenwich, Connecticut, JAI Press.
Heathcote, R. L. (1967) 'The effects of past droughts on the national economy', paper presented to ANZAAS Conference, Melbourne.
Heathcote, R. L. (1969) 'Drought in Australia: A problem of perception', *Geographical Review*, 59: 175–94.
Heathcote, R. L. (1985) 'Extreme event analysis', in Kates, R. W., Ausubel, J. H., and Berberian, M. (eds) *Climate Impact Assessment: Studies of the Interaction of Climate and Society*, SCOPE 27, New York, Wiley, 369–402.
Hentshel, G. (1964) 'Sports and climate', in Light, S. and Kamenetz, H. L. (eds) *Medical Climatology*, Baltimore, Maryland, Waverly Press.
Hewitt, K. (ed.) (1983) *Interpretations of Calamity: From the Viewpoint of Human Ecology*, London, Edward Arnold.
Hickman, J. S. (1982) Editorial, *Weather and Climate* 3 (August).
Hill, H. W. (1984) 'Weather forecasts and the beekeeper', *The New Zealand Beekeeper*, Winter: 12–13.
Hipp, G. (1972) 'Sun or rain? Use of a European Centre for Medium-Term Weather Forecasting', *Euro Spectra* (Scientific and Technical Review of the European Communities), 11(3): 66–84.
Hoch, J. (1974) 'Wages, climate and the quality of life', *Journal of Environmental Economics and Management*, 1: 268–95.
Holdgate, M. W., Kassas, M., and White, G. F. (1983) *The World Environment 1972–1982* (A report by the United Nations Environment Programme), Dublin, Tycooly International Publishing for the United Nations Environment Programme.
Hoskins, W. G. (1964) 'Harvest fluctuations and English economic history, 1480–1619', *Agricultural History Review*, 12: 28–46.
Hosler, C. L. (1974) 'Overt weather modification', *Reviews of Geophysics and Space Physics*, 12: 523–7.
Hudson, E. A. and Jorgenson, D. W. (1974) 'US energy policy and economic growth, 1975–2000', *The Bell Journal of Economics*, 5(2): 461–514.
Hughes, B. B. (1982) *International Futures Simulation: User's Manual*, Denver, Colorado, University of Denver.
Hughes, P. (1983) 'Weather, climate, and the economy', *Weatherwise*, 35: 60–3.
Hulme, M. (1984) 'An exceptionally dry year in central Sudan', *Weather*, 39: 281, 310.
Hunt, R. D. (1982) 'BBC Television weather forecasts – audience research results', *Meteorological Magazine*, 111: 45–7.

Huntington, E. (1945) *Mainsprings of Civilization*, New York, Wiley.
Hurnard, S. M. (1979) 'Long-term climate changes: Their possible impact on New Zealand horticulture', *New Zealand Commercial Grower*, May: 27–31.

Idso, S. B. (1980) 'The climatological significance of a doubling of the earth's atmospheric carbon dioxide concentration', *Science*, 207: 1462–3.
Idso, S. B. (1984) 'The Co_2 climate controversy: An issue of global concern', *New Zealand Geographer*, 40: 110–12.
Ingram, M. J., Farmer, G., and Wigley, T. M. L. (1981) 'Past climates and their impact on man: A review', in Wigley, T. N. L., Ingram, M. J., and Farmer, G. (eds) *Climate and History*, Cambridge, Cambridge University Press, 3–50.
Islam, M. A. and Kunreuther, H. (1973) 'The challenge of long term recovery from natural disasters: Implications for Bangladesh', *Oriental Geographer*, 17 (2): 51–63.

Jäger, J. (1983) *Climate and Energy Systems*, Energy Systems Program, Laxenburg, Austria, International Institute for Applied Systems Analysis.
Jäger, J. (1985) 'Energy sources', in Kates, R. W., Ausubel, J. H., and Berberian, M. (eds) *Climate Impact Assessment: Studies of the Interaction of Climate and Society*, SCOPE 27, New York, Wiley, 215–45.
Jodha, N. S. (1975) 'Famine and famine policies: Some empirical evidence', *Economic and Political Weekly*, 10(41), 1609–23.
Jodha, N. S. and Mascarenhas, A. C. (1985) 'Adjustment in self-provisioning societies', in Kates, R. W., Ausubel, J. H., and Berberian, M. (eds) *Climate Impact Assessment: Studies of the Interaction of Climate and Society*, SCOPE 27, New York, Wiley, 437–64.
Johnson, D. G. (1983) *The World Grain Economy and Climate Change to the Year 2000: Implications for Policy*, Washington, DC, National Defense University Press.
Johnson, D. G. and Sumner, D. (1976) 'An optimization approach to grain reserves for developing countries', in Eaton, D. J. and Steele, W. S. (eds) *Analysis of Grain Reserves: A Proceedings*, USDA Economic Research Service Report no. 634, Washington, DC, US Department of Agriculture.
Johnson, S. R. and Haigh, A. (1970) 'Agricultural land price differentials and their relationship to potentially modifiable aspects of the climate', *The Review of Economics and Statistics*, 52: 173–80.
Johnson, S. R. and McQuigg, J. D. (1972) 'Application of linear probability models in using weather forecasts to plan construction activities', unpublished paper, Department of Atmospheric Science, University of Missouri.

Johnson, S. R. and McQuigg, J. D. (1974) 'Some useful approaches to the measurement of economic relationships which include climatic variables', in Taylor, J. A. (ed.) *Climatic Resources and Economic Activity*, New York, Wiley.
Johnson, S. R., McQuigg, J. D., and Rothrock, T. P. (1969) 'Temperature modification and cost of electric power generation', *Journal of Applied Meteorology*, 8: 919–26.
Jordan, J. (1975) 'Summary of hearings on the Climate Act of 1975', *Bulletin of the American Meteorological Society*, 58: 235–40.
Joyce, J. (1982) 'Climate and energy', *The Economic Impact of Climate*, vol. 10, Norman, Oklahoma Climatological Survey.

Kahneman, D., Slovic, P., and Tversky, A. (eds) (1982) *Judgement Under Uncertainty: Heuristics and Biases*, New York, Cambridge University Press.
Kahneman, D. and Tversky, A. (1979) 'Prospect theory; An analysis of decision under risk', *Econometrica*, 47(2): 263–92.
Kalkstein, L. S. (1979) 'A synoptic climatological approach to environmental analysis', *Proceedings of the Middle States Division, Association of American Geographers*, 13: 68–75.
Karl, T. R. and Riebsame, W. E. (1984) 'The identification of 10 to 20-year temperature and precipitation fluctuations in the contiguous United States', *Journal of Climate and Applied Meteorology*, 23: 950–66.
Kates, R. W. (1979) 'The Australian experience: Summary and prospect', in Heathcote, R. L. and Thom, B. G. (eds) *Natural Hazards in Australia*, Canberra, Australian Academy of Science, 511–20.
Kates, R. W. (1981a) 'Drought in the Sahel: Competing views as to what really happened in 1910–14 and 1968–74', *Mazingira*, 5: 72–83.
Kates, R. W. (1981b) *Drought Impact in the Sahelian-Sudanic Zone of West Africa: A Comparative Analysis of 1910–15 and 1968–74*, Worcester, Massachusetts, CENTED, Clark University.
Kates, R. W. (1983) 'Part and apart; Issues in human kind's relationship to the natural world', in Hare, F. K. (ed.) *The Experiment of Life*, Toronto, University of Toronto Press.
Kates, R. W. (1985) 'The interaction of climate and society', in Kates, R. W., Ausubel, J. H., and Berberian, M. (eds) *Climate Impact Assessment: Studies of the Interaction of Climate and Society*, SCOPE 27 New York, Wiley, 3–36.
Kates, R. W., Ausubel, J. H., and Berberian, M. (eds) (1985) *Climate Impact Assessment: Studies of the Interaction of Climate and Society*, SCOPE 27, New York, Wiley.
Kates, R. W., Changnon, S. A., Jr, Karl, T. R., Riebsame, W., and Easterling, W. E. (1984) *The Climate Impact, Perception, and Adjustment Experiment (CLIMPAX): A Proposal for Collaborative Research*,

Worcester, Massachusetts, Climate and Society Research Group, Center for Technology, Environment, and Development, Clark University.
Katz, R. W. (1979) 'Sensitivity analysis of statistical crop-weather models', *Agricultural Meteorology*, 20, 291–300.
Katz R. W., Murphy, A. H., and Winkler, R. L (1982) 'Assessing the value of frost forecasts to orchardists: A dynamic decision-making approach', *Journal of Applied Meteorology*, 21: 519–31.
Kawasaki, T. (1985) 'Fisheries', in Kates, R. W., Ausubel, J. H., and Berberian, M. (eds) *Climate Impact Assessment: Studies of the Interaction of Climate and Society*, SCOPE 27, New York, Wiley, 131–54.
Kellogg, W. W. (1977) *Effects of Human Activities on Global Climate*, WMO Technical Note no. 156, Geneva, World Meteorological Organization.
Kellogg, W. W. and Schware, R. (1981) *Climate Change and Society: Consequences to Increasing Atmospheric Carbon Dioxide*, Boulder, Colorado, Westview Press, 24–8.
Kidd, D. A. (1984) 'Weather lore in Aratus' *Phaenomena*', *Weather and Climate*, 4: 32–5.
Kirkby, A. V. (1974) 'Individual and community responses to rainfall variability in Oaxaca, Mexico', in White, G. F. (ed.) *Natural Hazards*, New York, Oxford University Press, 119–28.
Kneese, A. V. (1977) *Economics and the Environment*, New York, Penguin Books.
Kogan, F. N. (1982) 'Perspectives for grain production in the USSR', *Agricultural Meteorology*, 28: 213–27.
Kolb, L. L. and Rapp, R. R. (1962) 'The utility of weather forecasts to the raisin industry', *Journal of Applied Meteorology*, 1: 8–12.
Kraemer, R. S. (1978) 'Meeting reviews: Session on climatic futures at the annual meeting of the AAAS, 17 February 1978', *Bulletin of the American Meteorological Society*, 59: 822–3.

Labys, W. C. (1978) 'Commodity markets and models: The range of experience', in Adams, F. G. and Klein, S. A. (eds) *Stabilizing World Commodity Markets*, Lexington, Massachusetts, Lexington Books.
Lamb, H. H. (1979) *Climate: Present, Past, and Future*. Vol. 2: *Climatic History and the Future*, London, Methuen.
Lamb, H. H. (1982) *Climate, History and the Modern World*, London, Methuen.
Lambert, L. D. (1975) 'The role of climate in the economic development of nations', *Land Economics*, 47: 339.
Landsberg, H. E. (1946) 'Climate as a natural resource', *The Scientific Monthly*, 63: 293–8.
Landsberg, H. E. (1970) 'Man-made climatic chänges', *Science*, 170: 1265 –74.

Landsberg, H. E. (1975) 'Sahel drought: change of climate or part of climate?', *Arch. Met. Geoph. Biokl.*, B, 23: 193–200.

Landsberg, H. E. (1981) *The Urban Climate*, New York, Academic Press.

Landsberg, H. E. (1984) 'Climate and health', in Biswas, A. K. (ed.) *Climate and Development*, Dublin, Tycooly International Publishing.

Landsberg, H. E. (1985) 'The value and challenge of climatic predictions', in *Scientific Lectures Presented at the Ninth World Meteorological Congress*, WMO no. 614, Geneva, World Meteorological Organization, 19–32.

Laur, T. L. (1976) *E.O.S. Transactions of the American Geophysical Union*, 57: 189–95.

Laurmann, J. A. (1976) 'Variance estimates in seasonal climate forecasting and related food reserve requirements', *Journal of Applied Meteorology*, 15: 529–34.

Lave, L. B. (1963) 'The value of better weather information to the raisin industry', *Econometrica*, 31: 151–64.

Lave, L. B. and Epple, D. (1985) 'Scenario analysis', in Kates, R. W., Ausubel, J. H., and Berberian, M. (eds) *Climate Impact Assessment: Studies of the Interaction of Climate and Society*, SCOPE 27, New York, Wiley, 511–28.

Lawford, R. G. (1981) 'Impacts of the climate of 1980 on the energy demand/supply cycle', in Phillips, D. W. and McKay, G. A. (eds) *Canadian Climate in Review 1980*, Ottawa, Environment Canada, Atmospheric Environment Service.

Le Comte, D. M. and Warren, H. E. (1981) 'Modelling the impact of summer temperatures on national electricity consumption', *Journal of Applied Meteorology*, 20: 1415–19.

Leech, L. S. (1985) 'A provisional assessment of the recreational quality of weather in summer, in terms of thermal comfort and the adverse effect of rainfall', *Meteorological Service of Ireland Technical Note*, no. 47.

Le Houérou, H. N. (1985) 'Pastoralism', in Kates, R. W., Ausubel, J. H., and Berberian, M. (eds) *Climate Impact Assessment: Studies of the Interaction of Climate and Society*, SCOPE 27, New York, Wiley, 155–86.

Le Houérou, H. N. and Popov, G. F. (1981) *An Ecoclimatic Classification of Intertropical Africa*, Rome, AGPE, FAO.

Leontief, W. (1977a) 'Structure of the world economy: Outline of a simple input-output formulation', *American Economic Review*, December: 823–34.

Leontief, W. (1977b) *The Future of the World Economy: A United Nations Study*, New York, Oxford University Press.

Le Roy Ladurie, E. (1971) *Times of Feast, Times of Famine: A History of Climate since the Year 1000* (translated by B. Bray), Garden City, New York, Doubleday.

Liebhardt, K. (1981) 'The socioeconomic impact of climate', *Enquiry*, 1(3), Spring. (Research at the University of Delaware, Newark, Delaware.)

Liljas, E. (1984) 'Benefit resulting from tailored very-short-range forecasts in Sweden', in *Proceedings of the Nowcasting II Symposium*, Norrköping, Sweden.

Linden, F. (1959) 'Weather in business', *The Conference Board Business Record*, 16: 90–4, 101.

Linden, F. (1962) 'Merchandising weather', *The Conference Board Business Record*, 19(6): 15–16.

Liverman, Diana M. (1983) *The Use of a Simulation Model in Assessing the Impacts of Climate on the World Food System*, University of California, Los Angeles–National Center for Atmospheric Research Cooperative Thesis no. 7, Boulder, Colorado, National Center for Atmospheric Research.

Logothetti, T. J. (1974) 'Social systems', in Weisbecker, L. W. (ed.) *The Impacts of Snow Enhancement*, Norman, Oklahoma, University of Oklahoma Press, 353–88.

Lorenz, E. (1968) 'Climatic determinism', *Meteorological Monographs*, 8(30): 1–3.

Lovell, C. A. K. and Smith, V. K. (1985) 'Microeconomic analysis', in Kates, R. W., Ausubel, J. H., and Berberian, M. (eds) *Climate Impact Assessment: Studies of the Interaction of Climate and Society*, SCOPE 27, New York, Wiley, 293–321.

McCracken, M. C. (1983) *Nuclear War, Preliminary Estimates of the Climatic Effects of a Nuclear Exchange*, paper presented at the Third International Seminar on Nuclear War, Erice, Sicily, 19–23.

McCracken, M. C. and Luther, F. M. (eds) (1985) *Detecting the Climatic Effects of Increasing Carbon Dioxide*, DOE/ER-0235, Washington, DC, US Department of Energy.

McFadden, D. (1984) 'Welfare analysis of incomplete adjustment to climatic change', in Smith, V. K. and Witte, A. D. (eds) *Advances in Applied Micro-economics*, vol. 3, Greenwich, Connecticut, JAI Press.

McHale, M. C. (1981) *Ominous Trends and Valid Hopes: A Comparison of Five World Reports*, Minneapolis, Minnesota, Hubert H. Humphrey Institute of Public Affairs, University of Minnesota.

McIntyre, A. J. (1973) 'Effects of drought in the economy', in Lovett, J. V. (ed.) *The Environmental, Economic and Social Significance of Drought*, Sydney, Angus & Robertson.

McKay, G. A. (1979) Editorial in *Climatic Change*, 2(1).

McKay, G. A. (1980) 'Socioeconomic impacts of climate', in Powell, J. M. (compiler) *Socioeconomic Impacts of Climate*, Proceedings of the Workshop and Annual Meeting of the Alberta Climatological Association, Northern Forest Research Centre, Canadian Forestry Service, Information Report NOR-X-217, 1–6.

McKay, G. A. and Allsopp, T. (1977) 'Climate and climate variability', in

Climatic Variability in Relation to Agricultural Productivity and Practices, Ottawa, Committee on Agrometeorology, Department of Agriculture Research Branch.

McKay, G. A. and Allsopp, T. (1980) 'The role of climate in affecting energy demand/supply', in Bach, W., Pankrath, J., and Williams, J. (eds) *Interactions of Energy and Climate*, Dordrecht, Holland, Reidel, 53–72.

McQuigg, J. D. (1964) 'The economic value of weather information', PhD dissertation, University of Missouri, Columbia, Missouri.

McQuigg, J. D. (1967) 'Some brief comments on the use of simulation models as a tool to study the relationships of weather and human activity', unpublished paper, Department of Atmospheric Science, University of Missouri.

McQuigg, J. D. (1970a) Foreword to W. J. Maunder *The Value of the Weather*, London, Methuen.

McQuigg, J. D. (1970b) 'Some attempt to estimate the economic response of weather information', *WMO Bulletin*, 19: 72–8.

McQuigg, J. D. (1972a) 'The use of meteorological information in economic development', unpublished paper prepared for WMO Executive Committee Panel on Economic Development, May.

McQuigg, J. D. (1972b) *Training in the Applications of Meteorology to Economic Development*, unpublished paper prepared for the Sixth Session of the WMO Executive Panel of Experts on Meteorological Education and Training, Cairo, April.

McQuigg, J. D. (1972c) 'Report on Agricultural Meteorology Committee', *Bulletin of the American Meteorological Society*, 53: 352.

McQuigg, J. D. (1974) 'World without hunger? The weather factor grows more critical', *Economic Impact*, 8: 34–8.

McQuigg, J. D. (1975a) 'Climatic change and world food production', address given at University of Florida, Frontier of Science Series, 23 April 1975.

McQuigg, J. D. (1975b) *Economic Impacts of Weather Variability*, Columbia, Missouri, Department of Atmospheric Science, University of Missouri.

McQuigg, J. D. (1979) 'Climatic variability and agriculture in the temperate regions', in *Proceedings of the World Climate Conference*, Publication no. 537, Geneva, World Meteorological Organization, 406–25.

McQuigg Consultants, Inc. (1981) 'The weighted weather index', in *McQuigg Crop/Weather News*, 5 June.

McQuigg, J. and Decker, W. (1962) 'The probability of completion of outdoor work', *Journal of Applied Meteorology*, 1: 178–82.

McQuigg, J. D. and Johnson, S. R. (1973) 'Increased precision of weather-load models for long term planning of large interconnected systems', paper presented at American Power Conference, Chicago, 10 May.

McQuigg, J. D., Johnson, S. R., and Tudor, J. R. (1972) 'Meteorological diversity/load diversity, a fresh look at an old problem', *Journal of Applied Meteorology*, 11: 561–6.

McQuigg, J. D. and Thompson, R. G. (1966) 'Economic value of improved methods of translating weather information into operational terms', *Monthly Weather Review*, 94: 83–7.

Maler, K. G. (1974) *Environmental Economics: A Theoretical Inquiry*, Baltimore, Maryland, Johns Hopkins University Press.

Manabe, S. and Wetherald, R. T. (1980) 'On the distribution of climatic change resulting from an increase in CO_2 content of the atmosphere', *Journal of the Atmospheric Sciences*, 37: 99–118.

Marchuk, G. I. (1979) 'Modelling of climatic changes and the problem of long-range weather forecasting', in *Proceedings of the World Climate Conference*, Publication no. 537, Geneva, World Meteorological Organization, 132–53.

Margolis, M. (1980) 'Natural disaster and socioeconomic change: Post-frost adjustments in Parana, Brazil', *Disasters*, 4(2): 231–5.

Marsh and McLennan Companies (1980) *Risk in a Complex Society: A Marsh and McLennan Public Opinion Survey*. Conducted by L. Harris and Associates for Marsh and McLennan, New York.

Martin, B. (1979) *The Bias of Science*, O'Connor, Australia, Society for Social Responsibility in Science.

Martin, D. B. (1970) 'Construction and seasonality: The new federal program', *Construction Review*, 16: 4–7.

Mascarenhas, A. C. (ed.) (1973) 'Studies in famines and food shortages', *Journal of the Geographical Association of Tanzania*, no. 8.

Mason, B. J. (1966) 'The role of meteorology in the national economy', *Weather*, 21: 382–93.

Mason, B. J. (1976a) 'The nature and prediction of climatic changes', *Endeavour*, 35: 51–7.

Mason, B. J. (1976b) 'Towards the understanding and prediction of climatic variations', *Quarterly Journal of the Royal Meteorological Society*, 102: 473–98.

Mason, B. J. (1981) Review of *Climatic Constraints and Human Activities* (Ausubel and Biswas, 1980), *Quarterly Journal of the Royal Meteorological Society*, 107: 743–4.

Massachusetts Institute of Technology, Center for Advanced Engineering Study (1980) *Climate and Risk*, MTR-80W322-01, McLean, Virginia, Mitre Corporation.

Mather, J. R. (1974) *Climatology: Fundamentals and Applications*, New York, McGraw-Hill.

Mather, J. R. (1978) *The Climatic Water Balance in Environmental Analysis*, Lexington, Massachusetts, D. C. Heath.

Mather, J. R., Field, R. T., Kalkstein, L. S., Willmott, C. J., and Maunder,

W. J. (1981) 'Climatology: The impact of the seventies and the challenge for the eighties', *Weather and Climate*, 1: 69–76.

Mattei, F. (1979) 'Climatic variability in agriculture in the semi-arid tropics', in *Proceedings of the World Climate Conference*, Publication no. 537, Geneva, World Meteorological Organization, 475–509.

Matthews, S. W. (1976) 'What's happening to our climate?' *National Geographic*, 50 (5): 576–615.

Maunder, W. J. (1963) 'The climates of the pastoral production areas of the world', *Proceedings of the New Zealand Institute of Agricultural Science*, 9: 25–40.

Maunder, W. J. (1965) 'The effect of climatic variations on some aspects of agricultural production in New Zealand, and an assessment of their significance in the national agricultural income', unpublished PhD dissertation, University of Otago.

Maunder, W. J. (1966a) 'Climatic variations and dairy production in New Zealand, a review', *New Zealand Science Review*, 24: 69–73.

Maunder, W. J. (1966b) 'Climatic variations and agricultural production in New Zealand', *New Zealand Geographer*, 22: 55–69.

Maunder, W. J. (1967a) 'Climatic variations and wool production: a New Zealand review', *New Zealand Science Review*, 25(4): 35–9.

Maunder, W. J. (1967b) 'Climatic variations and meat production: a New Zealand review', *New Zealand Science Review*, 25(5): 9–12.

Maunder, W. J. (1968a) 'The effect of significant climatic factors on agricultural production and incomes: a New Zealand example', *Monthly Weather Review*, 96: 39–46.

Maunder, W. J. (1968b) 'An econoclimatic model for Canada: problems and prospects', paper presented at the Conference and Workshop on Applied Climatology of the American Meteorological Society, Ashville, North Carolina.

Maunder, W. J. (1969) 'The consumer and the weather forecast', *Atmosphere*, 7: 15–22.

Maunder, W. J. (1970a) *The Value of the Weather*, London, Methuen.

Maunder, W. J. (1970b) 'Weather forecasting and society', in Sewell, W. R. D. and Foster, H. D. (eds) *The Geographer and Society*, Western Geographical Series, vol. 1, University of Victoria, BC, 134–59.

Maunder, W. J. (1971a) 'The value and use of weather information', *Transactions of the Electric Supply Authority Engineers' Institute of New Zealand Inc.*, 10–20.

Maunder, W. J. (1971b) *The Economic Consequences of Drought: With Particular Reference to the 1969/70 Drought in New Zealand*, New Zealand Meteorological Service Technical Note, no. 192.

Maunder, W. J. (1971c) *Temperature Forecasts and the Assessment of Electric Power Demand in New Zealand*, New Zealand Meteorological Service Technical Note, no. 195.

Maunder, W. J. (1972a) 'The formulation and use of weather indices weighted according to the significance of areas: a New Zealand example', *New Zealand Geographer*, 28: 130–50.

Maunder, W. J. (1972b) 'A review of research into the economic impact of weather on the building and construction industry', *Symposium on Meteorology and the Building Industry*, Wellington, New Zealand Meteorological Service, 8/1–8/33.

Maunder, W. J. (1972c) 'National economic analyses of responses to weather variations', *Proceedings of the Seventh New Zealand Geography Conference*, Hamilton, 207–16.

Maunder, W. J. (1972d) *National Econoclimatic Models: Problems and Applications*, New Zealand Meteorological Service Technical Note, no. 208.

Maunder, W. J. (1972e) 'Assessing the value of weather information – with particular reference to the New Zealand Meteorological Service', unpublished paper, Wellington, New Zealand Meteorological Service.

Maunder, W. J. (1973a) 'Weekly weather and economic activities on a national scale: An example using United States retail trade data', *Weather*, 28: 2–18.

Maunder, W. J. (1973b) 'Climate and climatic resources', in *Waikato –Coromandel–King Country Region*, National Resources Survey, Part VIII, Wellington, Government Printer, 23–49.

Maunder, W. J. (1973c) 'Application of meteorology to industry', in *Proceedings of the WMO/ECAFE Regional Conference on the Role of Meteorological Services in the Economic Development of Asia and the South-west Pacific*, Bangkok, 51–65.

Maunder, W. J. (1973d) 'Meteorology in relation to economic development plans', in *Proceedings of the WMO/ECAFE Regional Conference on the Role of Meteorological Services in the Economic Development of Asia and the South-west Pacific*, Bangkok, 110–30.

Maunder, W. J. (1974a) 'National econoclimatic models: problems and applications', in Taylor, J. A. (ed.) *Climatic Resources and Economic Activity*, Aberystwyth Symposium XV (1974), London, David & Charles, 237–57.

Maunder, W. J. (1974b) *The Prediction of Monthly Dairy Production in New Zealand through the Use of Weighted Indices of Water Deficit*, New Zealand Meteorological Service Technical Note, no. 227.

Maunder, W. J. (1974c) 'Climate and climatic resources', in Johnson, R. J. (ed.) *Society and Environment in New Zealand*, Christchurch, Whitcombe & Tombs, 47–63.

Maunder, W. J. (1975) 'Weather, food and agriculture: plans and programmes of the World Meteorological Organization', *Symposium on Meteorology and Food Production*, Wellington, New Zealand Meteorological Service, 197–212.

Maunder, W. J. (1977a) 'Weather and operational decision-making: The challenge', *Quarterly Predictions* (New Zealand Institute of Economic Research), September: 21–6.

Maunder, W. J. (1977b) 'Weather and climate as factors in forecasting national dairy production', in *Proceedings of the Symposium on the Management of Dynamic Systems in New Zealand Agriculture*, Lower Hutt, NZ, DSIR, 101–26.

Maunder, W. J. (1977c) 'National economic planning: The value of weather information and the role of the meteorologist', *Proceedings of the Ninth New Zealand Geography Conference*, Dunedin, 116–20.

Maunder, W. J. (1977d) 'Climatic constraints to agricultural production: Prediction and planning in the New Zealand setting', *New Zealand Agricultural Science*, 11: 110–19.

Maunder, W. J. (1978a) 'Economic and political issues', in Pittock, A. B. *et al.* (eds) *Climatic Change and Variability: A Southern Perspective*, Cambridge, Cambridge University Press, 327–34.

Maunder, W. J. (1978b) 'Forecasting pastoral production: The use and value of weather based forecasts and the implications to the transportation industry and the nation', *Symposium on Meteorology and Transport*, Wellington, New Zealand Meteorological Service, 107–27.

Maunder, W. J. (1979) 'New Zealand's real economic climate – evaluation and prospects', *Symposium on the Value of Meteorology in Economic Planning*, Wellington, New Zealand Meteorological Service, 203–45.

Maunder, W. J. (1980a) 'The use of econoclimatic models in national supply forecasting: New Zealand wool production', *Proceedings of the Tenth New Zealand Geography Conference*, 22–8.

Maunder, W. J. (1980b) 'Weather and climatic variations and their significance to man', *Proceedings of the Tenth New Zealand Geography Conference*, 315–19.

Maunder, W. J. (1981a) 'The economic climate: fact or fiction?' *Proceedings of the Eleventh New Zealand Geography Conference*, 187–92.

Maunder, W. J. (1981b) *National Econoclimate Studies: A Case Study Using United States Housing Starts*, unpublished paper, Columbia, Missouri, Center for Environmental Assessment Services, National Oceanic and Atmospheric Administration.

Maunder, W. J. (1982a) 'National economic indicators: The importance of the weather', in Heenan, L. D. B. and Kearsley, G. W. (eds) *Essays in Honour of Ronald Lister*, Dunedin, NZ, University of Otago, 41–60.

Maunder, W. J. (1982b) 'Canterbury weather patterns: Past, present, future', in Crabb, D. (ed.) *Drought and Drought Strategies*, Rural Development and Extension Centre, Studies in Agricultural Extension no. 1, Canterbury, NZ, Lincoln College.

Maunder, W. J. (1983) 'The weather game', *Weather and Climate*, 3: 2–10.

Maunder, W. J. (1984a) 'Weather information and weather forecasting', in *Proceedings of the 1984 Fertiliser Seminar: Utilizing New Technology*, Napier, East Coast Fertiliser Co., 12–21.

Maunder, W. J. (1984b) 'Climatology: Past, present, future . . . a personal view', *Weather and Climate*, 4: 2–10.

Maunder, W. J. (1985a) 'Climate and socio-economics', in *Scientific Lectures Presented at the Ninth World Meteorological Congress, 1983*, WMO no. 614, Geneva, World Meteorological Organization, 33–59.

Maunder, W. J. (1985b) 'Matching wits with the weather', *Straight Furrow* (NZ), 43 (50), 7.

Maunder, W. J. (1986a) 'Weather impacts and weather management', in Slater, F. (ed.) *People and Environments*, London, Collins, 441–52.

Maunder, W. J. (1986b) 'Economic aspects of agricultural climatology', in Monteith, J. L., Elston, J. F., and Mount, L. E. (eds) *Agricultural Meteorology*, London, Academic Press (forthcoming).

Maunder, W. J. and Ausubel, J. H. (1985) 'Identifying climate sensitivity', in Kates, R. W., Ausubel, J. H., and Berberian, M. (eds) *Climate Impact Assessment: Studies of the Interaction of Climate and Society*, SCOPE 27, New York, Wiley, 85–104.

Maunder, W. J., Johnson, S. R., and McQuigg, J. D. (1971a) 'A study of the effect of weather on road construction: A simulation model', *Monthly Weather Review*, 99: 939–45.

Maunder, W. J., Johnson, S. R., and McQuigg, J. D. (1971b) 'The effect of weather on road construction: Applications of a simulation model', *Monthly Weather Review*, 99: 946–53.

Maunder, W. J. and Sewell, W. R. D. (1968) 'Adjustments to the weather – choice or chance', *Atmosphere*, 6: 93–6, 105–8.

Maunder, W. J., Sewell, W. R. D., and Kates, R. W. (1968) 'Measuring the economic impact of weather and weather modification: A review of techniques of analysis', in Sewell, W. R. D. *et al. Human Dimensions of the Atmosphere*, Washington, DC, National Science Foundation, 103–12.

Mayer, J. (1976) 'The dimensions of human hunger', *Scientific American*, 235 (September): 40–9.

Meadows, D. H., Meadows, D. L., Randers, J., and Behrens, W. W. III (1972) *The Limits to Growth*, New York, Universe Books.

Meier, W. J. Jr (1977) 'Identification of economic and societal impacts of water shortages', in National Research Council, *Climate, Climatic Change, and Water Supply*, Washington, DC, National Academy of Sciences, 85–95.

Meillassoux, C. (1974) 'Development or exploitation: Is the famine good business?', *Review of African Political Economy*, 1: 27–33.

Meyer-Abich, K. M. (1980) 'Chalk on the white wall? On the transformation of climatological facts into political facts', in Ausubel, J. and Biswas,

A. K. (eds) *Climatic Constraints and Human Activities*, Oxford, Pergamon Press, 61–74.
Miller, A. A. (1956) 'The use and misuse of climatic resources', *Advancement of Science*, 13: 56–66.
Miller, J. M. (1984) 'Acid rain', *Weatherwise*, 37: 233–9.
Mitchell, J. M., Felch, R. E., Gilman, D. L., Quinlan, F. T., and Rotty, R. M. (1973) *Variability of Seasonal Total Heating Fuel Demand in the United States*, Report to the Energy Policy Office, Washington, DC.
Moisseiev, N. N., Svirezhev, Yu. M., Krapivin, V. F., and Tarko, A. M. (1985) 'Biosphere models', in Kates, R. W., Ausubel, J. H., and Berberian, M. (eds) *Climate Impact Assessment: Studies of the Interaction of Climate and Society*, SCOPE 27, New York, Wiley, 493–510.
Monteith, J. L. (1977) 'Climate and the efficiency of crop production in Britain', *Royal Society of London Philosophical Transactions*, Series B281: 277–94.
Monteith, J. L. (1981) 'Climate variation and the growth of crops', *Quarterly Journal of the Royal Meteorological Society*, 107: 749–74.
Moodie, D. W. and Catchpole, A. J. W. (1976) 'Valid climatological data from historical sources by content analysis', *Science*, 193: 51–3.
Morentz, J. W. (1980) 'Communication in the Sahel drought: Comparing the mass media with other channels of international communication', in National Academy of Sciences, *Disasters and the Mass Media*, Washington, DC, National Academy of Sciences, 158–86.
Morris, L. R. (1961) 'Periodicity of seasonal rhythm of wool growth in sheep', *Nature*, 190: 102–5.
Mostek, A. and Walsh, J. E. (1981) 'Corn yield variability and weather patterns in the USA', *Agricultural Meteorology*, 25: 111–24.
Murphy, A. H. (1976) 'Decision-making models in the cost-loss ratio situation and measures of the value of probability forecasts', *Monthly Weather Review*, 104: 1058–65.
Murphy, A. H. (1977) 'The value of climatological, categorical and probabilistic forecasts in the cost-loss ratio situation', *Monthly Weather Review*, 105: 803–16.
Murphy, A. H. and Brown, B. G. (1983) 'Forecast terminology: Composition and interpretation of public weather forecasts', *Bulletin of the American Meteorological Society*, 64: 13–22.
Murphy, A. H. and Brown, B. G. (1984a) 'A comparative evaluation of objective and subjective weather forecasts in the United States', *Journal of Forecasting*, 3: 369–93.
Murphy, A. H. and Brown, B. G. (1984b) 'Short-range weather forecasts and current weather information: User requirements and economic value', *Proceedings of the Nowcasting II Symposium*, Norrköping, Sweden.
Murphy, A. H. and Katz, R. W. (eds) (1985) *Probability, Statistics, and*

Decision Making in Atmospheric Sciences, Boulder, Colorado, Westview Press.

Murphy, B. L., Schulman, L. L., and Mahoney, J. R. (1977) 'Potential use of applied meteorology in energy conservation programs', *Bulletin of the American Meteorological Society*, 58: 304–17.

Murray, T., Le Duc, S., and Ingham, M. (1983) 'Impact of climate on early life stages of Atlantic mackerel *Scomber scombrus*, L.: An application of meteorological data to a fishery problem', *Journal of Climate and Applied Meteorology*, 22: 57–68.

Musgrave, J. C. (1968) 'Measuring the influence of weather on housing starts', *Construction Review*, 14(8): 4–7.

National Academy of Sciences (1975a) *Long-term Worldwide Effects of Nuclear Weapon Detonations*, Washington, DC, National Academy of Sciences.

National Academy of Sciences (1975b) *Environmental Impact of Stratospheric Flight*, Washington, DC, National Academy of Sciences.

National Academy of Sciences (1975c) *Understanding Climatic Change*, Washington, DC, National Academy of Sciences.

National Academy of Sciences (1976) *Climate and Food – Climatic Fluctuation and US Agricultural Production*, A Report of the Committee on Climate and Weather Fluctuations and Agricultural Production, Washington, DC, National Academy of Sciences.

National Defense University (1980) *Crop Yields and Climate Change to the Year 2000*, Report on the Second Phase of a Climate Impact Assessment, Fort Lesley J. McNair, Washington, DC, National Defense University.

National Defense University (1983) *World Grain Economy and Climate Change to the Year 2000: Implications for Policy*, Fort Lesley J. McNair, Washington, DC, National Defense University.

National Research Council (1979) *Carbon Dioxide and Climate: A Scientific Assessment*, Washington, DC, National Academy Press.

National Research Council (1981) *Managing Climatic Resources and Risks*, Washington, DC, National Academy Press.

National Research Council (1982) *Carbon Dioxide and Climate: A Second Assessment*, Washington, DC, National Academy Press.

National Research Council, Board on Atmospheric Sciences and Climate (1983), *Changing Climate* (Report of the Carbon Dioxide Assessment Committee), Washington, DC, National Academy Press.

Neale, A. A. (1985) 'Weather forecasting: Magic, art, science and hypnosis', *Weather and Climate*, 5: 2–5.

Nelson, J. P. (1976) 'Climate and energy demand: Fossil fuels', in Ferrar, T. (ed.) *The Urban Costs of Climate Modification*, New York, Wiley.

Newell, R. K. and Dopplick, T. G. (1979) 'Questions concerning the possible influence of anthropogenic CO_2 on atmospheric temperature', *Journal of Applied Meteorology*, 18: 822–5.

Newman, J. E. (1978) 'Drought impacts on American agricultural productivity', in Rosenberg, N. J. (ed.) *North American Droughts*, Boulder, Colorado, Westview Press.

Newman, J. E. (1980) 'Climate change impacts on the growing season of the North America "cornbelt" ', *Biometeorology*, 7(2): 128–42.

Newsweek (1977) 14 March.

Newsweek (1983a) 11 April.

Newsweek (1983b) 9 May.

Newsweek (1983c) 29 August.

Newsweek (1984) 2 January.

New Zealand Meteorological Service (1979) *Symposium on the Value of Meteorology in Economic Planning*, Wellington, New Zealand Meteorological Service.

Nicholls, N. (1985) 'Impact of the southern oscillation on Australian crops', *Journal of Climatology*, 5: 1–8.

Nicodemus, M. L. and McQuigg, J. D. (1969) 'A simulation model for studying possible modification of surface temperature', *Journal of Applied Meteorology*, 8: 199–204.

Nix, H. A. (1980) 'Strategies for crop research', *Proceedings of the Agronomic Society of New Zealand*, 10: 107–10.

Nix, H. A. (1985) 'Agriculture', in Kates, R. W., Ausubel, J. H., and Berberian, M. (eds) *Climate Impact Assessment: Studies of the Interaction of Climate and Society*, SCOPE 27, New York, Wiley, 105–30.

Nováky, B., Pachner, C., Szesztay, K., and Miller, D. (1985) 'Water resources', in Kates, R. W., Ausubel, J. H., and Berberian, M. (eds) *Climate Impact Assessment: Studies of the Interaction of Climate and Society*, SCOPE 27, New York, Wiley, 187–214.

Nye, R. H. (1965) 'The value of services provided by the Bureau of Meteorology in planning within the State Electricity Commission of Victoria', in *'What is Weather Worth?'*, Melbourne, Bureau of Meteorology, 87–91.

Office of Technology Assessment (OTA) (1983) *World Futures and Public Policy*, Washington, DC, Technology Assessment Board, OTA.

Ogilvie, A. E. J. (1981a) 'Climate and economy in eighteenth century Iceland', in Delano Smith, C. and Parry, M. L. (eds), *Consequences of Climatic Change*, Nottingham, Department of Geography, University of Nottingham, 54–69.

Ogilvie, A. E. J. (1981b) 'Climate and Society in Iceland from the Medieval Period to the Late Eighteenth Century', unpublished PhD dissertation, University of East Anglia, Norwich.

Oguntoyinbo, J. A. and Odingo, R. S. (1979) 'Climatic variability and land use: An African perspective', *Proceedings of the World Climate Conference*, Publication no. 537, Geneva, World Meteorological Organization, 552–80.

Oguntoyinbo, J. A. and Richards, P. (1978) 'Drought and the Nigerian farmer', *Journal of Arid Environments*, 1, 165–94.

Oliver, J. E. (1973) *Climate and Man's Environment*, New York, Wiley.

Oliver, J. E. *et al.* (1975) 'Recollection of past weather by the elderly in Terre Haute, Indiana', *Weatherwise* (August), 161–71.

Olsson, L. E. (1979) 'Meteorology and wind energy', in *Scientific Lectures Presented at the Eighth World Meteorological Congress*, WMO no. 568, Geneva, World Meteorological Organization, 106–12.

Oram, P. A. (1982) 'The economic cost of climatic variation', in Blaxter, K. and Fowden, L. (eds) *Food, Nutrition and Climate*, London, Applied Science Publishers, 355–92.

Oury, B. (1965) 'Allowing for weather in crop production model building', *Journal of Farm Economics*, 47: 270–83.

Oury, B. (1969) 'Weather and economic development', *Finance and Development*, 6: 24–9.

Palutikof, J. (1983a) 'The impact of weather and climate on industrial production in Great Britain', *Journal of Climatology*, 3: 65–79.

Palutikof, J. (1983b) 'Drought without water', *Nature*, 203, 635.

Parker, A. (1983) 'Met. office looks ahead to ensure economic sunshine', *National Business Review* (New Zealand), 31 January 1983: 17.

Parker, J. (1978) 'Meteorological office participation in Prestel – the Post Office viewdata system', *Meteorological Magazine*, 107: 249–54.

Parry, M. L. (1975) 'Secular climatic change and marginal agriculture', *Transactions of the Institute of British Geographers*, 64: 1–13.

Parry, M. L. (1976) 'The significance of the variability of summer warmth in upland Britain', *Weather*, 31: 212–17.

Parry, M. L. (1978) *Climatic Change, Agriculture and Settlement*, Hamden, Connecticut, Archon Books.

Parry, M. L. (1981a) 'Climatic change and the agricultural frontier: A research strategy', in Wigley, T. M. L., Ingram, M. J., and Farmer, G. (eds) *Climate and History*, Cambridge, Cambridge University Press, 319–36.

Parry, M. L. (1981b) 'Evaluating the impact of climatic change', in Delano Smith, C. and Parry, M. L. (eds) *Consequences of Climatic Change*, Nottingham, Department of Geography, University of Nottingham, 3–16.

Parry, M. L. (1983) *Climate Impact Assessment in Cold Areas*, Report of working group on cold margins, WMO/UNEP Workshop on Impacts of Climate Change, Villach, Austria, September 1983, Laxenburg, Austria, International Institute for Applied Systems Analysis.

Parry, M. L. (1985a) 'The impact of climatic variations on agricultural margins', in Kates, R. W., Ausubel, J. H., and Berberian, M. (eds) *Climate Impact Assessment: Studies of the Interaction of Climate and Society*, SCOPE 27, New York, Wiley, 351–68.

Parry, M. L. (ed.) (1985b) *The Sensitivity of National Ecosystems and Agriculture to Climatic Change*, Laxenburg, Austria, International Institute for Applied Systems Analysis. (Reprinted from *Climatic Change*, 1985, 7: 1–152.)

Parry, M. L. and Carter, T. (1983a) *Assessing Impacts of Climatic Change in Marginal Areas: The Search for Appropriate Methodology*, IIASA Working Paper WP-83-77. Laxenburg, Austria, International Institute for Applied Systems Analysis.

Parry, M. L. and Carter, T. R. (1983b) 'Assessing the impact of climatic change in cold regions', *Summary Report SR-84-1, International Institute for Applied Systems Analysis*, Laxenburg, Austria.

Parthasarthy, B. and Mooley, D. A. (1978) 'Some features of a long homogeneous series of Indian summer rainfall', *Monthly Weather Review*, 106: 771–81.

Paul, A. H. (1972) 'Weather and the daily use of outdoor recreation areas in Canada', in Taylor, J. A. (ed.) *Weather Forecasting for Agriculture and Industry*, Newton Abbot, David & Charles.

Pearman, G. I. (ed.) (1980) *Carbon Dioxide and Climate: Australian Research*, Canberra, Australian Academy of Science.

Pearson, R. (1978) *Climate and Evolution*, London, Academic Press.

Perkey, D. J., Young, K. N., and Kreitzberg, C. W. (1983) 'The 1980–81 drought in Eastern Pennsylvania', *Bulletin of the American Meteorological Society*, 64: 140–7.

Perry, A. H. (1971) 'Econoclimate – a new direction for climatology', *Area* (Institute of British Geographers), 3: 178–9.

Perry, A. H. (1972) 'Weather, climate, and tourism', *Weather*, 27: 199–203.

Perry, R. A. (1962) 'Notes on the Alice Springs area following rain in early 1962', *Arid Zone Newsletter*, 85–91.

Petty, M. T. (1963) 'Weather and consumer sales', *Bulletin of the American Meteorological Society*, 44: 68–71.

Pfister, C. (1978) 'Climate and economy in eighteenth century Switzerland', *Journal of Interdisciplinary History*, 9: 223–43.

Philander, S. G. H. (1983) 'El Niño southern oscillation phenomenon', *Nature*, 302: 295–301.

Phillips, D. W. (1985) 'Developing a National Climate Program. A decade of Canadian experience', lecture presented to the Ninth Session of the WMO Commission for Climatology, Geneva, December 1985.

Phillips, D. W. and McKay, G. A. (eds) (1980) *Canadian Climate in Review – 1980*, Ottawa, Environment Canada, Atmospheric Environment Service.

Pimentel, D. (1981) 'Food energy and climate change', in Bach, W., Pankratl, J., and Schneider, S. H. (eds) *Food – Climate Interactions*, Dordrecht, Holland, Reidel.
Pippard, Sir Brian (1982) 'Instability and chaos – Physical models of everyday life', *Interdisciplinary Science Reviews*, 7: 92–101.
Pittock, A. B., Fraser, P. J. B., Galbally, I. E., Hyson, P., Kulkarni, R. N., Pearman, G. I., Bigg, E. K., and Hunt, B. G. (1981) 'Human impact on the global atmosphere: Implications for Australia', *Search*, 12: 260–72.
Post, J. D. (1980) 'The impact of climate on political, social, and economic change', *Journal of Interdisciplinary History*, 10(4): 719–23.
Press (1981) 10 October.
Primack, J. and von Hippel, F. (1972) 'Scientists, politics and SST: A critical review', *Bulletin of the Atomic Scientists*, April: 24–30.
Prowse, T. D., Owens, I. F., and McGregor, G. R. (1981) 'Adjustment to avalanche hazard in New Zealand', *New Zealand Geographer*, 37: 25–31.

Quirk, W. J. (1981a) 'Climate and energy emergencies', *Bulletin of the American Meteorological Society*, 62: 623–31.
Quirk, W. J. (1981b) *Energy Supply Interruptions and Climate*, UCRL-86254, Rev. 1, Livermore, California, Lawrence Livermore National Laboratory.
Quirk, W. J. and Moriarty, J. E. (1980) 'Prospects for using improved climate information to better manage energy systems', in Bach, W., Pankrath, J., and Williams, J. (eds) *Interactions of Energy and Climate*, Dordrecht, Holland, Reidel, 89–99.
Quiroz, R. S. (1983) 'The climate of the El-Niño winter of 1982–83 – A season of extraordinary climatic anomalies', *Monthly Weather Review*, 111: 1685–706.

Radcliffe, J. E. (1974) 'Seasonal distribution of pasture production in New Zealand. I. Method of measurement', *New Zealand Journal Experimental Agriculture*, 2: 337–40.
Rapp, R. R. and Huschke, R. E. (1964) *Weather Information: Its Uses Actual and Potential*, Memo. RM4083 (US Weather Bureau), Santa Monica, California, The Rand Corporation.
Rasmusson, E. M. and Wallace, J. M. (1983) 'Meteorological aspects of the El Niño/Southern Oscillation', *Science*, 222: 1195–202.
Ravelo, A. C. and Decker, W. L. (1981) 'An iterative regression model for estimating soybean yields from environmental data', *Journal of Applied Meteorology*, 20: 1284–9.
Revelle, R. P. and Waggoner, P. E. (1983) 'Effects of a carbon dioxide induced climatic change on water supplies in the Western United States', in Carbon Dioxide Assessment Committee, *Changing Climate*, Washington, DC, National Academy Press.

Rhodes, S. L. and Middleton, P. (1983) 'The complex challenge of controlling acid rain', *Environment*, 25: 7–9, 31–8.

Rich, M. M. (1977) *Wool Clip Forecasting – An Econoclimatic Model*, New Zealand Meat and Wool Boards' Economic Service Paper no. 130.

Rich, M. M. and Taylor, N. W. (1977) 'Econoclimatic models for forecasting the national tailing percentage and wool clip', *New Zealand Agricultural Science*, 11: 65–72.

Richardson, C. W. (1984) *WGEN: A Model for Generating Daily Weather Variables*, USDA, ARS-8, Washington, DC, US Department of Agriculture, Agricultural Research Service.

Riebsame, W. E. (1981) 'Adjustments to drought in the spring wheat area of North Dakota: A case study of climate impacts on agriculture', PhD dissertation, Department of Geography, Clark University, Worcester, Massachusetts.

Riebsame, W. E. (1983) 'News media coverage of seasonal forecasts: The case of winter 1982–83', *Bulletin of the American Meteorological Society*, 64: 1351–6.

Riebsame, W. E. (1985) 'Research in climate–society interaction', in Kates, R. W., Ausubel, J. H., and Berberian, M. (eds) *Climate Impact Assessment: Studies of the Interaction of Climate and Society*, SCOPE 27, New York, Wiley, 69–84.

Roberts, W. O. and Lansford, H. (1979) *The Climate Mandate*, San Francisco, W. H. Freeman.

Robertson, G. W. (1973) 'Development of simplified agroclimate procedures for assessing temperature effects on crop development', in Slatyer, R. O. (ed.) *Plant Response to Climatic Factors, Proceedings of the Uppsala Symposium*, Paris, UNESCO, 327–41.

Robertson, G. W. (1974) 'World Weather Watch and wheat', *WMO Bulletin*, 23: 149–54.

Robinson, J. (1985) 'Global modeling and simulations', in Kates, R. W., Ausubel, J. H., and Berberian, M. (eds) *Climate Impact Assessment: Studies of the Interaction of Climate and Society*, SCOPE 27, New York, Wiley, 469–92.

Robinson, J. and Ausubel, J. H. (1983) 'A game framework for scenario generation for the CO_2 issue', *Simulation and Games*, 14(3): 317–44.

Robinson, P. (1974) 'Evaluation of air conditioning energy costs', *The Building Services Engineer*, 42: 195–8.

Rojko, A. S. and Schwartz, M. W. (1976) 'Modelling the world grains, oilseeds and livestock economy to assess the world food prospects', *Agricultural Economics Research*, 28: 89–98.

Rosenberg, N. J. (ed.) (1978) *North American Droughts*, AAAS Selected Symposium Series, 15, Boulder, Colorado, Westview Press.

Rosenberg, N. J. (1982) 'The increasing CO_2 concentration in the atmosphere and its implications on agricultural productivity II. Effects through CO_2-induced climatic change', *Climatic Change*, 4: 239–54.

Rotberg, R. I. and Raab, T. K. (eds) (1981) *Climate and History: Studies in Interdisciplinary History*, Princeton, New Jersey, Princeton University Press.

Royal Society of New Zealand (1985) *The Threat of Nuclear War: A New Zealand Perspective*, The Royal Society of New Zealand Miscellaneous Series no. 11.

Russo, J. A. (1966) 'The economic impact of weather on the construction industry of the United States', *Bulletin of the American Meteorological Society*, 47: 967–71.

Ruttenberg, S. (1981) 'Climate, food and society', in Slater, L. E. and Levin, S. K. (eds) *Climate's Impact on Food Supplies*, Boulder, Colorado, Westview Press, 23–37.

Sakamoto, C. (1978) 'The Z-index as a variable for crop yield estimation', *Agricultural Meteorology*, 19: 305–13.

Sakamoto, C., Le Duc, S., Strommen, N., and Steyaert, L. (1980) 'Climate and global grain yield variability', *Climatic Change*, 2: 349–61.

Sakamoto, C., Strommen, N., and Yao, A. (1979) 'Assessment with agroclimatological information', *Climatic Change*, 2: 7–20.

Santer, B. (1984) 'Impacts on the agricultural sector', in Meinl, H. and Bach, W. *Socioeconomic Impacts of Climatic Changes Due to a Doubling of Atmospheric CO_2 Content*, Contract No. CL1-063-D, Brussels, Commission of the European Communities.

Sassone, P. (1975) 'Public sector costs of climate change', in *Economic and Social Measures of Climatic Change*, CIAP Monograph no. 6, Washington, DC, US Department of Transportation.

Sawyer, J. S. (1967) 'Weather forecasting and its future. Part 1 – Forecasting from 1850 to the present', *Weather*, 22: 350–60; 'Part II – Forecasting in the years to come', *Weather*, 22: 400–6.

Sawyer, J. S. (1980) 'Climatic change and temperature extremes', *Weather*, 35: 353–7.

Schelling, T. C. (1983) 'Climatic change: Implications for welfare and policy', in National Research Council, *Changing Climate* (Report of the Carbon Dioxide Assessment Committee), Washington, DC, National Academy Press, 449–82.

Schneider, R., McQuigg, J. D., Means, L. L., and Klyukin, N. K. (1974) *Applications of Meteorology to Economic and Social Development*, WMO Technical Note no. 132, Geneva, World Meteorological Organization.

Schneider, S. H. (with L. E. Mesirow) (1976) *The Genesis Strategy*, New York, Plenum Press.

Schneider, S. H. and Mesirow, L. E. (1976) 'Genesis strategy: Climate and global survival', *The Futurist*, August: 192–7.

Schneider, S. H. and Temkin, R. L. (1978) 'Climatic changes and human affairs, in Gribbin, J. (ed.) *Climatic Change*, Cambridge, Cambridge University Press, 228–46.

Schramm, C. J. (1975) 'The perils of wheat trading without a grain policy', *Challenge*, March/April: 36–41.

Science Council of Canada (1976) *Living with Climatic Change*, proceedings of a conference held in Toronto in November 1975, chaired by P. D. McTaggart-Cowan, Science Council of Canada.

SCOPE (1978) *Report of the Workshop on Climate/Society Interface, Toronto, 10–14 December 1978*, Paris, SCOPE Secretariat.

Seaborg, G. T. (1972) 'Science, culture, universities and government', *Impact of Science on Society*, 22: 111–22.

Sears, P. D. (1961) 'The potential of New Zealand grassland farming', *Proceedings of the Third New Zealand Geography Conference*, 65–70.

Seifert, W. W. and Kamrany, N. M. (1974) *A Framework for Evaluating Long-Term Strategies for the Development of the Sahel-Sudan Region.* Vol. 1: *Summary Report: Project Objectives, Methodologies, and Major Findings*, Cambridge, Massachusetts, MIT Center for Policy Alternatives.

Sewell, W. R. D. (ed.) (1966) *Human Dimensions of Weather Modification*, Research Paper no. 105, Department of Geography, University of Chicago.

Sewell, W. R. D. (1968) 'Emerging problems in the management of atmosphere resources: The role of social science research', *Bulletin of the American Meteorological Society*, 49: 326–36.

Sewell, W. R. D. *et al.* (1968) *Human Dimensions of the Atmosphere*, Washington, DC, National Science Foundation.

Sewell, W. R. D. *et al.* (1973) *Modifying the Weather: A Social Assessment*, Western Geographical Series, vol. 9, Department of Geography, University of Victoria, BC, Canada.

Sewell, W. R. D. and MacDonald-McGee, D. (1983) 'Climatic Variability and Change in Canada: Towards an Acceleration of the Social Science Research Effort', discussion paper prepared for the Canadian Climate Planning Board, August 1983.

Sewell, W. R. D. and Maunder, W. J. (1985) 'A day in the life of the weather administrator: 15 March 1994', *Weather and Climate*, 5: 26–31.

Shapley, D. (1976) 'Crops and climatic change: USDA's forecasts criticized', *Science*, 193 (September 24): 1222–4.

Sherretz, L. A. and Farhar, B. C. (1978) 'An analysis of the relationship between rainfall and the occurrence of traffic accidents', *Journal of Applied Meteorology*, 17: 711–15.

Shor, M. (1964) 'Exploratory work in measurement of the effect of weather factors on retail sales', *Proceedings of the American Statistical Association*, Business and Economics Section, 54–8.

Slovic, P., Fischoff, B., and Lichtenstein, S. (1979) 'Rating the risks', *Environment*, 21(3): 14–20, 36–9.

Slovic, P., Fischoff, B., and Lichtenstein, S. (1982) 'Facts versus fears: Understanding perceived risk', in Kahneman, D., Slovic, P., and Tversky, A. (eds) *Judgement under Uncertainty: Heuristics and Biases*, New York, Cambridge University Press.

Slovic, P., Kunreuther, H., and White, G. F. (1974) 'Decision processes, rationality, and adjustment to natural hazards', in White, G. F. (ed.) *Natural Hazard: Local, National, Global*, New York, Oxford University Press.

Smith, J. R. (1971) *Augmenting Power Generation through Weather Modification*, Paper 2.2-197, The Eighth World Energy Conference, Bucharest.

Smith, L. P. (1961) 'Measuring the effects of climatic changes', *New Scientist*, 12: 608–11.

Smith, V. K. (1981) 'Economic impact analysis and climate change: A conceptual introduction, *Climatic Change*, 3: 5–22.

Smith, V. K. and Krutilla, J. V. (1982) 'Toward a restructuring of the treatment of natural resources in economic models', in Smith, V. K. and Krutilla, J. V. (eds) *Explorations in Natural Resource Economics*, Baltimore, Johns Hopkins University Press.

Spitz, P. (1980) 'Drought and self-provisioning', in Ausubel, J. and Biswas, A. K. *Climatic Constraints and Human Activities*, New York, Pergamon, 125–47.

Stanhill, G. (1977) 'Quantifying weather-crop relations', in Landsberg, J. J. and Cutting, C. V. (eds), *Environmental Effects on Crop Physiology*, New York, Academic Press.

Starr, C. (1972) 'Benefit-cost studies in sociotechnical systems', in *Perspectives on Benefit-Risk Decision Making*, Washington, DC, National Academy of Engineering, 17–42.

Starr, T. B. and Kostrow, P. I. (1978) 'The response of spring wheat yield to anomalous climate sequences in the United States', *Journal of Applied Meteorology*, 17: 1101–15.

Steele, A. T. (1951) 'Weather's effect on the sales of a department store', *Journal of Marketing*, 15: 436–43.

Stephens, F. B. (1951) 'A method of analysing weather effects on electrical power consumption', *Bulletin American Meteorological Society*, 32: 16–20.

Stewart, R. B. (1983) 'Production potential for spring wheat in the Canadian prairie provinces – an estimate', *Agriculture, Ecosystems and Environment*, 11: 1–13.

Stewart T. R. and Glantz, M. H. (1985) 'Expert judgment and climate forecasting: A methodological critique of "climate change to the year 2000"', *Climatic Change*, 7: 159–83.

Stewart, T. R., Katz, R. W., and Murphy, A. H. (1984) 'Value of weather information: A descriptive study of the fruit-frost problem', *Bulletin of the American Meteorological Society*, 65: 126–36.

Steyaert, L. T., Le Duc, S. K., and McQuigg, J. D. (1978) 'Atmospheric pressure and wheat yield modelling', *Agricultural Meteorology*, 19: 23–4.

Stockton, C. W. and Meko, D. M. (1983) 'Drought recurrence in the Great Plains as reconstructed from long-term tree-ring records', *Journal of Climate and Applied Meteorology*, 22: 17–29.

Studt, F. E. (1971) 'Economic problems of scientific growth in New Zealand science – profitable investment or bad debt', *New Zealand Science Review*, 29: 143–51.

Sugg, A. L. (1967) 'Economic aspects of hurricanes', *Monthly Weather Review*, 95: 143–6.

Susman, P., O'Keefe, P., and Wisner, B. (1983) 'Global disasters, a radical interpretation', in Hewett, K. (ed.) *Interpretations of Calamity from the Viewpoint of Human Ecology*, London, Allen & Unwin, 263–83.

Szarec, S. R. (1979) 'Primary production in four North American desert communities: Indices of efficiency', *Journal of the Arid Environment*, 2: 187–209.

Tapp, M. C. (1984) 'Weather, climate and predictability', *Weather*, 39: 271–5.

Tarko, A. M. (1977) 'Global importance of the "atmosphere-plants-soil" system in compensation of impacts on the biosphere', *Docl. of the U.S.S.R. Academy of Sciences*, 237: 234–7 (in Russian).

Tarpley, J. D., Schneider, S. R., and Money, R. L. (1984) 'Global vegetation indices from the NOAA-7 meteorological satellite', *Journal of Climatology and Applied Meteorology*, 23: 491–4.

Tass (Soviet News Agency) (1982) 28 January.

Taylor, B. F., Dini, P. W., and Kidson, J. W. (1985) 'Determination of seasonal and interannual variation in New Zealand pasture growth from NOAA-7 data', *Remote Sensing of the Environment*, 18: 177–92.

Taylor, J. A. (ed.) (1970) *Weather Economics*, University College of Wales Memo. 11, 1968, Oxford, Pergamon Press.

Taylor, J. A. (1971) 'Curbing the cost of bad weather', *New Scientist and Science Journal*, 3 June, 560–3.

Taylor, J. A. (ed.) (1974) *Climatic Resources and Economic Activity*, Newton Abbot, David & Charles.

Taylor, J. A. (1979) *Recreation Weather and Climate*, Aberystwyth, Sports Council.

Taylor, N. W. (1977a) 'Short term forecasting in the meat industry', in *Management of Dynamic Systems in New Zealand Agriculture*, DSIR Information Series, 129: 79–87.

Taylor, N. W. (1977b) *The Trend in Farm Costs and their Effect on Farm*

Output, New Zealand Meat and Wool Boards' Economic Service Paper 1775.

Taylor, N. W. (1978) *The Value of Incentives and Subsidies for Increasing Livestock Production*, New Zealand Meat and Wool Boards' Economic Service Paper 1778.

Terjung, W. H. (1976) 'Climatology for geographers', *Annals of the Association of American Geographers*, 66: 199–222.

Terjung, W. H., Liverman, D. M., and Hayes, J. T. (1984) 'Climatic change and water requirements for grain corn in the North American great plains', *Climatic Change*, 6: 193–220.

Theil, H. (1966) *Applied Economic Forecasting*, Amsterdam, Holland Publishing Co. (See pp. 283–301.)

Thomas, A. and Drummond, J. J. (1953) 'The role of weather correction in load forecasting', *Edison Electronic Institute Bulletin*, August, 306.

Thompson, J. C. (1966) *The Potential Economic and Associated Values of World Weather Watch*, World Weather Watch Planning Report no. 4, Geneva, World Meteorological Organization.

Thompson, J. C. (1967) *Assessing the Economic Value of a National Meteorological Service*, World Weather Watch Planning Report no. 17, Geneva, World Meteorological Organization.

Thompson, J. C. (1969) 'The value of weather forecasts', *Science Journal*, 5: 62–6.

Thompson, J. C. (1977) in Aspen Institute, *Living with Climatic Change*, McLean, Virginia, MITRE Corporation.

Thompson, J. D. (1981) 'Climate, upwelling and biological productivity', in Glantz, M. H. and Thompson, J. D. (eds) *Resource Management and Environmental Uncertainty*, New York, Wiley, 13–33.

Thompson, L. M. (1962) 'Evaluation of weather factors in the production of wheat', *Journal of Soil and Water Conservation*, 17: 149–56.

Thompson, L. M. (1969) 'Weather and technology in the production of corn in the US corn belt', *Agronomy Journal*, 61: 453–6.

Thompson, L. M. (1970) 'Weather and technology in the production of soybeans in the Central United States', *Agronomy Journal*, 62: 232–6.

Thompson, L. M. (1975) 'Weather variability, climatic change, and grain production', *Science*, 188: 535–41.

Thomson, R. T. (1975) *An Application of Regression Analysis to the Forecasting of National Lambing Percentage and per head Wool Clip*, New Zealand Meat and Wool Boards' Economic Service Paper no. T4.

Thomson, R. T. and Taylor, N. W. (1975) 'The application of meteorological data to supply forecasting in the meat and wool industry', in *Symposium on Meteorology and Food Production*, Wellington, New Zealand Meteorological Service, 217–29.

Thornthwaite, C. W. (1953) 'Topoclimatology', in Wormell, T. W. *et al.*

(eds) *Proceedings of the Toronto Meteorological Conference*, London, Royal Meteorological Society, 227–32.
Time (1976) 9 August.
Time (1977a) 21 February.
Time (1977b) 27 June.
Time (1980) 17 March.
Time (1981) 31 August.
Time (1983) 28 March.
Timmerman, P. (1981) *Vulnerability, Resilience and the Collapse of Society*, Environmental Monograph no. 1, Institute for Environmental Studies, University of Toronto.
Todorov, A. V. (1984) 'Sahel' The changing rainfall regime and the "normals" used for its assessment', *Journal of Climate and Applied Meteorology*, 24: 97–107.
Tolstikov, B. I. (1968) 'Benefits of meteorological services in the U.S.S.R.', in *Economic Benefits of Meteorology*, WMO Bulletin, 17: 181–6.
Tooze, S. (1984) 'Sahel drought – call for action', *Nature*, 307: 497.
Trager, J. (1975) *The Great Grain Robbery*, New York, Ballantyne Books (revised edition of *The Amber Waves of Grain* published in 1973).
Tremayne, T. (1967) 'Manitobans can take cold weather – but it costs', *Winnipeg Tribune*, 8 May, pp. 1, 7.
Troughton, J. H. (1977) 'Preface to Proceedings of a Symposium on the Management of Dynamic Systems in New Zealand Agriculture', D.S.I.R. Information Series, 129: 5–6.
Tucker, G. B. (1976) 'Research and services: Differing attitudes within the science of meteorology', *Weather*, 31: 104–12.
Tucker, G. B. (1985) 'The global CO_2 problem', *Clean Air*, 19: 5–8.
Tucker, C. J., Fung, I. Y., Keeling, C. D., and Gammon, R. H. (1986) 'Relationship between atmospheric CO_2 variations and a satellite-derived vegetation index', Nature, 319: 195–9.
Tudge, C. (1979) *The Famine Business*, New York, Penguin Books.
Turco, P. P., Toon, O. B., Ackerman, T. P., Pollock, J. B., and Sagan, C. (1983) 'Nuclear winter consequences of multiple nuclear explosions', *Science*, 222: 1283–92.
Tyndall, J. (1861) 'On the absorption and radiation of heat and gases and vapours, and on the physical connection of radiation, absorption and conduction', *Philosophical Magazine Series*, 4, 22: 169–94, 273–85.
Tyson, P. D. (1977) 'The enigma of changing world climates', *The South African Geographical Journal*, 59: 77–116.

Uchijima, Z. (1978) 'Long-term change and variability of air temperature above 10°C in relation to crop production', in Takahashi, K., and Yoshino, M. M. (eds) *Climatic Change and Food Production*, Tokyo, University of Tokyo Press, 217–29.

United Nations (1974) *The World Food Problem – Proposals for National and International Action*, Paper E/CONF. 65/4 of the World Food Conference, Rome.

United Nations Environment Programme (UNEP) (1985) *Joint UNEP/WMO/ICSU Statement on 1985 Villach Conference* on 'An assessment of the role of carbon dioxide and of other greenhouse gases in climate variations and associated impacts'.

Unninayar, S. (1983) 'Climate system monitoring', unpublished paper, World Climate Programme Department, Geneva, World Meteorological Organization.

US Conference Board (1977) *Across the Board*, June issue.

US Council on Environmental Quality (1980) *The Global 2000 Report to the President*, Washington, DC, US Government Printing Office.

US Department of Agriculture, Forest Service, and US Army Corps of Engineers Waterways, Experiment Station (1959) *Development and Testing of Some Average Relations for Predicting Soil Moisture*, USDA Technical Memorandum, 3-331, Report no. 5, Vicksburg, Miss.

US Department of Commerce (1966) *Weather and the Construction Industry*, Washington, DC, US Department of Commerce, ESSA.

US Department of Energy (1980) *Environmental and Societal Consequences of a Possible CO_2-Induced Climate Change: A Research Agenda*, vol. 1, DOE/EV/10019-01, Springfield, Virginia, National Technical Information Service.

US Environmental Protection Agency (1983) *Can We Delay a Greenhouse Warming?* (Seidel, S. and Keyes, D.). Washington, DC, US Government Printing Office.

US National Research Council (1981) *Managing Climatic Resources and Risks*, Washington, DC, National Academy Press.

US News and World Report (1976) 5 July.

US Weather Bureau (1964) *The National Research Effort on Improved Weather Description and Prediction for Social and Economic Purposes*, Federal Council for Science and Technology, Interdepartmental Committee on Atmospheric Sciences.

Vlachos, E. (1977) 'The use of scenarios for social impact assessment', in Finsterbusch, K. and Wolf, C. P. (eds) *Methodology of Social Impact Assessment*, Stroudsberg, Pennsylvania, Hutchinson Ross.

Waddell, E. (1983) 'Coping with frosts, governments and disaster experts: Some reflections based on a New Guinea experience and a perusal of the relevant literature', in Hewitt, K. (ed.) *Interpretations of Calamity from the Viewpoint of Human Ecology*, London, Allen & Unwin, 33–43.

Waggoner, P. E. (1975) 'What to do?' (Editorial), *Agricultural Meteorology*, 15: 161–4.

Waggoner, P. E. (1979) 'Variability of annual wheat yields since 1909 and among nations', *Journal of Agricultural Meteorology*, 20: 42.

Waggoner, P. E. (1983) 'Agriculture and a climate changed by more carbon dioxide', in National Research Council, *Changing Climate* (Report of the Carbon Dioxide Assessment Committee), Washington, DC, National Academy Press, 383–418.

Wahlibin, C. (1984) 'Use of short-range weather forecasts at construction sites: Some results from a user and market test', *Proceedings of the Nowcasting II Symposium*, Norrköping, Sweden.

Wall Street Journal (1970) 23 January.

Wall Street Journal (1981) January–March period.

Wallén, C. C. (1968) 'Definitions and scales in climatology as applied to agriculture', in *WMO Regional Training Seminar on Agrometeorology*, Wageningen, The Netherlands, University of Wageningen, 207–12.

Wallén, C. C. (1984) 'Present century climate fluctuations in the northern hemisphere and examples of their impact', *World Climate Programme Series Publication*, WCP – 37, Geneva, World Meteorological Organization.

Wallén, C. C. and Gwynne, M. D. (1978) 'Drought – a challenge to rangeland management', *Proceedings of the First International Rangeland Congress*, 21–31.

Walsh, M. J. (1981) 'Farm income production and exports: Part II', in Deane, R. S., Nicholl, P. W. E., and Walsh, M. J. *External Economic Structure and Policy – An Analysis of New Zealand's Balance of Payments*, Wellington, Reserve Bank of New Zealand.

Ward, Barbara (1977) Article in *The Economist*, 13 August.

Ward, F. L. (1978) *Growth and Development of the N.Z. Meat and Wool Industry*, New Zealand Meat and Wool Boards' Economic Service Paper 1786.

Ward, G. F. A. (1985) 'The southern oscillation and its effects on New Zealand weather', *New Zealand Agricultural Science*, 19: 34–8.

Warren, H. E. and Le Duc, Sharon K. (1981) 'Impact of climate on energy sector in economic analysis', *Journal of Applied Meteorology*, 20: 1431–9.

Warrick, R. A. (1980) 'Drought in the Great Plains', in Ausubel, J. H. and Biswas, A. K. (eds) *Climatic Constraints and Human Activities*, Oxford, Pergamon Press, 93–123.

Warrick, R. A. and Bowden, M. J. (1979) 'The changing impacts of droughts in the Great Plains', in Lawson, M. P. and Baker, M. E. (eds) *The Great Plains: Perspectives and Prospects*, Lincoln, Nebraska, Center for Great Plains Study, University of Nebraska.

Warrick, R. A. and Bowden, M. J. (1981) 'Changing impacts of droughts in the Great Plains', in Lawson, M. and Baker, M. (eds) *The Great Plains, Perspectives and Prospect*, Lincoln, Nebraska, Center for Great Plains Studies, University of Nebraska.

Warrick, R. A. and Riebsame, W. E. (1981) 'Societal response to CO_2-induced climate change: opportunities for research', *Climatic Change*, 3: 387–428.

Watson, D. J. (1963) 'Climate, weather and plant yield', in L. T. Evans (ed.) *Environmental Control of Plant Growth*, New York, Academic Press.

Weatherwise (1984) Special section on acid rain, October, 37: 233–52.

Weihe, W. (1979) 'Climate, health and disease', in *Proceedings of the World Climate Conference*, Publication no. 527, Geneva, World Meteorological Organization, 313–68.

Weisbecker, L. W. (1974) *The Impacts of Snow Enhancement*, Norman, Oklahoma, University of Oklahoma Press.

White, G. F. and Haas, J. E. (1975) *Assessment of Research on Natural Hazards*, Cambridge, Massachusetts, MIT Press.

White, R. M. (1964) 'New weather discoveries will serve you', *Nations' Business*, November.

White, R. M. (1979) 'Climate at the millennium', in *Proceedings of the World Climate Conference*, Publication no. 527, Geneva, World Meteorological Organization, 1–14.

White, R. M. (1982) 'Science, politics, and international atmospheric and oceanic programs', *Bulletin of the American Meteorological Society*, 63: 924–33.

Whyte, A. V. T. (1983) 'Probabilities, consequences and values in the perception of risk', in *Risk: Proceedings of a Symposium on the Assessment and Perception of Risk to Human Health in Canada*, Ottawa, The Royal Society of Canada, 121–34.

Whyte, A. V. T. (1985) 'Perception', in Kates, R. W., Ausubel, J. H., and Berberian, M. (eds) *Climate Impact Assessment: Studies of the Interaction of Climate and Society*, SCOPE, 27, New York, Wiley, 403–36.

Whyte, A. V. T. and Harrison, M. R. (1981) *Public Perception of Weather and Climatic Change: Report on a Pilot Study in Ontario, Canada*, Ottawa, Atmospheric Environment Service.

Wigley, T. M. L. (1983) 'The role of statistics in climate impact analysis', in *Proceedings of the Second International Meeting on Statistical Climatology*, Lisbon, 8.1.1–8.1.10. Lisbon, Instituto Nacional de Meteorologia e Geofisica.

Wigley, T. M. L., Huckstep, N. J., Ogilvie, A. E. J., Farmer, G., Mortimer, R., and Ingram, M. J. (1985) 'Historical climate impact assessments', in Kates, R. W., Ausubel, J. H., and Berberian, M. (eds) *Climate Impact Assessment: Studies of the Interaction of Climate and Society*, SCOPE, 27, New York, Wiley, 529–64.

Wigley, T. M. L., Ingram, M. J., and Farmer, G. (eds) (1981) *Climate and History*, Cambridge, Cambridge University Press.

Wigley, T. M. L. and Tu Oipu (1983) 'Crop-climate modelling using spatial patterns of yield and climate: Part 1, Background and an example from

Australia', *Journal of Climate and Applied Meteorology*, 22: 1831–41.
Wilks, D. S. and Murphy, A. H. (1985) 'The value of seasonal precipitation forecasts in a hay/pasturing problem in Western Oregon', *Monthly Weather Review*, 113: 1738–45.
Willett, J. W. (1981) Preface to *Climate Change and the World Grain Economy to the Year 2000: Some Implications for Domestic and International Agricultural Policy Report on the Third Phase of the NDU Climate Impact Assessment* (D. G. Johnson). Draft manuscript dated April 1979, cleared for release June 1981. Washington, DC, National Defense University.
Williams, G. D. V. and Oakes, W. T. (1978) 'Climatic resources for maturing barley and wheat in Canada', in Hage, K. D. and Reinelt, E. R. (eds) *Essays on Meteorology and Climatology: In Honour of Richard W. Longley*, Studies in Geography, Monograph 3, Edmonton, University of Alberta.
Wilson, A. (1966) 'The impact of climate on industrial growth; Tucson, Arizona: A case study', in Sewell, W. R. D. (ed.) *Human Dimensions of Weather Modification*, Research Paper no. 105, Department of Geography, University of Chicago, 249–60.
Winstanley, D. (1985) 'Africa in drought – a change of climate?', *Weatherwise*, 38: 75–81.
Wintle, B. J. (1960) 'Railways versus the weather', *Weather*, 15: 137–9.
Wisner, B. G., Jr (1977) 'The human ecology of drought in eastern Kenya', PhD dissertation, Department of Geography, Clark University, Worcester, Massachusetts.
Won, T. K. (1980) 'Climate and energy', in *Socioeconomic Impacts of Climate*, Alberta, Northern Forest Research Center.
Wooster, W. S. and Guillen, O. (1974) 'Characteristics of El Niño in 1972', *Journal of Marine Research*, 32: 387–404.
World Climate Programme (1980–5) Reports published in the World Climate Programme Series, WCP-1 to WCP-100 (continuing). Unpublished reports. Geneva, World Meteorological Organization.
World Climate Programme (1981) *On the Assessment of the Role of CO_2 on Climate Variations and their Impact* (Villach, Austria, Workshop, November 1980), Geneva, WMO, UNEP, and ICSU.
World Meteorological Organization (1968) 'Economic benefits of meteorology', *WMO Bulletin*, 17: 181–6.
World Meteorological Organization (1975) *The Physical Basis of Climate and Climate Modelling*, GARP Publication Series no. 16, Geneva, World Meteorological Organization.
World Meteorological Organization (1979) *Proceedings of the World Climate Conference: A Conference of Experts on Climate and Mankind*, Publication no. 5371, Geneva, World Meteorological Organization.
World Meteorological Organization (1980) *Outline Plan and Basis for the*

World Climate Programme 1980–1983, WMO no. 540, Geneva, World Meteorological Organization.

World Meteorological Organization (1982) *Report of the Meeting of Experts on Potential Climatic Effects of Ozone and Other Minor Trace Gases, Boulder, Colorado, 13–17 September 1982*, WMO Report no. 14, Geneva, World Meteorological Organization.

World Meteorological Organization (1983) *First WMO Long-Term Plan, Part I: Overall Policy and Strategy (1984–1993)*, Geneva, World Meteorological Organization.

World Meteorological Organization (1985a) *Meteorology and Society: Statements Presented at the Ninth Congress*, WMO Publication no. 643, Geneva, World Meteorological Organization.

World Meteorological Organization (1985b) *The Global Climate System – A Critical Review of the Climate System during 1982–84*, a contribution to the Global Environmental Monitoring System (GEMS), World Climate Data Programme, Geneva, World Meteorological Organization.

Wright, P. B. (1978) 'The southern oscillation', in Pittock, A. B., Frakes, L. A., Jenssen, D., Peterson, J. A., and Zillman, J. W. (eds) *Climatic Change and Variability: A Southern Perspective*, Cambridge, Cambridge University Press, 180–5.

Wright, S. M. (1976) 'Barton Blount: Climatic or economic change?' *Medieval Archaeology*, 20: 148–52.

Yared, R. (1984) 'Coping with climatic change', *Options*, 4: 1–5, Laxenburg, Austria, International Institute for Applied Systems Analysis.

Yates, H. W., Tarpley, J. D., Schneider, S. R., McGinnis, D. F., and Schofield, R. A. (1984) 'The role of meteorological satellites in agricultural remote sensing', *Remote Sensing of the Environment*, 14: 219–33.

Young, M. D. and Wilson, A. D. (1978) 'The influence of climatic variability on Australian society's economy', in *Phillip Island Conference, Victoria, 27–30 November 1978*, Canberra, CSIRO.

Zeisel, H. (1950) 'How temperature affects sales', *Printers' Ink*, 223: 40–2.

Indexes

AUTHOR INDEX

GEOGRAPHICAL INDEX

ORGANIZATIONAL INDEX

SUBJECT INDEX

Note The subject index, or more correctly the topic index, lists most of the subjects discussed in the book, *except* those subjects which are basically meteorological or climatological. For example, subjects such as rain, temperature, cloud, meteorology, anticyclone, thermometer are not specifically listed, since they can be assumed to be directly or indirectly associated with the wide range of topics that *are* listed, such as agriculture, construction, energy, population, umbrellas, welfare.